ASPHALT
SCIENCE
AND
TECHNOLOGY

ASPHALT SCIENCE AND TECHNOLOGY

edited by

ARTHUR M. USMANI

Usmani Development Company
Indianapolis, Indiana

CRC Press
Taylor & Francis Group
Boca Raton London New York

CRC Press is an imprint of the
Taylor & Francis Group, an **informa** business

CRC Press
Taylor & Francis Group
6000 Broken Sound Parkway NW, Suite 300
Boca Raton, FL 33487-2742

First issued in paperback 2019

© 1997 by Taylor & Francis Group, LLC
CRC Press is an imprint of Taylor & Francis Group, an Informa business

No claim to original U.S. Government works

ISBN-13: 978-0-8247-9998-4 (hbk)
ISBN-13: 978-0-367-40089-7 (pbk)

Library of Congress Cataloging-in-Publication Data

Asphalt science and technology/edited by Arthur M. Usmani
     p. cm.
    Includes index.
    ISBN 0-8247-9998-4 (hardcover: alk. paper)
    1. Asphalt. I. Usmani, Arthur M.
TP692.4.A8A68   1997
665'.4—dc21

                          97-22475
                            CIP

Visit the Taylor & Francis Web site at
http://www.taylorandfrancis.com

and the CRC Press Web site at
http://www.crcpress.com

Preface

Asphalt is an old, low-cost, thermoplastic material. Petroleum asphalt became available from petroleum refining about 100 years ago. Today, about 70 billion pounds of asphalt are used annually in the United States alone, and asphalt usage will grow dramatically in Asia during the next 10 years. Asphalt-containing materials find application in paving, road construction, roofing, coatings, adhesives, and batteries.

Despite asphalt's wide use, it is not a well-characterized chemical. Polymer modifications invariably mitigate asphalt's shortcomings. This has not been thoroughly understood until now. To sustain the current usage of asphalt and develop newer applications, there is a pressing need to revitalize research, development, and engineering in asphalt materials. The objective of this book is to initiate this revitalization.

Organization of this book follows characterization, mechanical and rheological aspects, polymer modifications, and performance-modification relationships of asphalt. I am most appreciative of the leading experts who contributed to this volume.

Chemists, chemical engineers, and civil engineers involved in asphalt research, development, and engineering will find this book useful. Professionals in highway departments should also benefit. Students of civil engineering and construction technology particularly will find this book helpful.

With the help of asphalt experts, I am most pleased to bring this authoritative book to the readership.

Arthur M. Usmani

Contents

Preface   *iii*

Contributors   *ix*

Characterization of Asphalt

1. Asphalt Chemistry: An NMR Investigation of the Benzylic
   Hydrogens and Oxidation   1
   P. W. Jennings, Jacqueline Fonnesbeck, Jennifer Smith,
   and J. A. S. Pribanic

2. Molecular Motions and Rheological Properties of Asphalts:
   An NMR Study   11
   Daniel A. Netzel, Francis P. Miknis, J. Calhoun Wallace, Jr.,
   Clinton H. Butcher, and Kenneth P. Thomas

3. DSC Studies of Asphalts and Asphalt Components   59
   Thomas F. Turner and Jan F. Branthaver

4. Detection of Strongly Acidic Compounds in Extensively Aged
   Asphalts   103
   Sang-Soo Kim, Gerald W. Gardner, John F. McKay,
   Raymond E. Robertson, and Jan F. Branthaver

5. Oxidation of Asphalt Fractions   119
   Jinmo Huang

6. Reduction of Odor from Hot Asphalt   135
   Peter Kalinger, Robert J. Booth, and Ralph M. Paroli

Mechanical Rheological Aspects

7. Propagation and Healing of Microcracks in Asphalt Concrete
    and Their Contributions to Fatigue   149
    Dallas N. Little, Robert L. Lytton, Devon Williams,
    and Y. Richard Kim

8. Polymer Modifiers for Improved Performance of Asphaltic
    Mixtures   197
    Heshmet A. Aglan

9. Effect of SBS Polymer Modification on the Low-Temperature
    Cracking of Asphalt Pavements   217
    Robert Q. Kluttz and Raj Dongré

10. Rheological Properties of Polymer-Modified Asphalt Binders   235
    Peter E. Sebaaly

Polymer Modifications

11. Current and Potential Applications of Asphalt-Containing
    Materials   249
    Arthur M. Usmani

12. Amorphous Polyalphaolefins (APAOs) as Performance Improvers
    in Asphalt-Containing Materials   259
    Andrés Sustic

13. Chemistry and Technology of SBS Modifiers   279
    Elio Diani, Mauro Da Via, and Maria Grazia Cavaliere

14. Polychloroprene-Modified Aqueous Asphalt Emulsion:
    Use in Roofs and Roads   297
    S. S. Newaz and M. Matner

15. Polymer Science and Technology   307
    Arthur M. Usmani

16. Factors Influencing Compounding of Constitutents in
    Bitumen–Polymer Compositions   337
    Janusz Zielinski

17. Development and Characterization of Asphalt Modifiers from
Agrobased Resin   349
Sangita, I. R. Arya, R. Chandra, and Sandeep

18. Polymer Network Formation in Asphalt Modification   369
Arthur M. Usmani

19. Production, Identification, and Application of Crumb Rubber
Modifiers for Asphalt Pavements   385
Michael Wm. Rouse

20. Asphalt Coatings and Mastics for Roofing and Waterproofing   443
Eileen M. Dutton

Performance–Modification Relationships

21. Polymer-Modified Asphalt Cement Used in the Road
Construction Industry: Basic Principles   463
Bernard Brûlé

22. Performance Characteristics of Asphalt Concrete Mixes
Containing Conventional and Modified Asphalt Binders   479
*Akhtarhusein A. Tayebali, John Harvey, Alejandro J. Tanco,
Robert N. Doty, and Carl L. Monismith*

23. Analytical Approach to Bitumen–Polymer Blends   505
Patrick Cogneau

Index   523

Contributors

Heshmet A. Aglan, Ph.D.   Mechanical Engineering Department, Tuskegee University, Tuskegee, Alabama

I. R. Arya   Flexible Pavements Division, Central Road Research Institute, New Delhi, India

Robert J. Booth, Ph.D.   Department of Building Science and Technology, Hansed-Booth, Dalkeith, Ontario, Canada

Jan F. Branthaver, Ph.D.   Asphalt and Heavy Oil Division, Western Research Institute, Laramie, Wyoming

Bernard Brûlé, Ph.D.   Department of Technical Management, Entreprise Jean Lefebvre, Neuilly sur Seine, France

Clinton H. Butcher   Chemical Engineering Department, University of Wyoming, Laramie, Wyoming

Maria Grazia Cavaliere, Ph.D.   Elastomers Division, EniChem, Milan, Italy

R. Chandra, Ph.D.   Department of Applied Chemistry and Polymer Technology, Delhi College of Engineering, Delhi, India

Patrick Cogneau, Ph.D.   Parc Industriel, Perwez, Belgium

Mauro Da Via, Ph.D.   Elastomers Division, EniChem, Milan, Italy

Elio Diani, Ph.D.   Elastomers Division, EniChem, Milan, Italy

Raj Dongré, Ph.D.   EBA Engineering, Turner-Fairbank Highway Research Center, McLean, Virginia

Robert N. Doty   California Department of Transportation, Sacramento, California

Eileen M. Dutton, B.A., M.S.   Research and Development, Quality Control, Karnak Corporation, Clark, New Jersey

Jacqueline Fonnesbeck   Department of Chemistry, Montana State University, Bozeman, Montana

Gerald W. Gardner   Western Research Institute, Laramie, Wyoming

John Harvey, P.E., Ph.D.   Institute of Transport Studies, University of California, Berkeley, Richmond Field Station, Richmond, California

Jinmo Huang, Ph.D.   Chemistry Department, The College of New Jersey, Trenton, New Jersey

P. W. Jennings, Ph.D.   National Science Foundation, Arlington, Virginia, and Montana State University, Bozeman, Montana

Peter Kalinger, B.A., M.A.   Canadian Roofing Contractors' Association, Ottawa, Ontario, Canada

Sang-Soo Kim   Western Research Institute, Laramie, Wyoming

Y. Richard Kim, P.E., Ph.D.   Department of Civil Engineering, North Carolina State University, Raleigh, North Carolina

Robert Q. Kluttz, Ph.D.   Elastomers Department, Shell Chemical Company, Houston, Texas

Dallas N. Little, Ph.D.   Texas Transportation Institute, Texas A&M University, College Station, Texas

Robert L. Lytton, Ph.D.   Department of Civil Engineering, Texas A&M University, College Station, Texas

M. Matner   Bayer AG, Dormagen, Germany

John F. McKay   Western Research Institute, Laramie, Wyoming

Francis P. Miknis, Ph.D.   Division of Advanced Technology, Western Research Institute, Laramie, Wyoming

Carl L. Monismith, Ph.D.   Institute of Transport Studies, University of California, Berkeley, Richmond Field Station, Richmond, California

Daniel A. Netzel, Ph.D.   Division of Advanced Technology, Western Research Institute, Laramie, Wyoming

S. S. Newaz, Ph.D.   Bayer Corporation, Houston, Texas

Ralph M. Paroli, Ph.D., C.Chem   Institute for Research in Construction, National Research Council of Canada, Ottawa, Ontario, Canada

*J. A. S. Pribanic   Department of Chemistry, Montana State University, Bozeman, Montana

Raymond E. Robertson, Ph.D.   Asphalt and Heavy Oil Division, Western Research Institute, Laramie, Wyoming

Michael Wm. Rouse, B.Sc.   Rouse Rubber Industries, Inc., Vicksburg, Mississippi

Sandeep   Department of Applied Chemistry and Polymer Technology, Delhi College of Engineering, Delhi, India

Sangita, Ph.D.   Flexible Pavements Division, Central Road Research Institute, New Delhi, India

Peter E. Sebaaly, Ph.D.   Department of Civil Engineering, University of Nevada, Reno, Nevada

Jennifer Smith   Department of Chemistry, Montana State University, Bozeman, Montana

Andrés Sustic, Ph.D.   Department of Materials Development, Rexene Corporation, Odessa, Texas

Alejandro J. Tanco   University of Córdoba, Córdoba, Argentina

Akhtarhusein A. Tayebali, P.E., Ph.D.   Department of Civil Engineering, North Carolina State University, Raleigh, North Carolina

Kenneth P. Thomas, Ph.D.   Division of Advanced Technology, Western Research Institute, Laramie, Wyoming

Thomas F. Turner   Asphalt and Heavy Oil Division, Western Research Institute, Laramie, Wyoming

Arthur M. Usmani, Ph.D.   Usmani Development Company, Indianapolis, Indiana

J. Calhoun Wallace, Jr.   Chemical Engineering Department, University of Wyoming, Laramie, Wyoming

Devon Williams   Texas Transportation Institute, Texas A&M University, College Station, Texas

Janusz Zielinski, D.Sc.   Institute of Chemistry in Płock, Warsaw University of Technology, Płock, Poland

---

* Current affiliation: Bozeman, Montana.

1

Asphalt Chemistry: An NMR Investigation of the Benzylic Hydrogens and Oxidation

P. W. Jennings

National Science Foundation, Arlington, Virginia, and
Montana State University, Bozeman, Montana

Jacqueline Fonnesbeck,
Jennifer Smith, and *J.A.S. Pribanic

Montana State University, Bozeman, Montana

As all of you know, asphaltic materials are very important to the quality of our lives. A day does not pass in which we do not intersect with some form of these materials. In most cases, these interactions are beneficial. Thus, it is worrisome that the public is not more aware of the overall significance of asphalt and its many uses. Beyond potholes, there is not much public concern. However, more germane to this conference are the challenges that we face in understanding this very complex material and in finding ways to improve its performance [1–5].

Over the past few years, with the support of the asphalt industry, the state highway departments, and the Strategic Highway Research Program (SHRP), many of us have done some in-depth investigations of the chemistry of these materials. I would like to tell you about the results of our investigations on the oxidation of asphalt. More specifically, this chapter is about our efforts to detect and then to measure the benzylic groups that may participate in asphalt oxidation. Furthermore, I will discuss an experiment that was designed to elaborate the

---

* Current affiliation: Bozeman, Montana.

mechanism of molecular self-assembly that is observed in the LMS region of the HP-GPC profile.

In structure 1, Figure 1, the benzylic groups are identified by the letters a–d, and the possible consequences of oxidation are shown in structure 2. Benzylic carbon (a) is referred to as a primary benzylic group (i.e., three hydrogens) that upon vigorous oxidation could yield the carboxylic acid moiety (a) in structure 2. Benzylic group (b) is a secondary group (two hydrogens) and can yield either a ketone, under mild oxidation conditions, or another carboxylic acid with concomitant chain cleavage under harsh conditions. Tertiary benzylic groups (c) and (d) can be transformed into a variety of products on oxidation. However, it is likely that oxidation at these two positions will lead to an aromatic residue as shown in structure 2. There is a considerable driving force for such a transformation as aromatization is a stabilizing process. Furthermore, in this case an aromatic link would be made between two formerly isolated aromatic units (i.e., rings B and D). An interesting feature here is that by aromatizing ring C, the substituent groups on ring C ($-CH_2-CH_3$ and $-CH_3$) become benzylic groups. The point of this discussion is that if this reaction (1 to 2) occurred, there would be a loss of seven benzylic hydrogens (a = 3, b = 2, c = 1, d = 1) and a gain of five (e = 2, f = 3). If this reaction was being monitored by nuclear magnetic resonance (NMR) spectroscopy, the observed result would be a loss of two benzylic hydrogens or the net from

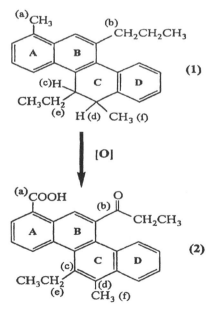

Figure 1

above. With this introductory discussion, it would be easy to develop scenarios wherein there could be larger losses of benzylic hydrogens or even a net gain.

Thus, the challenge we undertook was (a) to find a way to observe the benzylic groups unambiguously, (b) to attempt to quantify the benzylic hydrogens, and finally (c) to make sense of the results.

Nuclear magnetic resonance spectroscopy was chosen for these measurements because it, better than infrared (IR) spectroscopy, should provide more detail and be less ambiguous. Simple one-dimensional NMR experiments proved to be inadequate as both the benzylic hydrogens and benzylic carbons resonate at chemical shift values that are among a complex group of resonances from other groups such as alkyl ethers. Therefore, a two-dimensional experiment referred to as COLOC was chosen. In this experiment, the linkage between the benzylic hydrogens and the IPSO carbon may be observed (see Figure 2). Although this is a difficult and a time-consuming experiment, it can provide unambiguous identity of the benzylic groups.

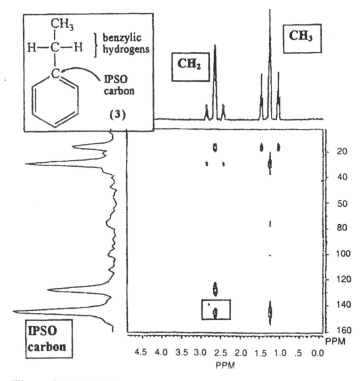

Figure 2 NMR COLOC experiment on model compound, ethylbenzene.

The model system, ethylbenzene (structure <u>3</u>), which was chosen to illustrate the experiment, and the results derived therefrom are shown in Figure 2. The hydrogen (H) spectrum is shown on the upper horizontal axis and the carbon (C) spectrum is shown on the left vertical axis. In the graph, there are several spots that are in fact peaks coming out of the plane of the paper. The two-dimensional resonance spot or peak of importance in this context is circumscribed by the box. This resonance peak at 2.63 ppm (H) and 144 ppm (C) represents the coupling (J^2) between the benzylic hydrogens (Hs) of the CH_2 group and the IPSO carbon of the aromatic ring. The other peak at 2.63 ppm and 128 ppm represents the coupling (J^3) between the CH_2 group and the ortho aromatic carbon. Finally, the peak at 1.23 ppm and 144 ppm is the J^3 coupling between the CH_3 group and the IPSO carbon. J^3 coupling is usually a bit more intense and easier to obtain than J^2 coupling, but it is not the specific target for the present study. Applying this concept to a real asphalt system results in the graph shown as Figure 3. The peak at 2.1/138 ppm is a benzylic group, which we believe to be a methyl (CH_3) group. The other benzylic peaks are assigned as CH_2, 2.4, and

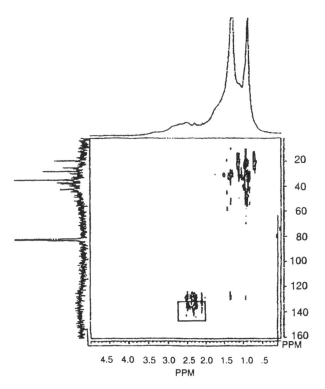

Figure 3   NMR COLOC experiment on a real asphalt sample.

CH, 2.6 ppm on the hydrogen scale and 138 ppm on the carbon scale. The peaks lying outside and just above the box are the J^3 coupling peaks. These peaks are definitely benzylic groups, but the specific assignments (CH, CH_2, CH_3) must be regarded as tentative.

Thus, the first question that was asked earlier has been answered: Benzylic moieties can be observed by two-dimensional NMR spectroscopy in real asphalt systems.

The next question dealt with whether or not these peaks could be integrated and thereby quantified. For this measurement, the hydrogen axis is used to project the two-dimensional peaks. Once projected, the area under the curves can be integrated. The initial results for model compounds quickly revealed that we could not obtain absolute concentrations. This is demonstrated in Figure 4, where the two-dimensional spectrum of an equal mixture of 9,10-dimethylanthracene (A) and 2,6-dimethylphenol (B) is shown. It is clear that the relative intensities

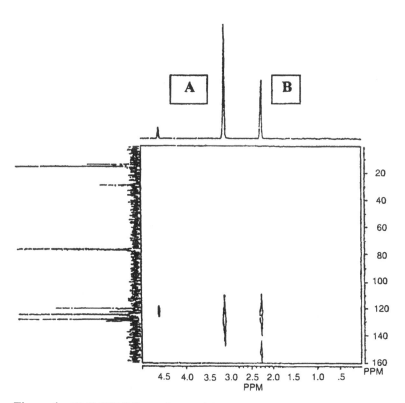

Figure 4   NMR COLOC experiment with equal concentrations of two model compounds A and B.

of the methyl groups in the proton projection (upper horizontal axis) do not have equivalent peak areas. This is probably due to the fact that there are several variable delay settings on the NMR spectrometer in this measurement and not all samples will respond to a given setting to the same degree.

Subsequently, we tried to develop a relative concentration methodology by using the internal standard, 9,10-dimethylanthracene (DMA). This appears to have been successful. For this measurement, the internal standard was placed in the inner cell of a coaxial tube and the asphalt sample was placed in the outer cell. An example of the results of this experiment is displayed in Figure 5. This technique appears to be useful for making measurements on the same sample under different conditions (e.g., before and after a chemical reaction).

In a series of NMR experiments, we compared the original asphalt sample with the same asphalt that had been oxidized at WRI using the thin-film oven/pressure oxygen vessel (TFO/POV) procedure. Data for the eight SHRP core asphalts are shown in Table 1. Also listed in this table are the carbonyl concentrations of the same asphalts as garnered by integrating the carbonyl region (1540–1820 cm^{-1}) of the infrared spectrum. In all cases, the carbonyl concentration increased by a factor of 2 to 4 on oxidation. The benzylic hydrogen concentration increased in samples A, C, D, F, G, and K and decreased in two samples, B and M. These data suggest a couple of tentative conclusions. First, the TFO/POV

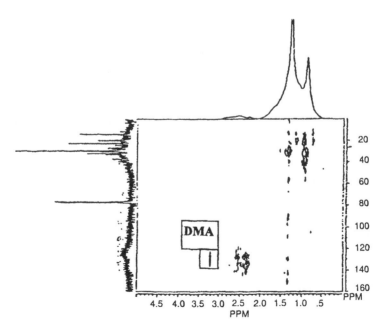

Figure 5 NMR COLOC experiment showing the DMA standard peak.

Table 1 NMR COLOC and Infrared Results for Asphalts

Asphalt	Aging	NMR COLOC[a]	Carbonyl concentration[b]
AAA-1	Unaged	2.80	2.86×10^{-6}
	TFO/POV	3.04	7.00×10^{-5}
	H_2O_2	1.15	3.09×10^{-4}
AAB-1	Unaged	1.87	2.86×10^{-4}
	TFO/POV	1.17	3.20×10^{-4}
AAC-1	Unaged	1.76	1.11×10^{-4}
	TFO/POV	1.99	4.16×10^{-4}
AAD-1	Unaged	1.99	2.09×10^{-4}
	TFO/POV	2.30	8.41×10^{-4}
AAF-1	Unaged	1.37	1.51×10^{-4}
	TFO/POV	1.59	3.39×10^{-4}
AAG-1	Unaged	2.24	2.99×10^{-4}
	TFO/POV	2.93	5.64×10^{-4}
AAK-1	Unaged	2.02	1.88×10^{-4}
	TFO/POV	2.84	4.32×10^{-4}
AAM-1	Unaged	2.10	1.58×10^{-4}
	TFO/POV	1.65	3.64×10^{-4}
Big Timber	Unaged	3.20	6.66×10^{-5}
G-18	Field aged	2.40	2.52×10^{-4}
Chemcrete	Field aged	0	3.61×10^{-4}

[a]Benzylic peak area relative to standard, 9,10-dimethylanthracene.
[b]Moles of carbonyl per gram of asphalt.

oxidation methodology appears to be creating an excess of benzylic hydrogens, which are likely to be derived from aromatizing saturated rings that have incipient benzylic hydrogens. Second, because the carbonyl concentration is increasing in all samples along with the benzylic hydrogens in most samples, it is likely that there are functionalities other than benzylic groups leading to carbonyl formation. More standardization work needs to be done here to relate carbonyl peak area to benzylic group oxidation.

In one sample, AAA-1, 30% H_2O_2 was used as the oxidant to determine whether this reagent was more selective for benzylic hydrogen oxidation. In this case, the benzylic hydrogen content decreased relative to the unaged as well as TFO/POV samples and the carbonyl concentration increased considerably. The infrared spectra for these three asphalts are shown in Figure 6. It is evident that this reagent is harsher than TFO/POV and will oxidize benzylic groups.

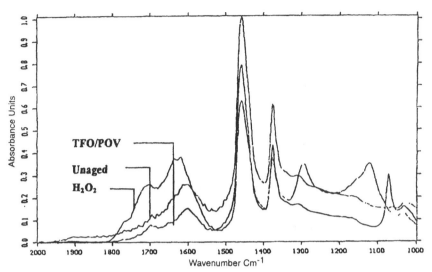

Figure 6   Infrared spectra of asphalt AAA-1.

In addition to NMR and infrared spectroscopy, HP-GPC analyses of the three AAA-1 samples were conducted. The graphs are shown in Figure 7. As can be seen, TFO/POV oxidation results in a slight increase in the LMS region of the profile and indicates an enhancement of self-assembly of molecular components. A much more dramatic increase in the LMS region results from peroxide oxidation. Thus, hydrogen peroxide appears to have a more pronounced effect

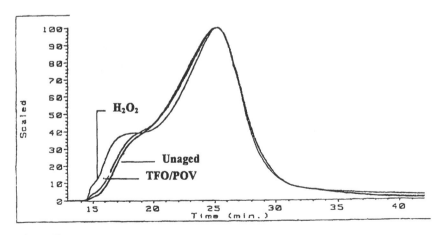

Figure 7   HP-GPC profiles of asphalt AAA-1.

than TFO/POV with regard to loss of benzylic hydrogens, increase in carbonyl concentration, and enhanced molecular self-assembly.

The last entry in Table 1 shows the results for real roadway samples. In this case, two samples from a test site in Montana (Big Timber Test Sections) were analyzed versus the original asphalt sample, which had been sealed in a container for 12 years. The unaged sample had HP-GPC characteristics similar to those of AAG-1 (i.e., low concentrations of LMS). In the field-aged sample, the benzylic hydrogens decreased slightly and the carbonyl concentration increased significantly. The second sample was the same original asphalt that had been treated with Chemcrete. In this case, no benzylic hydrogens were observed in the road-aged sample, suggesting rather harsh oxidizing conditions. The carbonyl content increased significantly over that of the unaged as well as the field-aged samples.

Finally, I want to report one unusual experiment that we performed recently. The experiment was designed to interrupt π-π stacking in the asphalt self-assembled phase as detected by the LMS region of the HP-GPC analysis technique. We believe that π-π stacking between aromatic units is partially responsible for the LMS region of the HP-GPC. To probe this concept, we decided to alkylate the aromatic residues using Lewis acid catalysis and various alkyl halides. The Lewis acid of interest in this specific case is ferric chloride, $FeCl_3$. After conducting the experiment with ferric chloride, methyl chloride, and a standard work-up procedure, the HP-GPC analysis was conducted. As anticipated, the LMS region for asphalt K, which is about 25% of the area of an untreated sample, was nearly depleted. Obviously, the concept was correct and the alkylation had broken up the π-π stacking. However, we were concerned that because a number of other reagents were generated in this reaction, each should be tested for individual effects. After considerable testing of water, hydrochloric acid, and methyl chloride, we tested ferric chloride by itself. The result is that the observed loss of self-assembly that was found earlier was reproduced with just ferric chloride. This is interesting, as ferric chloride is used in some processes for the oxidation of asphalt for the roofing industry. However, that oxidation is usually done at elevated temperatures, whereas our experiment was conducted at room temperature, approximately 25°C. We have not had time to elaborate this result into a valid explanation. However, because it is well known that ferric chloride, like other Lewis acid reagents, is prone to complex with any Lewis base, I presume that the reagent is complexing with all of the oxygen, sulfur, nitrogen, and π bases in asphalt. Although it is not clear what the mode of action is for $FeCl_3$, it is destroying the molecular self-assembly of asphalt K as measured by the LMS region of the HP-GPC profile.

REFERENCES

1. P. W. Jennings, et al., NMR Spectroscopy in the Characterization of Eight Selected Asphalts, *Fuel Sci. Technol. Int.*, **10**: 887 (1992).

2.   P. W. Jennings, et al., High Performance Gel Permeation Chromatography in the Characterization of Self-Assemblies in Asphalt, *Fuel Sci. Technol. Int.*, **10**: 809, (1992).

3.   P. W. Jennings, et al., HP-GPC Analysis of Asphalt Fractions in the Study of Molecular Self-Assembly in Asphalt, *Prepr. ACS Div. Fuel Chem.*, **37**: 1312 (1992).

4.   P. W. Jennings, et al., Nuclear Magnetic Resonance Spectroscopy for the Characterization of Asphalt, *Prepr. ACS Div. Fuel Chem.*, **37**: 1320 (1992).

5.   P. W. Jennings, et al., Predicting the Performance of the Montana Test Sections by Physical and Chemical Testing, Transportation Research Record, No. 1171 (1989).

2

Molecular Motions and Rheological Properties of Asphalts: An NMR Study

Daniel A. Netzel, Francis P. Miknis, and Kenneth P. Thomas
Western Research Institute, Laramie, Wyoming

J. Calhoun Wallace, Jr. and Clinton H. Butcher
University of Wyoming, Laramie, Wyoming

I. INTRODUCTION

The road performance of an asphalt depends upon its susceptibility to changes in its physical and chemical properties with time, temperature, and exposure to moisture. Fundamentally, the response to changes is determined by its initial chemical composition, molecular structural configuration, and the extent of molecular motion.

The exact nature of asphalt structure is still a subject surrounded by considerable controversy. One model of asphalt structure assumes a complex molecular system consisting of polycondensed aromatic and heteroaromatic clusters with methyl and long-chain alkane hydrocarbon groups attached and surrounded by lower molecular weight, less polar aromatic and aliphatic components. A simplified "average" asphalt molecule is shown in Figure 1. Generally, there are ~30% aromatic carbons and ~70% aliphatic carbons in an average "asphaltic" molecule. However, many asphalt constituents vary considerably from this "typical" molecule. Not shown in the figure are the heteroatoms, which are also of major importance to asphalt chemistry.

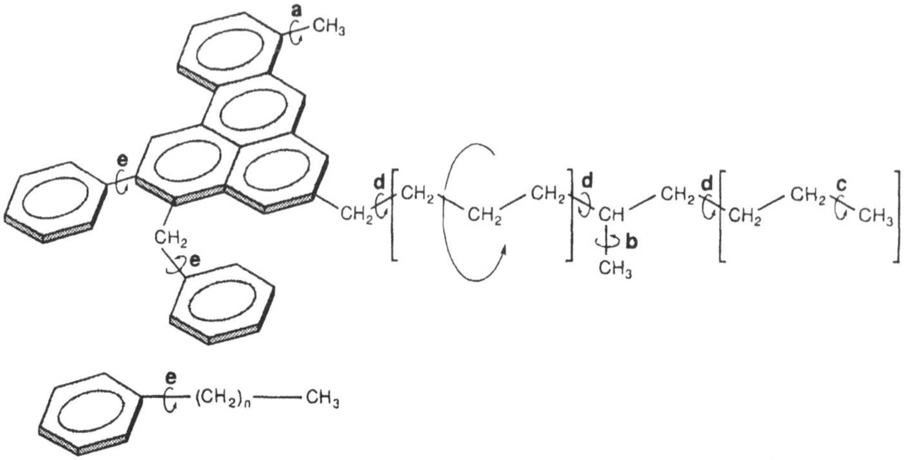

Figure 1 (a) Aromatic methyl rotation, (b) branched methyl rotation, (c) terminal methyl rotation, (d) segmental (rotation) motion, and (e) phenyl rotation (twisting) in a complex "asphalt-like" molecule.

The types and magnitudes of the molecular interactions in a macromolecular system such as an asphalt are extensive. These interactions may include molecular entanglement of the long-chain paraffinic material and weak noncovalent associative forces (hydrogen bonding, acid-base interactions, π-π interactions, etc.). At any given temperature, these interactions control the molecular motion of the constituents in asphalt (in addition to intramolecular configuration, the size of the substituted aromatic cluster, and the conformation of the methyls and long-chain alkane substituents on the aromatic clusters). Changes in structure and composition due to enhanced molecular association and/or bonding of the molecules brought about by chemical reactions also strongly affect the molecular motions. These changes, in turn, affect the overall long-term road performance of the asphalts. Thus, the extent of molecular motion associated with the asphalt molecular components is one of the most fundamentally important properties of an asphalt, which dictates its temperature-dependent road performance. However, the task of defining and quantifying the mobile, semirigid, and rigid components of such a heterogeneous material as an asphalt is by no means an unambiguous process.

On a macro scale the road performance behavior of asphalts is manifested in changes in the viscoelastic nature of the asphalt, that is, changes in its rheological properties. An understanding of the molecular dynamics (motion) in asphalts in relation to the rheological properties is paramount in achieving good road performance. It may be possible to enhance the performance of an asphalt by changing the extent of molecular mobility at a given temperature. The molecular mobility can be altered through chemical modification of the asphalt, physical blending of two or more asphalts, or the introduction of polymeric materials into asphalts.

The type and extent of molecular motions in a material can be obtained from carbon and hydrogen nuclear magnetic resonance (NMR) relaxation time measurements. These measurements can provide quantitative information on the molecular structure and degree of molecular motions associated with the carbon types in asphalts. NMR relaxation techniques have been used extensively in the study of molecular motions in polymeric and elastomeric systems [1–4], and in some cases the NMR relaxation parameters of polymeric systems have been correlated with viscoelastic and mechanical properties. Specifically, the segmental motion of the carbon main chain, as determined by carbon-13 (^{13}C) spin-lattice relaxation time, T_1^C, has been shown to obey the Williams, Landel, and Ferry (WLF) equation [5]. Schaefer et al. [6], using data on the ^{13}C spin-lattice relaxation time in the rotating frame, $T_{1\rho}^C$, for polystyrene, have assigned the broad, intense β-transition in the tan δ versus temperature plot to cooperative ring- and main-chain restricted oscillations. The mechanical impact strength and ^{13}C NMR relaxation times of several glassy polymers were also studied by Schaefer et al. [7]. In their studies, they found that the impact strength correlates specifically with the ratio of ^{13}C cross-polarization relaxation time, T_{CH}, to $T_{1\rho}^C$. This ratio ($T_{CH}/T_{1\rho}^C$) is in a sense a measure of the ratio of polymer segments that can absorb and dissipate the energy of an impact to the polymer chain segments unable to respond to the impact.

Before NMR polymer studies, the glass transition temperature (T_g) was associated with the cessation of long-range motions. However, NMR relaxation investigations of polymers have revealed that some local motions, such as methyl rotations and the segmental motion of the carbon main chain and methylene side groups, do occur below T_g. Measurements of the hydrogen spin-lattice relaxation time in the rotating frame, $T_{1\rho}^C$, and $T_{1\rho}^C$ suggest that motions of progressively smaller amplitude and segment size are frozen out as the temperature is lowered.

The use of NMR to study the molecular dynamics throughout the service temperature range for asphalts and asphalt-related materials has not received much attention [8–11]. Gülsün [12] investigated the ^{1}H (hydrogen-1) and ^{13}C spin-lattice relaxation times of asphalt suspensions and found that the correlation time is dependent on concentration as well as temperature. With increasing concentration, the molecular diffusion slows down as a consequence of increased friction among the solute molecules. VanderHart et al. [13] studied the structural inhomogeneity and physical aging of asphalts by solid-state NMR. Using ^{1}H line shapes and NMR spin-echo decay data, they observed no detectable changes in the molecular mobility during aging. Also, the presence of aggregate did not change the average molecular mobility of the asphalt cement at ambient temperature. These authors, using ^1H spin diffusion data as a probe for molecular size, showed that there was no evidence for micellar or gel structure in asphalts at ambient temperatures. In addition, VanderHart [14] has shown that the molecular motions in some asphalts are in the same frequency range as those of glassy polymers and suggested that these asphalts will, thus, have reduced impact strength and be prone to low-temperature cracking. Kohno et al. [8] measured the hydrogen spin-lattice relaxation time, T_1^H, and $T_{1\rho}^H$ as a function of temperature for 10 samples of pitch (a material similar to asphalt). They found that the $T_{1\rho}^H$ minimum occurred at approximately the same temperature as the softening point and concluded that the softening phenomenon of pitch can be detected at the molecular level by $T_{1\rho}^H$ measurement.

Except for the study by VanderHart et al. [13], few or no NMR data exist on the mobility of asphaltic molecules at or near the glass transition temperature for various types of asphalts. In addition, there do not exist in the literature any correlations of NMR relaxation parameters with the rheological properties of asphalts. However, recent studies by Netzel ([11] and reported here) have shown that the relaxation times of the amount of flexible aliphatic carbons in asphalts can be correlated quite well with their rheological properties.

II. AN OVERVIEW OF NMR CONCEPTS

Nuclear magnetic resonance chemical shifts, coupling constants, and integrated intensities of resonance lines are some of the most useful and powerful diagnostic parameters measured by the chemist to determine the structure, conformation, and composition of molecular systems. In addition, the nuclear spin relaxation time has become still another NMR parameter with both practical and theoretical significance. Relaxation time measurements have been used as a probe to investigate the structure and conformation of molecules, molecular motion, molecular interactions, kinetics of molecular systems, adsorption of molecular species, the quantitative analysis of natural and industrial products, and the mechanism of the relaxation phenomenon itself.

A. Chemical Shift and Quantitative Analysis

Carbon and hydrogen nuclei devoid of any surrounding electrons have specific resonance frequencies dictated by the static magnetic field, H_0, of the superconducting magnet used in the experiment. However, in reality the carbon and hydrogen nuclei are surrounded by electrons. These electrons shield the nuclei so that the resonance frequencies are shifted. The extent of the shift depends upon the electron density, which differs for the different carbon and hydrogen functionalities in a molecule. Because the electron density surrounding the different carbon and hydrogen types depends on the compound and the compound's molecular structure, the shift in resonance frequency is referred to as the chemical shift. The chemical shift values are dependent on the strength of the magnetic field. To remove the field dependence of the chemical shift for a given carbon or hydrogen type, chemical shifts are referenced to a standard compound and the shifts are reported in the dimensionless units of parts per million (ppm). Tetramethylsilane (TMS) is the most commonly used chemical shift reference compound for both carbon and hydrogen nuclei. The chemical shifts for different carbon types range from 0 to 215 ppm relative to TMS and for different hydrogen types range from 0 to 15 ppm. Nuclei that are less shielded than the reference compound are assigned positive values of the chemical shift. Aromatic carbons and hydrogens are less shielded by electrons than aliphatic carbons and hydrogens and, thus, have larger positive chemical shift values.

The integrated intensity for the resonance line of any given type of carbon or hydrogen nucleus can be related to the number of carbons or hydrogens. Thus, from the integration of the resonance intensities of several different carbon types in the same NMR spectrum, the relative amounts can be determined. The relative amounts of carbon types can be used to construct the molecular structure of a single compound or the average molecular structure of asphalt.

B. Spin Relaxation

When a sample containing ^1H and ^{13}C each with a nuclear spin, I, of $1/2$ is placed in a magnetic field, the degenerate nuclear spin energy states are split. The lower energy state for the case $I = +1/2$ contains a greater proportion of the nuclear spins than the upper energy state ($I = -1/2$) as prescribed by the Boltzmann distribution function. If a nuclear spin system is perturbed by a short but finite radiofrequency (rf) pulse, there is a redistribution of spins. The upper energy state may become populated to a greater extent than the lower energy state. After the perturbation, the spin system must revert to its original equilibrium condition. The recovery process by which the spin system returns to the equilibrium state is referred to as relaxation. The relaxation process occurs through the coupling interaction of the nuclear spin moments with resonance frequencies generated by fluctuating local magnetic and electric fields, which in turn are produced by thermal motions of the molecules. The extent of the coupling depends upon the magnitude and rate of fluctuation of the local fields. The time dependence of the fluctuating local magnetic fields is related to the molecular dynamics (motion) of a molecule.

The NMR signal intensity depends upon the difference in the spin population of the nuclear spin energy states. Changes in the population of the nuclear spin states can best be described by a vector representation of the net magnetization of the nuclear spins in a rotating-frame coordinate system (Figure 2). Specifically, the rotating-frame coordinate system rotates at the Larmor frequency of the nuclei

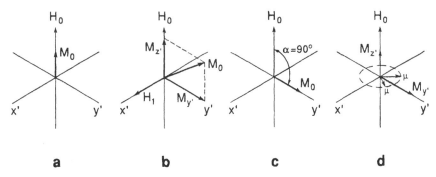

Figure 2 Effects of an intense rf pulse on the macroscopic magnetization vector, M_0, in a rotating frame coordinate system. (a) At equilibrium M_0 is colinear with H_0; (b) during a pulse M_0 rotates about the x' axis; (c) rotation of the magnetization vector with a 90° pulse; (d) after the pulse with loss of phase coherence of spins in an inhomogeneous field. (From Ref. 32.)

of interest (100 MHz for ^{1}H and 25 MHz for ^{13}C) in a static field of 2.3 tesla). Thus, the net spin magnetization vector $\mathbf{M}_0$ (vector sum of all the nuclear moments) at equilibrium is stationary and parallel to the direction of the external (static) magnetic field, $\mathbf{H}_0$ (Figure 2a). When a spin system is perturbed with an rf pulse (of magnetic field strength, $\mathbf{H}_1$, applied perpendicular to $\mathbf{H}_0$), the magnetization vector $\mathbf{M}_0$ is rotated toward the $x'y'$ plane, giving rise to magnetization components $\mathbf{M}'_y$ and $\mathbf{M}'_z$ in the y' and z' directions, respectively (Figure 2b). If the intensity of the pulse, $\mathbf{H}_1$, and its duration or pulse length, t_ω, are properly chosen, $\mathbf{M}_0$ can be rotated away from the z' direction through any angle, α. A 90° rotation ($\pi/2$ pulse angle) is shown in Figure 2c.

1. Spin-Lattice Relaxation Time (T_1)

The spin-lattice relaxation time constant, T_1, governs the return of the $\mathbf{z}'$ magnetization vector to its equilibrium Boltzmann distribution value, $\mathbf{M}_0$, after an rf perturbation. The spin-lattice relaxation time has also been defined as the average lifetime of the nuclear spin in the excited or higher energy state and the time necessary for the spins to come to equilibrium with the lattice. The term "lattice" represents the translational and rotational degrees of freedom of the molecular system. Vibrational degrees of freedom are too fast on an NMR time scale to provide a mechanism for relaxation.

2. Spin-Spin Relaxation Time (T_2)

The spin-spin relaxation time, T_2, governs the return of the x' and y' spin magnetization vectors to their equilibrium value of zero after an rf pulse perturbation. After a strong perturbation of the nuclear spin system, the nuclear spins of the nuclei are in phase (coherent) with each other and have a net spin magnetization component in the y' direction within the $x'y'$ plane (Figure 2c). After termination of the pulse, spin-spin interactions (among others) cause the nuclear spins to become out of phase with one another. Thus, some spins rotate faster and some slower than the rotating reference frame, resulting in a loss of phase coherence or a "fanning out" of the individual nuclear spins (Figure 2d). After a certain amount of time, the nuclear spins are again randomly distributed about the z' axes of the external magnetic field, resulting in a zero y' spin magnetization component. The time constant for this loss of phase coherence is the spin-spin relaxation time, T_2. Both spin-lattice and spin-spin relaxation processes are affected by high-frequency fluctuations (near the Lamor frequency) as a result of fluctuating local magnetic fields due to molecular motion, whereas low-frequency fluctuations (ms to s time scales) resulting from fluctuating local magnetic fields due to slow molecular motion affect only spin-spin relaxation processes.

C. Cross-Polarization ^{13}C NMR with Magic Angle Spinning (CP/MAS)

For solid materials, the ^{13}C NMR instrumentation and techniques used for liquids have been modified because problems associated with spectral resolution and sensitivity for solids are considerably different than for liquids. The more prominent problems are (a) ^{1}H-^{13}C dipole-dipole interactions, which cause excessive spectral line broadening in solids; (b) chemical shift anisotropy, which also causes spectral line broadening and unsymmetrical spectral line shapes because of the many different orientations the molecules in an amorphous state can assume in a magnetic field; and (c) long spin-lattice relaxation times, T_1, for ^{13}C in solids. An NMR technique known as cross-polarization (CP) with magic-angle spinning (MAS) overcomes many of these problems. This technique involves (a) high-power proton decoupling, which reduces the spectral line broadening due to ^{1}H-^{13}C dipolar coupling; (b) rapid rotation of the sample at an angle of 54.7°, relative to the external magnetic field, which reduces the spectral line broadening due to the chemical shift anisotropy; and (c) ^{1}H-^{13}C cross-polarization to overcome the problems of long carbon spin-lattice relaxation times, T_1^C, and consequent limited sensitivity.

The cross-polarization NMR experiment relies on the presence of abundant nuclear spins (^{1}H) in order to observe the NMR signal for a dilute nuclear spin (^{13}C, natural abundance = 1.1%). The experiment consists of four basic timed sequences of rf pulses and is illustrated in Figure 3. The four-part procedure consists of (1) polarization of the ^{1}H spin system, (2) spin locking in the rotating frame, (3) establishing ^{13}C-^{1}H contact, and (4) observation of the ^{13}C free induction decay. The left portions of Figure 3 represent the experimental timing

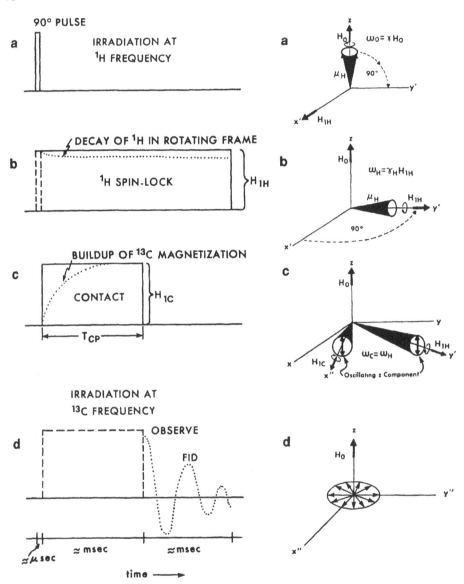

Figure 3 Timing sequence for cross-polarization experiment: (a) polarization of ¹H, (b) spin-locking of ¹H in rotating frame, (c) ¹³C-¹H contact under Hartman-Hahn conditions, and (d) observation of ¹³C free induction decay signal. (From Ref. 32.)

sequence in the CP experiment, and the right portions show what happens to the nuclei in the sample under the action of the rf pulses.

1. Polarization of ^{1}H

The sample is "prepared" by applying a 90° rf pulse, $\mathbf{H}_{1H}$, at the ^{1}H resonance frequency for a static magnetic field, $\mathbf{H}_0$. The action of this pulse rotates the spin vector away from its equilibrium position along $\mathbf{H}_0$ (in the z direction) and into the $x'y'$ plane, along y' (Figure 3a). It is assumed that the coordinate systems from which the ^{1}H and ^{13}C spins are viewed are rotating about $\mathbf{H}_0$ at their respective Larmor frequencies.

2. Spin-Locking

Immediately after the 90° pulse (<1 µs), a 90° phase shift is applied to the field, $\mathbf{H}_{1H}$, to redirect it along the y' axis, colinear with the ^{1}H spin magnetization, μ_H (Figure 3b). At resonance, and in the rotating frame, the only field that the ^{1}H spins now see is the $\mathbf{H}_{1H}$ field. This field causes the ^{1}H spin magnetic moments to process about $\mathbf{H}_{1H}$ at the corresponding frequency of $\omega_H = \gamma_H \mathbf{H}_{1H}$ ($\gamma_H =$ magnetogyric ratio) and the net ^{1}H spin magnetization vector to be spin locked along the $\mathbf{H}_{1H}$ field. It is called a "spin lock" because any spin that starts to dephase under the influence of a local field is driven back to the bundle by the precession about the rf vector. An important fact is that the precession of individual ^{1}H magnetic moments about the $\mathbf{H}_{1H}$ field causes oscillating components of the ^{1}H magnetic moments along the z axis. The so-called spin-lock condition (Figure 3b) cannot be maintained indefinitely because of the strong $\mathbf{H}_0$ field. As a result, the projection of ^{1}H spin magnetization along y' eventually decays to zero, even in the presence of the $\mathbf{H}_{1H}$ field. The time constant for this to occur is called the hydrogen spin-lattice relaxation time in the rotating frame and is generally denoted by $T_{1\rho}{}^H$. The dotted line in Figure 3b represents the rate of decay of the ^{1}H spins with this relaxation time constant.

3. ^{13}C-^{1}H Contact

Immediately following the 90° ^{1}H pulse and during the ^{1}H spin-lock condition, a radio frequency field ($\mathbf{H}_{1C}$) at the ^{13}C resonance frequency is switched on and applied along the direction, x'' axis, and maintained for a time, T_{CP}, the contact time (Figure 3c). Although there may not be substantial ^{13}C spin magnetization along the $\mathbf{H}_0$ direction at the start of the pulse, the ^{13}C nuclei in the sample will still precess about $\mathbf{H}_{1C}$ at the frequency $\omega_C = \gamma_C \mathbf{H}_{1C}$. Note also that this precession gives rise to an oscillating component of ^{13}C spin magnetization in the z direction. At this stage there are two oscillating components along the z axis, one at frequency $\omega_H = \gamma_H \mathbf{H}_{1H}$, the other at frequency $\omega_C = \gamma_C \mathbf{H}_{1C}$. If the rf power levels for hydrogen and carbon nuclei are adjusted so that $\gamma_H \mathbf{H}_{1H} = \gamma_C \mathbf{H}_{1C}$, then the frequencies of the oscillating components will be the same, $\omega_H = \omega_C$. This condition, known as the Hartmann-Hahn condition, is the key feature of the cross-polarization experiment. By making $\omega_H = \omega_C$, an efficient means is established for the transfer of magnetization from the abundant ^1H spin system to the dilute ^{13}C spin system. This transfer is called cross-polarization. The Hartmann-Hahn condition is maintained for a time, T_{CP}, the mixing or contact time, so that cross-polarization can occur, causing a significant buildup in ^{13}C spin magnetization (dotted line, Figure 3c) available for detection as a ^{13}C free induction decay (FID) signal (dotted line, Figure 3d). The time constant associated with the rate of buildup in ^{13}C spin magnetization, that is, the polarization of the carbon spins by the hydrogen spins, is the cross-polarization relaxation time, T_{CH}. For a solid sample the ^1H-^{13}C cross-polarization process is far more efficient for signal averaging than ordinary ^{13}C spin-lattice relaxation processes, so that a CP ^{13}C NMR experiment can be repeated at intervals shorter than the usual 3–5 T_1 values based upon 3–5 T_1^H of hydrogen instead of 3–5 T_1^C of carbon. Hydrogen T_1^H is, in general, much shorter than carbon T_1^C.

4. Observation of ^{13}C

The fourth part of the CP experiment is to terminate the ^{13}C pulse and observe the resultant free induction decay while the ^1H field is maintained for decoupling (Figure 3d). The entire sequence is repeated many times until a suitable signal-to-noise (S/N) ratio for ^{13}C is achieved. The resultant free induction decay is then Fourier transformed to give the more common frequency domain NMR spectrum.

D. Relaxation Time and Molecular Motion

The time a molecule or segment of a molecule remains in any given position before it changes its state of motion due to rotation or molecular collision is the correlation time, τ_C. Depending upon the physical state of the molecule, the correlation time describes the translational, rotational, or segmental motions of the molecule. The correlation time depends on the size and symmetry of the molecule, the intermolecular interactions, the viscosity, and the temperature.

The relationship between the correlation time and the various NMR relaxation times is shown qualitatively in Figure 4. The correlation time is directly proportional to viscosity, η, and inversely proportional to temperature, T (Equation 1).

$$\tau_C \propto \eta/T \tag{1}$$

The correlation time for fast and slow motions is related to the measured NMR relaxation times as shown in Equations 2 and 3, respectively.

$$1/T_x{}^y = n_i(\tau_C) \tag{2}$$

$$1/T_x{}^y = n_i(1/\tau_C) \tag{3}$$

where $x = 1$, 1_ρ, or 2; $y = $ C (carbon) or H (hydrogen); and n_i is the proportionality constant. In the liquid region (to the left of the T_1 minimum), as the temperature increases and the viscosity decreases for a molecular system, τ_C becomes shorter (fast molecular motion) and the spin-lattice relaxation time increases (Equation 2). To the right of the T_1 minimum (the solid region), the spin-lattice relaxation time increases with decreasing temperature and with increasing τ_C; that is, τ_C is long and molecular motion is slow.

The spin-spin relaxation time, T_2, is the time constant for expressing the rate of dephasing of the nuclear spins, and thus the exchange of energy is among the spins and not the lattice [15]. The dependence of T_2 on molecular motion differs from that of T_1. As shown in Figure 4, T_2 decreases monotonically with increasing τ_C throughout the liquid range (a and b) and reaches a limiting value that is characteristic of a completely rigid solid lattice.

Because $T_{1\rho}$ is measured at the strength of the rotating magnetic field rather than the static magnetic field, the magnitude of this time constant is considerably less than T_1 and, thus, can be used to describe much slower motions than T_1 measurements. The behavior of $T_{1\rho}$ is similar to that of T_2 in regard to temperature

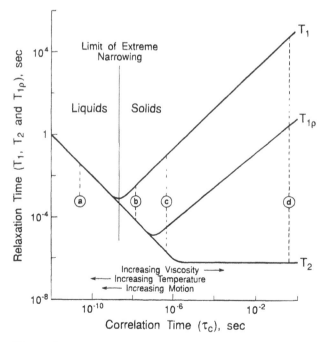

Figure 4 Schematic representation of the dependence of relaxation times on the correlation time for (a) nonviscous liquid, (b) viscous liquid, (c) nonrigid solid, and (d) rigid lattice.

and correlation time and, in fact, $T_{1\rho}$ was used to measure T_2 in liquids before the advent of commercial pulse NMR equipment.

III. NMR STUDY OF ASPHALTS

A. Spectral Assignments of Carbon Types

Before the motion of carbon types in asphalts can be studied, the carbon types need to be identified. The solid-state ^{13}C cross-polarization/magic angle spinning (CP/MAS) NMR spectra for the eight Strategic Highway Research Program (SHRP) core asphalts at 20°C are shown in Figure 5. The aromatic and aliphatic carbon chemical shift ranges are defined as 100–160 ppm and 0–60 ppm, respectively. A comparison of these spectra shows some differences in the line width, resolution, and concentration of the various carbon types in both the aromatic and aliphatic regions. From the integration values for the carbon groups (total

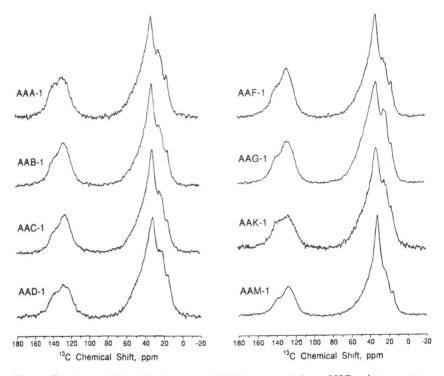

Figure 5   Carbon-13 CP/MAS spectra of SHRP core asphalts at 20°C and at a contact time of 500 μs.

carbon types in the aromatic and aliphatic regions) in these spectra, the relative percentage of aromatic carbons can be calculated (carbon aromaticity).

The CP/MAS spectra for all SHRP core asphalts at 20°C were obtained as a function of the carbon-hydrogen polarization contact time. Figure 6 shows the CP/MAS spectra of asphalt AAC-1 obtained using cross-polarization contact times of 100 μs, 300 μs, and 4 ms. As the contact time increases, the signal-to-noise ratio in the spectra decreases as a result of the decays of carbon and

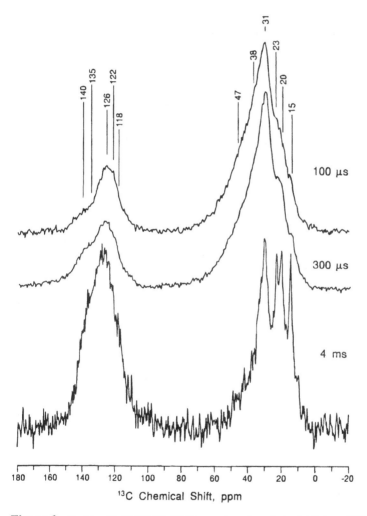

Figure 6 Carbon-13 CP/MAS NMR spectra of asphalt AAC-1 at 20°C for contact times of 100 μs, 300 μs, and 4 ms.

hydrogen spins to their ground state by relaxation. In addition, the resolution between carbon types in the aliphatic region (0–60 ppm) of the spectrum increases with increased contact time. The enhanced resolution is the result of signals for carbon types with short relaxation times (rigid) decaying faster than signals for carbon types with long relaxation times (mobile), which decay more slowly. Because of the greater resolution for spectra obtained at long contact times, these spectra in part were used to identify the carbon types of interest.

The aromatic region of the spectra shown in Figure 6 is very broad (100–160 ppm) with very little resolution because of the overlap of the chemical shifts for the various aromatic carbons and because of the line broadening that occurs when samples are in the solid state. Nevertheless, carbon type assignments can still be made with reference to liquid-state NMR spectra for the different aromatic carbon types. The carbons in the region of the carbon peak positions at 140 and 135 ppm are alkyl-substituted and bridgehead quaternary aromatic carbons and peak assignments at 126, 122, and 117 ppm are protonated (tertiary) aromatic carbons. The identification of the carbon types represented by the peaks in the aliphatic region (0–60 ppm, Figure 6) is also complicated by the overlap of the chemical shifts for the various carbons in the asphalt as well as the decrease in spectral resolution in the solid state due to line broadening effects.

The assignments of the aliphatic carbon types in the solid-state spectra of the asphalt were made from the high-resolution liquid-state spectrum of the asphalt. Figure 7 shows a comparison of the aliphatic region of the solid-state spectrum of asphalt AAC-1 with that of a high-resolution liquid-state spectrum of the asphalt [16]. In turn, the assignments of the carbon types in the liquid-state spectrum of an asphalt (Figure 8a) can be made by comparing the ^{13}C NMR spectrum with that of a mixture of farnesane and n-decane as shown in Figure 8b. Normal decane is a straight-chain alkane having eight methylenes and two terminal methyl groups. Farnesane is an isoprenoid with five methyl groups (CH_3), seven methylene carbons (CH_2), and three methine carbons (CH). Both compounds are representative of common compound types in petroleum. This mixture was used by Netzel et al. [17] to assign the carbon types in the saturate fraction of shale oil (Figure 8c). As shown in Figure 8, the saturate components in asphalt and shale oil are very similar, that is, a mixture of isoprenoids and n-alkanes. This fact is not surprising because the biogenetic origins of the two materials are the same. It is not to be construed that the saturate components of asphalts (or shale oil) contain only farnesane and n-decane, rather they contain a complex mixture of isoprenoids, other branched alkanes, and n-alkanes of different carbon numbers.

The aliphatic carbon assignment at 47 ppm (see Figure 6) is undefined but the region is characteristic of aliphatic quaternary and methine carbons and aliphatic carbons attached to heteroatoms such as oxygen and nitrogen. The carbon assignment at 38 ppm is assigned to the methylene carbons of types C_5,

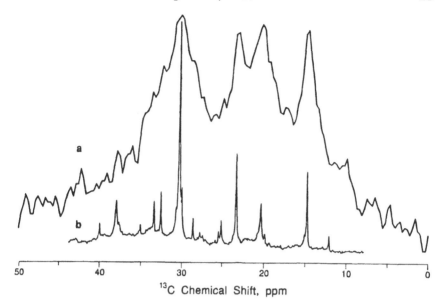

Figure 7 Carbon-13 NMR spectra of the aliphatic region for asphalt AAC-1: (a) solid-state CP/MAS and (b) liquid state. (From Ref. 16.)

C_7, and C_9 of farnesane (see Figure 8). The carbon resonance peak at 30 ppm is due to methylene carbons in long-chain hydrocarbons such as C_4 and C_5 of *n*-decane. The resonance at 23 ppm can be assigned to geminal methyl groups (C_1 of farnesane) and to a methylene carbon next to a terminal methyl group of a long-chain hydrocarbon (C_2 of *n*-decane). However, based on liquid-state NMR data, the *n*-alkane structures are more dominant than branched alkanes in an asphalt and, thus, the relaxation properties should reflect those of the methylene carbons. The carbon resonance signal at 20 ppm is that of branched methyl groups of an isoprenoid (for example, C_{13} and C_{14} of farnesane) and methyl carbons attached directly to an aromatic ring. The relative concentrations of the two types of methyl groups are not known and would be difficult to determine. Terminal methyl carbons of long-chain hydrocarbons (C_1 of *n*-decane) have high upfield signals and, thus, the resonance at 14.6 ppm in the ^{13}C NMR spectrum of asphalt AAC-1 is assigned to this carbon type.

B. Relaxation Times

The molecular motions associated with an average asphaltic molecule are methyl rotation, segmental motions of the carbon main chain, and phenyl twisting. Segmental motions correspond to changes in bond direction (conformational

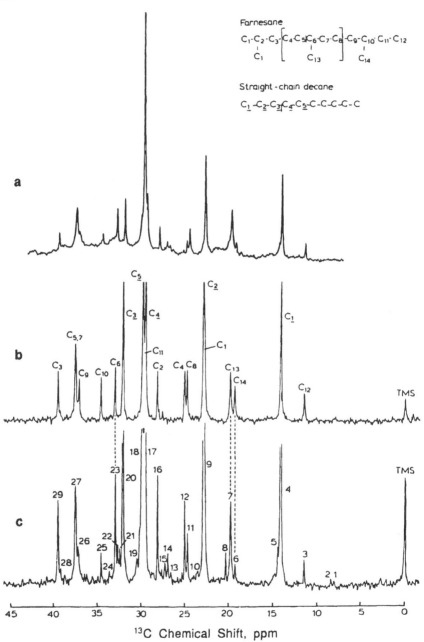

Figure 8 Carbon-13 NMR spectra of the aliphatic region for (a) an asphalt, (b) a mixture of farnesane and *n*-decane, and (c) a saturate fraction from a shale oil. (From Ref. 17.)

changes) for one or several backbone units. Figure 1 illustrates the different kinds of motions in a complex asphaltic molecule. Molecular dynamic processes in complex systems such as asphalts can be studied using several different pulsed NMR techniques to measure the ^{13}C and ^1H nuclear spin relaxation times. The molecular motions of interest in asphalts are low in frequency, on the order of 1–50 kHz.

1. Dipolar Relaxation Times (T_{CH}, $T_{1\rho}^H$, $T_{1\rho}^C$)

Cross-Polarization Relaxation Time (T_{CH}). The cross-polarization rate, $1/T_{CH}$, describes the rate of transfer of magnetization from the hydrogens to the carbon spins. The time constant, T_{CH}, is proportional to the strength of the dipolar C−H coupling. This C−H dipolar interaction is averaged by molecular motion. As the molecular motion increases, the value of T_{CH} increases. In the limit of no motion, T_{CH} is also dependent upon the number and proximity of hydrogens. That is, methyl carbon would be expected to have a much shorter cross-polarization relaxation time than a methine carbon. In terms of carbon types, T_{CH} values decrease in the following order: nonprotonated carbon > methyl carbon (rotating) > protonated aromatic (aliphatic) methine carbon > methylene carbon > methyl carbon (static). The cross-polarization process is further complicated in semicrystalline materials because the various phases (amorphous, crystalline) may be characterized by significant differences in the degree of molecular mobility. Thus, to interpret the cross-polarization dynamic data, the carbon type and the phase must be considered. Values of T_{CH} can provide information about molecular motions having frequencies on the order of 10^0 to 10^3 Hz [18].

A plot of the integrated areas for the aromatic and aliphatic carbon groups of asphalt AAA-1 at 0°C as a function of the carbon-hydrogen cross-polarization time (contact time) is shown in Figure 9. The data were fitted with a biexponential curve of the form given by Equation 4.

$$M_t = M_0 b^{-1} e^{-t/T_{1\rho}^H}(1 - e^{-bt/T_{CH}}) \qquad (4)$$

where $\quad M_t$ = carbon spin magnetization (integrated intensity) at t
$\qquad t$ = contact time
$\qquad b = 1 - T_{CH}/T_{1\rho}^H$
$\qquad M_0$ = carbon spin magnetization at $t = \infty$

The biexponential curve represents (a) the rate of buildup ($1/T_{CH}$) of the carbon spins through polarization by the hydrogen spins, which depends primarily upon the number of hydrogens attached to the carbon, and (b) the decrease in carbon signal as a result of the rate of decay ($1/T_{1\rho}^H$) of the number of hydrogen spins available for polarization of the carbons. Both rates depend on the type and extent of molecular motion and, thus, are functions of the temperature. The T_{CH} and $T_{1\rho}^H$ can also be determined for single carbon types using peak height

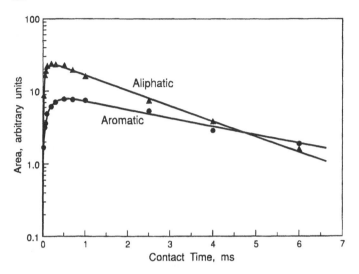

Figure 9 Plot of the integrated areas of the aromatic and aliphatic carbon groups in asphalt AAA-1 as a function of contact time at 0°C.

intensity instead of area measurements and Equation 4. The relaxation times T_{CH} and $T_{1\rho}{}^{H}$ for the 11 different carbon types for the SHRP core asphalts at 20°C are given in Tables 1 to 4.

Aromatic Carbons. The T_{CH} values for the aromatic carbons at five chemical shift positions for the different asphalts fall into two distinct classes as shown in Table 1. The cross-polarization relaxation times for the aromatic quaternary

Table 1 Cross-Polarization Relaxation Time, T_{CH} (μs), for the Aromatic Carbons in the Eight SHRP Core Asphalts at 20°C

Asphalt	Carbon-13 chemical shift (ppm)				
	139.9	134.9	126.1	122.0	117.8
AAA-1	486	366	69	52	35
AAB-1	299	210	59	50	48
AAC-1	364	203	58	—	—
AAD-1	213	154	50	54	43
AAF-1	319	232	67	54	38
AAG-1	443	346	89	79	91
AAK-1	138	93	35	99	5
AAM-1	229	202	60	49	50

Table 2   Cross-Polarization Relaxation Time, T_{CH} (μs), for the Aliphatic Carbons in the Eight SHRP Core Asphalts at 20°C

Asphalt	Carbon-13 chemical shift (ppm)					
	46.8	38.0	30.5	23.4	20.5	14.7
AAA-1	50	56	79	80	90	118
AAB-1	52	59	86	94	106	147
AAC-1	46	61	86	86	98	119
AAD-1	50	56	72	82	88	118
AAF-1	55	63	90	84	96	135
AAG-1	57	64	78	99	127	176
AAK-1	49	51	71	68	77	89
AAM-1	52	64	98	82	86	124

Table 3   Aromatic Hydrogen Spin-Lattice Relaxation Time in the Rotating Frame, $T_{1\rho}{}^{H}$ (ms), for the Eight SHRP Core Asphalts at 20°C

Asphalt	Carbon-13 chemical shift (ppm)				
	139.9	134.9	126.1	122.0	117.8
AAA-1	2.7	3.7	5.1	4.7	6.5
AAB-1	4.1	4.7	4.8	4.5	5.0
AAC-1	4.5	5.6	5.3	—	—
AAD-1	3.7	4.6	4.7	4.2	4.3
AAF-1	4.6	5.0	5.5	6.0	7.1
AAG-1	4.4	4.9	6.9	7.6	7.3
AAK-1	2.3	2.7	2.1	2.3	6.1
AAM-1	5.2	6.5	6.5	7.9	7.5

carbons (nonprotonated) at 139.9 and 134.9 ppm are longer than the T_{CH} of the tertiary (protonated) aromatic carbons (126.1, 122.0, and 117.8 ppm). The longer T_{CH} values observed for the quaternary aromatic carbons are due to ineffective coupling (cross-polarization) between the quaternary carbon and the nearest hydrogens, which are at least two bonds removed. Very effective coupling, resulting in very short T_{CH}, is observed between the tertiary aromatic carbons and the directly attached hydrogens.

It should be noted that for asphalt AAK-1, the aromatic and aliphatic T_{CH} values are much smaller than for the other asphalts except for the peak at 122 ppm. Most relaxation times are sensitive to paramagnetic impurities such as dissolved oxygen, unpaired electrons, and metals in organometallic compounds.

Table 4 Aliphatic Hydrogen Spin-Lattice Relaxation Time in the Rotating Frame, $T_{1\rho}{}^H$ (ms), for the Eight SHRP Core Asphalts at 20°C

Asphalt	Carbon-13 chemical shift (ppm)					
	46.8	38.0	30.5	23.4	20.5	14.7
AAA-1	1.3	1.5	1.8	2.3	2.7	4.3
AAB-1	1.3	1.6	2.2	2.6	3.0	4.5
AAC-1	1.7	1.7	2.2	2.9	3.3	5.2
AAD-1	1.3	1.4	1.9	2.2	2.7	3.2
AAF-1	1.8	1.9	2.4	3.3	3.6	6.9
AAG-1	2.3	2.3	2.6	3.4	3.7	5.2
AAK-1	0.8	0.8	1.0	1.2	1.3	1.6
AAM-1	1.5	1.7	2.4	3.1	3.4	7.9

However, T_{CH} is not unless $T_1{}^H$ is made exceedingly short by paramagnetic impurities. The smaller T_{CH} values for AAK-1 appear to be related to its metal content. The vanadium content alone is 1480 ppm. A plot of the cross-polarization relaxation rate ($1/T_{CH}$) for the tertiary aromatic carbon at 126.1 ppm for the eight asphalts versus the logarithm of the total metal concentration (V + Fe + Ni) in each asphalt is shown in Figure 10. Similar plots are obtained for the aromatic quaternary carbons at 139.9 and 134.9 ppm. Thus, the paramagnetic materials

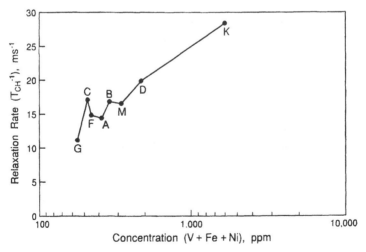

Figure 10 Carbon-hydrogen cross-polarization rate for aromatic tertiary carbon at 126 ppm as a function of the total metal content in the asphalts.

in the asphalts are affecting the relaxation rates of the hydrogens ($1/T_1^H$), which, in turn, affect the cross-polarization relaxation rates. It is of interest that the highest metal concentration in asphalt is only a factor of 10 less than the normal concentration of paramagnetic relaxation reagents used to shorten carbon relaxation times for quantitative analysis of carbon spins. The effect of metal content on the relaxation times is most noticeable for asphalts AAD-1 and AAK-1. The correlation of $1/T_{CH}$ versus the logarithm of the total metal concentration suggests that the differences in measured T_{CH} for the different asphalts are not totally related to differences in molecular mobility. Any real effect of mobility on the cross-polarization relaxation time must be extracted from NMR variable-temperature studies of the asphalts.

Aliphatic Carbons. Table 2 lists the T_{CH} values for the six selected aliphatic carbon types in each asphalt. It can be seen that the T_{CH} values for the methine and methylene aliphatic carbons at chemical shifts of 46.8 and 38.0 ppm are of the same magnitude as those of the tertiary aromatic carbons. The methylene carbons at 30.5 and 23.4 ppm (mixture of methylene and methyl carbons) and the methyl carbons at 20.5 and 14.7 ppm have progressively larger T_{CH} values. Under static conditions in which there is no molecular motion, one would expect shorter or at least equal T_{CH} values as the number of hydrogens attached to a carbon atom increases. The observed increase in the T_{CH} for the methylene carbons with decrease in chemical shift must be due to increased segmental (rotational) motion of the backbone carbons of the paraffinic and branched hydrocarbons. The farther the carbon type is from the aromatic cluster, the more molecular motion it has. However, for very fast molecular motions, the cross-polarization mechanism becomes less effective and the rate of polarization is slow. The much longer T_{CH} value observed for the terminal methyl carbon (14.7 ppm) for the different asphalts is due to the rapid rotation of the methyl group attenuating the dipolar coupling interaction between the methyl hydrogens and carbon. In contrast to the aliphatic terminal methyl carbon, the branched aliphatic and/or aromatic substituent methyl carbons (20.5 ppm) have a T_{CH} value less than that of the terminal methyl carbon. The shorter T_{CH} value for these methyl carbons suggests that methyl rotation is more hindered than the terminal methyl of a long-chain hydrocarbon. It should be noted that the high metal concentration in asphalt AAK-1 also affected the aliphatic carbons by reducing the T_{CH} values relative to most of the other asphalts. The methylene (30.5 ppm) and methine (38 and 46.8 ppm) carbons in asphalt AAA-1 and AAD-1 also appear to be affected by the metal concentration.

Hydrogen Spin-Lattice Relaxation Time in the Rotating Frame, $T_{1\rho}^H$. Because $T_{1\rho}^H$ is associated with a dipolar relaxation process, it will be affected by molecular motion in the range 10–100 kHz [19]. An increase in the molecular motion in solids will decrease the value of $T_{1\rho}^H$. If the motion in a solid is moderately fast ($T_{1\rho}{}^H$ is short), the cross-polarization transfer process can be reduced or entirely eliminated and, thus, no spectrum will be observed.

It has been observed that $T_{1\rho}{}^H$ values for the different hydrogen types in many solid materials are of the same magnitude. This effect is a result of an efficient rapid spin diffusion mechanism for strongly dipolar coupled hydrogens that tends to equalize the relaxation rate for spins in any part of the hydrogen spin system, yielding a single relaxation time. If spin diffusion is not present, differences in $T_{1\rho}{}^H$ values may be observed. These differences can then be related to differences in the degree of molecular motion.

Aromatic Hydrogens.   The aromatic $T_{1\rho}{}^H$ values for the core asphalts are given in Table 3. It can be seen that there is very little variation in the $T_{1\rho}{}^H$ for the aromatic hydrogens for any given asphalt. The similarity in $T_{1\rho}{}^H$ for a given asphalt is the result of spin diffusion such that all hydrogens in the aromatic structure effectively relax at the same rate. The only real difference among the asphalts is again seen with asphalt AAK-1, with the high paramagnetic metal concentration effectively reducing the $T_{1\rho}{}^H$ values.

Aliphatic Hydrogens.   The aliphatic $T_{1\rho}{}^H$ values for the core asphalts are shown in Table 4. For all asphalts the $T_{1\rho}{}^H$ values increase with decrease in ^{13}C chemical shift. The longer $T_{1\rho}{}^H$ values for the methyl hydrogens (at carbon-13 chemical shifts of 20.5 and 14.7 ppm) again show the effects of rapid methyl rotation relative to the segmental motion of the methylene group. Asphalt AAK-1 also has the smallest $T_{1\rho}{}^H$ as a result of the high metal concentration. The methine and methylene hydrogen $T_{1\rho}{}^H$ values for asphalt AAG-1 also differ from values for the other asphalts. The higher values may suggest a slightly more rigid methine and methylene structure for AAG-1 compared with the other asphalts. That is, the magnitude of the segmental motions of the backbone carbons of normal and branched alkanes is less. A comparison of the aromatic and aliphatic $T_{1\rho}{}^H$ values (Tables 3 and 4) shows that, in general, the aromatic $T_{1\rho}{}^H$ values are longer than the aliphatic $T_{1\rho}{}^H$ values. This suggests that the aromatic clusters in all asphalts are less mobile than the long-chain aliphatic components.

Carbon and Hydrogen Spin-Lattice Relaxation Times ($T_1{}^C$, $T_1{}^H$).   The spin-lattice relaxation time (T_1) is the time constant for the carbon or hydrogen nuclear spins to return to equilibrium after being perturbed by an rf pulse. The mechanism for this return for carbon spins in a solid is the coupling of carbon spins with the fluctuating magnetic fields generated by the segmental and rotational motions of the molecules in the solid state. The most efficient coupling occurs when T_1 is a minimum (Figure 4); that is, the frequency of the motions is about the same as the carbon resonance frequency in a static magnetic field. Thus, T_1 is generally sensitive to motions in the 1–1000 MHz frequency range. That is, carbons having very long T_1 values are immobilized (far right of the minimum in Figure 4), whereas the carbons having short values of T_1 are engaged in rapid internal rotations.

The carbon and hydrogen spin-lattice relaxation times T_1^C and T_1^H were determined at 20°C for the 11 carbon positions in asphalt AAA-1 using a CP inverse-recovery NMR technique. The values are given in Table 5. The carbon and hydrogen relaxation times were measured at frequencies of 50 and 200 MHz, respectively. The T_1^C has a single relaxation time for the aromatic carbons and two relaxation times (T_1^{CS} and T_1^{CL}) for the aliphatic carbons. It should be noted that the methylene carbons at 23.4 ppm and the methyl carbons at 20.5 and 14.7 ppm have significantly smaller relaxation times for T_1^{CL} (the long relaxation component) relative to the other aliphatic carbons. The smaller values reflect the rapid rotation of the methyl groups. The T_1^H values shown in Table 5 for asphalt AAA-1 are, in general, essentially the same for all hydrogen types.

Carbon Spin-Lattice Relaxation Time in the Rotating Frame ($T_{1\rho}^C$). As with $T_{1\rho}^H$, the $T_{1\rho}^C$ values of asphalts are also sensitive to (and indicative of) internal rotational and segmental motions within molecules in the frequency range of the applied rf field. However, $T_{1\rho}^C$ is less susceptible to homonuclear spin diffusion than $T_{1\rho}^H$. This is because the low natural abundance of carbon-13 reduces the likelihood of two ^{13}C atoms being bonded together within a molecule and hence reduces the spin diffusion or transfer rate. A decrease in molecular motion results in an increase in $T_{1\rho}^C$ value (long relaxation time constant).

Aromatic and Aliphatic Carbons. Semilogarithmic plots of the $T_{1\rho}^C$ data for the asphalts indicate a dispersion or distribution of relaxation times even for chemically unique carbons such as the aromatic tertiary carbons and the aliphatic methyl and methylene carbon types. A biexponential function was used to fit

Table 5 Carbon and Hydrogen Spin-Lattice Relaxation Times for the Aromatic and Aliphatic Carbons in Asphalt AAA-1 at 20°C

^{13}C chemical shift (ppm)	T_1^{CS} (s)	T_1^{CL} (s)	T_1^H (s)
139.9		4.64	0.39
134.9		5.84	0.43
126.1		6.99	0.52
122.0		8.01	0.54
117.8		—	0.59
46.8	0.22 (37)[a]	7.24 (63)	0.44
38.0	0.22 (69)	4.67 (31)	0.49
30.5	0.27 (86)	3.75 (14)	0.51
23.4	0.29 (67)	1.72 (33)	0.52
20.5	0.17 (26)	12.6 (74)	0.55
14.7	0.15 (16)	1.54 (84)	0.54

[a]Data in parentheses are the relative amounts of the components for a given relaxation time.

the data for each carbon type, resulting in relaxation time constants for short decay time (fast decay, more mobile), $T_{1\rho}^{CS}$, and long decay time (slow decay, less mobile), $T_{1\rho}^{CL}$, components. The aromatic and aliphatic $T_{1\rho}^{C}$ values for the eight core asphalts are given in Tables 6 and 7. The data in parentheses are the relative percentages of the carbon types having short and long relaxation times. As shown in Table 6, the $T_{1\rho}^{CS}$ values for the aromatic quaternary and tertiary carbons vary from about 0.5 to 3.0 ms and the $T_{1\rho}^{CL}$ values range from 5.0 to 25 ms. There do not appear to be any trends in the data in regard to carbon type. The $T_{1\rho}^{CS}$ values for the aliphatic carbons (Table 7) range from about 0.5 to 2 ms with the methyl carbons (at 20.5 and 14.7 ppm) having slightly higher values than the methine and methylene carbons (at 46.8, 38.0, and 30.5 ppm). With some exceptions, the values of $T_{1\rho}^{CL}$ for methine and methylene carbons range from about 3.0 to 6.0 ms and the $T_{1\rho}^{CL}$ values for the methyl carbons range from about 4.0 to 10 ms. The slightly higher $T_{1\rho}^{CL}$ values for the methyl carbons are the result of ineffective dipolar coupling (increasing the $T_{1\rho}^{CL}$ values) as a result of rapid rotation of the methyl groups.

The observed nonlinearity (two components) of the $T_{1\rho}^{C}$ data can be interpreted as due to (a) the presence of a multiplicity of relaxation times, which might be expected for mixed-phase systems in which the various component phases have significantly different motional properties resulting in different relaxation times but have sufficiently similar chemical properties that no net chemical shift differences are observed [20], or (b) the molecular dynamic heterogeneity of the glassy (amorphous) state. Asphalts are nominally homogeneous single-phase systems that contain no internal static domains of varying density [10,13]. Thus, the nonlinearity in the $T_{1\rho}^{C}$ plots is not due to static heterogeneity but arises from a time average over the complexity of the side groups and main-chain molecular motions (dynamic heterogeneity) [7]. That is, local motion of a main-chain carbon will depend not only upon the configuration and conformations of the side groups nearest to it in the chain but also, via connected pairwise interactions, upon more distant side groups in the chain. A given carbon in a main chain may be in any of a multiplicity of conformational environments. For solid asphalts the molecular conformations are not rapidly interconverting, so there is no averaging of these environments. Even though local segmental or twisting (torsional) motions are cooperatively coupled, these motions can be different in different parts of the sample, giving rise to $T_{1\rho}^{C}$ dispersion (multiplicity of the relaxation times) [7]. For asphalts, this dispersion could include the effects of inter- as well as intrachain steric interactions.

2. Ratio of Carbon-13 Relaxation Times and the Softening Point of Asphalts

It has been shown from mechanical spectroscopy data that low-frequency main-chain motions correlate with mechanical properties [7,21–23], such as impact

Table 6 Carbon Spin-Lattice Relaxation Time in the Rotating Frame, $T_{1\rho}^C$ (ms), for the Aromatic Carbons in the Eight SHRP Core Asphalts at 20°C

Asphalt	139.9		134.9		126.1		122.0		117.8	
	$T_{1\rho}^{CS}$	$T_{1\rho}^{CL}$	$T_{1\rho}^{CS}$	$T_{1\rho}^{CL}$	$T_{1\rho}^{CS}$	$T_{1\rho}^{CL}$	$T_{1\rho}^{CS}$	$T_{1\rho}^{CL}$	$T_{1\rho}^{CS}$	$T_{1\rho}^{CL}$
AAA-1	2.3 (14)[a]	18.4 (86)	0.49 (3)	13.1 (97)	0.86 (30)	15.8 (70)	3.2 (55)	23.3 (45)	0.45 (21)	10.4 (79)
AAB-1	14.0 (6)	14.7 (94)	1.4 (10)	16.2 (90)	2.2 (50)	24.7 (50)	1.9 (49)	24.5 (51)	0.44 (29)	11.0 (71)
AAC-1	4.3 (71)	119 (29)	1.8 (41)	25.0 (59)	2.6 (50)	23.3 (50)	—	—	—	—
AAD-1	0.42 (31)	14.6 (69)	2.9 (57)	37.6 (43)	0.60 (37)	14.2 (63)	1.9 (64)	58.8 (36)	0.49 (23)	13.6 (77)
AAF-1	0.92 (28)	20.1 (72)	0.89 (21)	15.1 (79)	1.3 (31)	12.7 (69)	1.7 (34)	13.9 (66)	0.59 (13)	11.4 (87)
AAG-1	2.6 (17)	20.0 (83)	0.80 (14)	17.0 (86)	1.2 (34)	17.0 (66)	3.1 (59)	33.9 (41)	3.8 (61)	79.3 (39)
AAK-1	0.48 (34)	6.6 (66)	1.5 (51)	10.5 (49)	1.4 (56)	14.3 (44)	0.14 (17)	5.3 (83)	7.3 (31)	7.7 (69)
AAM-1	1.1 (15)	18.1 (85)	0.58 (17)	16.4 (83)	1.4 (28)	14.8 (72)	2.3 (38)	20.4 (62)	0.5 (15)	11.8 (85)

Carbon-13 chemical shift (ppm)

[a] Data in parentheses are the relative amounts of the components for a given relaxation time.

Table 7 Carbon Spin-Lattice Relaxation Time in the Rotating Frame, $T_{1\rho}{}^{C}$ (ms), for the Aliphatic Carbons in the Eight SHRP Core Asphalts at 20°C

	Carbon-13 chemical shift (ppm)					
	46.8		38.0		30.5	
Asphalt	$T_{1\rho}{}^{CS}$	$T_{1\rho}{}^{CL}$	$T_{1\rho}{}^{CS}$	$T_{1\rho}{}^{CL}$	$T_{1\rho}{}^{CS}$	$T_{1\rho}{}^{CL}$
AAA-1	0.79 (83)[a]	13.4 (17)	0.72 (79)	6.8 (21)	0.85 (76)	5.3 (24)
AAB-1	0.93 (92)	669 (8)	0.67 (67)	3.0 (33)	1.0 (75)	3.9 (25)
AAC-1	0.94 (85)	5.8 (15)	0.65 (69)	3.3 (31)	1.1 (83)	5.5 (17)
AAD-1	0.67 (87)	4.0 (23)	0.59 (64)	2.7 (36)	0.75 (69)	3.4 (31)
AAF-1	0.76 (80)	8.5 (20)	0.89 (84)	5.8 (16)	0.64 (48)	2.6 (52)
AAG-1	0.76 (71)	6.1 (29)	0.77 (76)	4.6 (24)	0.97 (77)	4.5 (23)
AAK-1	0.53 (88)	433 (12)	0.29 (44)	1.5 (56)	0.57 (55)	1.8 (45)
AAM-1	0.87 (87)	16.2 (13)	1.0 (93)	15.8 (7)	0.85 (65)	3.6 (35)

	Carbon-13 chemical shift (ppm)					
	23.4		20.5		14.7	
Asphalt	$T_{1\rho}{}^{CS}$	$T_{1\rho}{}^{CL}$	$T_{1\rho}{}^{CS}$	$T_{1\rho}{}^{CL}$	$T_{1\rho}{}^{CS}$	$T_{1\rho}{}^{CL}$
AAA-1	1.2 (71)	10.4 (29)	1.9 (82)	22.9 (18)	1.0 (41)	9.3 (59)
AAB-1	0.93 (60)	7.0 (40)	0.87 (44)	6.0 (56)	1.1 (53)	12.0 (47)
AAC-1	1.3 (72)	8.5 (28)	1.2 (60)	9.9 (40)	2.0 (62)	18.2 (38)
AAD-1	0.79 (57)	5.5 (43)	0.90 (54)	7.0 (46)	0.91 (33)	6.3 (67)
AAF-1	1.2 (61)	7.9 (39)	1.1 (54)	8.8 (46)	1.0 (32)	8.8 (68)
AAG-1	1.2 (60)	9.6 (40)	1.5 (53)	10.8 (47)	1.3 (35)	10.6 (65)
AAK-1	0.67 (62)	4.2 (38)	0.71 (56)	4.1 (44)	1.4 (67)	9.7 (33)
AAM-1	0.88 (58)	6.7 (42)	0.99 (59)	8.8 (41)	1.7 (56)	18.2 (44)

[a]Data in parentheses are the relative amount of the components for a given relaxation time.

strength of glassy amorphous polymers. Schaefer et al. [7] studied several glassy polymers, focusing on the measurement of the ^{13}C relaxation parameters T_{CH}, the cross-polarization relaxation time, and $T_{1\rho}{}^{C}$, the carbon spin-lattice relaxation time in the rotating frame. These researchers established a correlation between the toughness or impact strength of glassy polymers and the ^{13}C relaxation times. In these studies, the impact strength correlates specifically with the ratios of ^{13}C relaxation rate parameters, $1/T_{1\rho}{}^{C}$ and $1/T_{CH}$. The ratio $(T_{1\rho}{}^{C})^{-1}/(T_{CH})^{-1}$ approximates the number of polymer chain segments that can absorb and dissipate the energy of an impact $(1/T_{1\rho}{}^{C})$ relative to the number of rigid polymer segments $(1/T_{CH})$ that are unable to respond to the impact. A short T_{CH} (reflecting considerable static character, that is, very low frequency motions) yields a small ratio and implies a rigid (hence brittle) polymer system. Such a system has a low impact strength. A short $T_{1\rho}{}^C$ in the denominator (reflecting substantial motion in the kHz range) yields a large ratio and implies that a sufficient number of suitable chain segments are present to dissipate the energy of impact as heat. Thus, a large ratio implies a large impact strength.

Even though the ratio of relaxation times correlates well with impact strength of polymers, serious limitations in applying the correlation to other materials exist [7]. Nevertheless, the ratios of $1/T_{1\rho}{}^{CS}$ to $1/T_{CH}$ were calculated for the SHRP asphalts. It is assumed that the softening point of the asphalt is, to a first approximation, an indicator of the impact strength of the asphalt. Figure 11 shows a plot of the ratio $T_{CH}/T_{1\rho}{}^{CS}$ versus the softening point temperature of the SHRP asphalts [24]. The ratios were calculated for the main-chain methylene carbons at 30.5 ppm and using the short (fast) relaxation component, $T_{1\rho}{}^{CS}$. As shown in the plot, a reasonably good correlation exists for asphalts AAA-1, AAC-1, AAD-1, AAF-1, and AAK-1. However, asphalts AAB-1, AAG-1, and AAM-1 do not fit the correlation curve. The saturate content of all asphalts equals 10 ± 2 mass percent except for AAK-1 (5.1%) and AAM-1 (1.9%) [24].

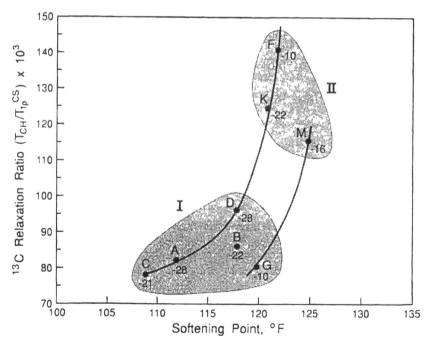

Figure 11   The carbon-13 $T_{CH}/T_{1\rho}{}^{CS}$ ratio at 20°C for the main-chain methylene carbon as a function of the softening point for the SHRP core asphalts: group I. SHRP grade 58; group II, SHRP grade 64. (Numbers in shaded areas represent the low-temperature rating.)

In addition, asphalts AAG-1 and AAM-1 have the highest polar contents ($\sim$50%) [24], whereas the polar contents of the other asphalts are nearly the same and range from 37.3 to 41.3%. Because of the higher polar contents, asphalts AAG-1 and AAM-1 may have a greater tendency to associate and raise the softening point of these asphalts relative to the other asphalts. No chemical explanation can be found for the position of asphalt AAB-1. However, the $T_{1\rho}{}^{CS}$ for this asphalt is slightly higher than for most of the other asphalts, resulting in a low value for the ratio.

The asphalts can also be separated into two groups. Asphalts in group I (AAA-1, AAB-1, AAC-1, AAD-1, and AAG-1) have a high-temperature performance grade (PG) of 58, whereas asphalts in group II (AAF-1, AAK-1, and AAM-1) have a performance grade of 64 (see Table 8). This separation of the SHRP core asphalts into two groups is also apparent in the discussion of aliphatic flexible carbons and their relationship to rheological properties.

3. Carbon-13 Dipolar Dephasing Relaxation Time (T_2')

The carbon spin-spin relaxation time, $T_2{}^C$, is the time constant for the coherent (aligned) carbon nuclear spins to return to phase equilibrium, that is, to become completely out of phase with each other (dephased) after being perturbed by an rf pulse. The dependence of $T_2{}^C$ on the correlation time, τ_C, is shown in Figure 4. For nonviscous liquids having short correlation times the carbon-13 spin-lattice ($T_1{}^C$) and spin-spin relaxation ($T_2{}^C$) times are essentially equal. However, for materials such as asphalts, which have slow molecular motions, that is, long τ_C, $T_2{}^C$ departs from $T_1{}^C$. The reason is that slow molecular motions cause inefficient transfer of the spin energy to the lattice, thus $T_1{}^C$ increases

Table 8 SHRP Core Asphalt Production Method and Grade

Asphalt	Production method	Grade MRL[a]	SHRP[b]
AAA-1	Distillation	150/200	58, -28
AAB-1	Distillation	AC-10	58, -22
AAC-1	Distillation	AC-8	58, -16
AAD-1	Distillation	AR-4000	58, -28
AAG-1	Distillation (lime-treated)	AR-4000	58, -10
AAF-1	Solvent-deasphaltened	AC-20	64, -10
AAK-1	Distillation (high vanadium content)	AC-30	64, -22
AAM-1	Solvent-deasphaltened	AC-20	64, -16

[a]Materials Reference Library Report on Asphalt Properties [24].
[b]Strategic Highway Research Program.

with slower motions, whereas the slow molecular motions (long τ_C) permit effective dipole-dipole spin interactions and thus T_2^C decreases with slower motions.

The carbon-13 spin-spin relaxation time can be measured using an NMR dipolar dephasing pulse technique (CP/MAS/dd). The relaxation time measured by this technique is referred to as T_2'. The rate at which the various carbons dephase depends upon the number of hydrogens attached to the carbon and the extent of molecular motion of the molecules. Carbons having attached hydrogens dephase more rapidly than quaternary carbons and carbons in a rigid network dephase more rapidly than carbons with segmental and rotational motions.

Figure 12 shows the conventional cross-polarization/magic angle spinning (CP/MAS) spectrum at 23°C of asphalt AAA-1 (a) and the dipolar dephasing spectrum of the same asphalt at 23°C using a 120-μs dephasing delay time (b). Comparing the two spectra, it is seen that in the dipolar dephased spectrum, the aromatic (100–160 ppm) and aliphatic (0–60 ppm) carbon resonances have narrower line widths and greater resolution of the aliphatic carbons than in the conventional CP/MAS spectrum. Also, in the dipolar dephased spectrum, the intensity of the aliphatic carbons is decreased relative to the aromatic carbons. This is because the signal for the more rigid aliphatic carbons has decayed. It should be noted that the center of the aromatic resonances shifted from about 127 ppm in the conventional CP/MAS spectrum to about 137 ppm in the dipolar dephased spectrum. The shift occurs because most of the signal for the aromatic

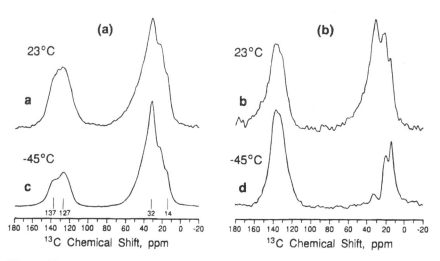

Figure 12   (a) Carbon-13 CP-MAS and (b) dipolar dephasing spectra (delay time = 120 μs) at 23 and −45°C for asphalt AAA-1.

tertiary carbons (~127 ppm), which have a short relaxation time, has decayed, leaving only the signal for the aromatic quaternary carbons (~137 ppm), which have relatively long relaxation times.

Figure 12c shows the CP/MAS NMR spectrum of asphalt AAA-1 at −45°C. At this temperature much of the motions of aromatic and aliphatic carbons is stopped except for methyl rotation. The CP/MAS spectrum at −45dgC shows that partially resolved aromatic quaternary and tertiary carbons and the aliphatic methylene carbons (32 ppm) have increased in intensity relative to the aromatic carbons (127 ppm) and the methyl carbons (20 and 14 ppm). This increase is due to stopping the segmental motions of methylene carbons in the long hydrocarbon chains. When the segmental motion is stopped, cross-polarization of the carbons is more effective, so the signal increases.

The dipolar dephased spectrum of asphalt AAA-1 at −45°C is shown in Figure 12d. Dramatic differences in the aromatic and aliphatic carbon chemical shift regions are observed. At the delay time of 120 μs the only resonances in the aromatic region are due to bridged or alkyl-substituted quaternary carbons. The spectrum also shows that essentially all of the resonance signal for the methylene carbons (32 ppm) has disappeared. The disappearance of the methylene signal at a delay time of 120 μs indicates that the segmental motions of the methylene carbons at −45°C have nearly stopped. The only signals remaining are those of the methyl resonances, because the methyl groups are still in rapid rotational motion. The alkane branched methyl groups and methyls attached to aromatic rings (both resonance signals at ~20 ppm) now have less intensity relative to the terminal methyl groups (14 ppm) of the long hydrocarbon chains. This is indicated by the change in the intensity ratio of the branched plus aromatic methyls to the terminal methyls in the CP/MAS (Figure 12c) and the dipolar dephased (Figure 12d) spectra. A temperature less than −160°C is needed to stop the rotational motion of non-sterically hindered methyl carbons. The lower intensity may be due to either the removal of the overlapping methylene carbons or slowing down of the rotation of these methyl groups (because of increased steric hindrance) relative to the terminal methyl carbons.

Two methods of data reduction were used to study the effects of temperature on the molecular motions in asphalts. These methods are the carbon type and carbon group analyses of the carbon-13 dipolar dephasing relaxation time data. For the carbon type analysis, the peak height intensities of the terminal methyl (14 ppm), methylene (32 ppm), tertiary aromatic (127), and quaternary aromatic (137) carbons were measured, whereas for the carbon group analysis the total integrated areas of the aliphatic and aromatic carbons were measured.

Figure 13 shows a plot of the normalized aliphatic carbon area as a function of the dipolar dephasing delay time for asphalt AAA-1 at 20°C. The solid line

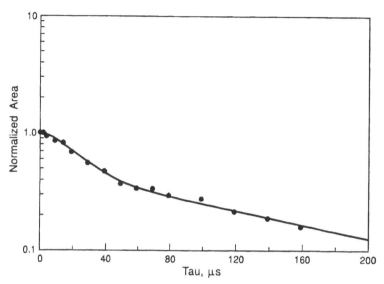

Figure 13 Dipolar dephasing spectral data for the aliphatic carbons in asphalt AAA-1 at 20°C.

represents a least-squares fit of either the aromatic or aliphatic carbon data to Equation 5.

$$M_0(\tau) = M_0{}^G e^{-0.5(\tau/T'_{2G})^2} + M_0{}^L e^{-\tau/T'_{2L}} \tag{5}$$

where $M_0{}^G$ = fractional amount of tertiary aromatic carbons (or rigid aliphatic carbons); G = Gaussian

$M_0{}^L$ = fractional amount of quaternary aromatic carbons (or flexible aliphatic carbons); L = Lorentzian

τ = dipolar dephasing delay time (μs)

T'_{2G} = carbon-13 spin-spin dipolar dephasing relaxation time for tertiary aromatic carbons (or rigid aliphatic carbons)

T'_{2L} = carbon-13 spin-spin dipolar dephasing relaxation time for quaternary aromatic carbons (or flexible aliphatic carbons)

Carbon Types (Resonance Peak Height Analysis). Figures 14 and 15 show the T'_2 data for the methyl, methylene, tertiary aromatic, and quaternary aromatic carbons at 23°C and −45°C for asphalts AAA-1, AAB-1, and AAM-1. Data from 160 to 200 μs in the plots at −45°C show the effect of rotational modulation due to sample spinning. Thus, the last two data points were not used for the

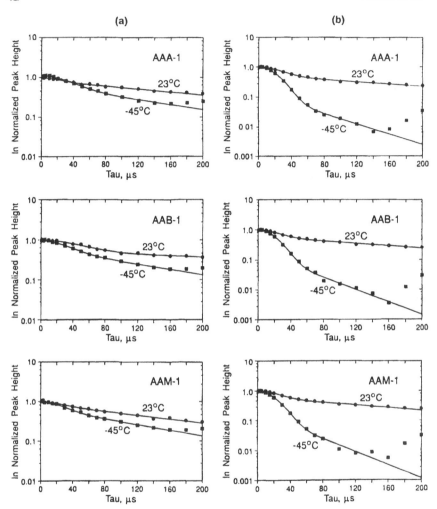

Figure 14   Effects of temperature on the carbon-13 NMR dipolar dephasing spectral data for (a) terminal carbons at 14 ppm methyl and (b) methylene carbons at 32 ppm in asphalts AAA-1, AAB-1, and AAM-1.

computation of T_2'. The differences in the relaxation data at 23°C and −45°C for the four carbon types are due to changes in the extent of molecular motions. Table 9 lists the carbon-13 dipolar dephasing relaxation times (T_2') for the four carbon types for asphalts AAA-1, AAB-1, and AAM-1 at 23°C and −45°C. For a simple, rigid, molecular system the T_2' value for any carbon type would be a single value. However, molecular motion strongly affects the T_2'. Thus, in a

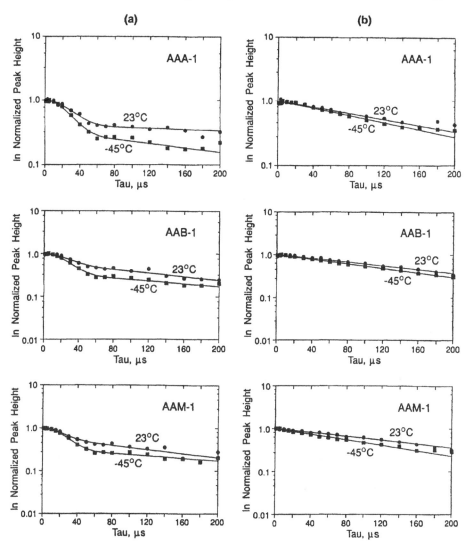

Figure 15 Effects of temperature on the carbon-13 NMR dipolar dephasing data for (a) tertiary and (b) quaternary aromatic carbons at 27 ppm in asphalts AAA-1, AAB-1, and AAM-1.

Table 9 Carbon-13 Dipolar Dephasing Relaxation Times (μs) for Carbon Types in Asphalts at 23°C and −45°C

Carbon type	Asphalt	23°C		−45°C	
		Rigid	Flexible	Rigid	Flexible
Methyl carbon (14 ppm)	AAA-1	21.9	231	32.6	140
	AAB-1	42.9	498	29.3	130
	AAM-1	26.3	181	25.3	122
Methylene carbon (32 ppm)	AAA-1	24.4	236	19.8	49.8
	AAB-1	22.6	222	19.2	41.7
	AAM-1	22.6	191	19.7	39.6
Tertiary aromatic carbon (127 ppm)	AAA-1	25.0	972	22.5	258
	AAB-1	24.2	211	22.7	276
	AAM-1	20.4	182	21.9	306
Quaternary aromatic carbon (137 ppm)	AAA-1	179		153	
	AAB-1	204		166	
	AAM-1	198		136	

complex molecular system such as an asphalt, any carbon type can have a range of molecular motions. The T_2' data for the carbon types in asphalts can best be represented by two types of motions broadly defined as motions related to flexible carbons (rotation, segmental, or twisting) and to carbons with a lesser degree of motion (relatively sluggish rotation, segmental, twisting, or completely rigid structures with no motion). For the purpose of simplifying the discussion, the terms flexible and rigid will be used to describe the relative differences in motions associated with the carbon types. As shown in Table 9, carbon atoms having at least one attached hydrogen have T_2' values on the order of 22 ± 2 μs. Quaternary aromatic carbon atoms without any attached hydrogens have T_2' values of 192 ± 12 μs. For terminal methyl, methylene, and tertiary aromatic carbons, the flexible components have T_2' values around 200 μs at 23°C. At −45°C, the T_2' values for the rotating methyl carbons for the three asphalts decrease to ∼130 ± 10 μs, whereas the T_2' values for the segmental motion of the methylene carbons of the long-chain hydrocarbons decrease to 45 ± 5 μs. This value, which approaches that of a rigid component, indicates that nearly all methylene segmental motion is stopped at −45°C, whereas considerable rotational motion is still present for the terminal methyl carbons. The T_2' values for the twisting of the tertiary aromatic carbons appear to increase at −45°C for reasons unknown at this time. The T_2' values of the rigid quaternary aromatic carbons show a small decrease at −45°C relative to 23°C, indicating that some motion does exist at room temperature for the aromatic cluster of the "asphaltic" molecules (see Figure 1).

Table 10 gives the relative percentages of the rigid and flexible aliphatic and aromatic carbons at 23°C and −45°C. The relative percentages of flexible methyl and methylene carbons decrease with decreasing temperature. The decrease is greater for the methylene carbons; only ~16% of all methylene carbons still have segmental motion (flexible) whereas ~62% of all the terminal methyl groups have considerable rotational motion.

The relative percentages of the rigid and flexible tertiary aromatic carbons show a temperature dependence; however, there is no change for the relative percentages of the rigid quaternary aromatic carbons. Current "average" molecular structural data suggest that all the tertiary and quaternary aromatic carbons are a part of the same aromatic cluster and, thus, would have nearly the same low-frequency molecular motions. However, the temperature dependence of the relaxation data is different for both the tertiary and quaternary aromatic carbons, indicating that there are tertiary carbons that are in motion in addition to rigid tertiary carbons. Table 10 shows that ~62% of the tertiary carbons are flexible at 23°C and this decreases to ~35% at −45°C. No change is assumed for the quaternary carbons. Figure 1 shows a model for the average structure of a simplified asphalt molecule based upon the relaxation data. In this model, phenyl groups are attached to the aromatic cluster either directly or through a methylene linkage. In addition, phenyl groups can be attached to an aliphatic long-chain hydrocarbon molecule. In both cases, these phenyl groups are "free" to rotate but are more likely to oscillate (twist) in their positions. This model allows for both flexible and rigid tertiary aromatic carbons and explains the observed small temperature dependence of the relaxation data.

Table 10 Relative Percent of Carbon Types in Asphalts at 23°C and −45°C

Carbon type	Asphalt	23°C Rigid	23°C Flexible	−45°C Rigid	−45°C Flexible
Terminal methyl carbon (14 ppm)	AAA-1	16	84	43	57
	AAB-1	45	55	39	61
	AAM-1	14	86	33	67
Methylene carbon (32 ppm)	AAA-1	47	53	86	14
	AAB-1	39	61	83	17
	AAM-1	41	59	82	18
Tertiary aromatic carbon (127 ppm)	AAA-1	58	42	67	33
	AAB-1	38	62	65	35
	AAM-1	39	61	67	33
Quaternary aromatic carbon (137 ppm)	AAA-1	100		100	
	AAB-1	100		100	
	AAM-1	100		100	

Carbon Groups (Aromatic and Aliphatic Area Analysis). The general method for determining the carbon-13 dipolar dephasing-relaxation times for complex solid carbonaceous materials (asphalts, coals, etc.) is to determine from total integrated area measurements the average relaxation times for all aromatic and all aliphatic carbons. The aromatic carbons can be subdivided into tertiary and quaternary carbons. This subdivision differs from the peak height analysis in that all tertiary and quaternary carbons over the whole aromatic chemical shift range (100–160 ppm) are measured rather than just the carbons at 127 and 137 ppm. Further subdivision of the tertiary and quaternary aromatic carbons into rigid and flexible carbons cannot be made as was done for the peak height method (carbon types). However, the advantage of the area analysis method is that the mass percentages of tertiary and quaternary carbons can be measured.

The dipolar dephasing relaxation time for the aliphatic carbons from area measurements is the average relaxation time of the three types of methyl groups (methyls attached to aromatic rings and the branched and terminal methyls of long-chain hydrocarbons) and the various methylene and methine carbons. However, with the area method of analysis, the aliphatic carbons can be identified only as rigid and flexible carbons, thus, the identity of the carbon types is lost. The advantage, however, is that the mass percent of rigid and flexible aliphatic carbons can be obtained for each asphalt. It is the mass percent of flexible aliphatic carbons in each asphalt that can be correlated with the rheological and binder (asphalt) performance parameters. The average relaxation times, T_2', for the aromatic and aliphatic carbon groups in the eight SHRP core asphalts at 23°C and −45°C are given in Table 11. At 23°C the tertiary aromatic and rigid aliphatic carbons for all asphalts have a T_2' value of 21 ± 2 μs, which is typical of carbons having at least one hydrogen attached. Lowering the temperature to −45°C does not affect the relaxation times of these carbon types.

The T_2' values for quaternary aromatic carbons at 23°C vary from 218 μs for asphalt AAC-1 to 361 μs for asphalt AAD-1. The variation in T_2' for the different asphalts is due to differences in the low-frequency motions of the carbons in the larger aromatic clusters. This conclusion is supported by the fact that upon reducing the temperature to −45°C, at which most of the motions of the quaternary carbons in the cluster are assumed to be stopped, the T_2' values (for asphalts AAA-1, AAB-1, and AAM-1) are significantly smaller and nearly the same (205 ± 10 μs).

The T_2' values for the flexible aliphatic carbons in the asphalts vary from 139 to 221 μs. This variation is due to the different degrees of segmental motion for the different asphalts. Reducing the temperature to −45°C reduces T_2' to a nearly constant value of 86 ± 3 μs. The magnitude of T_2' at −45°C suggests that some motion of the aliphatic carbons still exists. This motion is assumed to be mostly due to rotating methyl groups in the asphalts.

Table 11 Carbon Dipolar Dephasing Relaxation Times (μs) for Aromatic and Aliphatic Carbon Groups in Asphalts at 23°C and −45°C

Asphalt	Total tertiary aromatic carbons		Total quaternary aromatic carbons		Rigid aliphatic carbons		Flexible aliphatic carbons	
	23°C[a]	−45°C[b]	23°C	−45°C	23°C	−45°C	23°C	−45°C
AAA-1	21.3 (22.7)[c]	24.4	239 (480)[c]	215	20.6 (19.8)[c]	20.3	195 (195)[c]	86.1
AAB-1	24.9	23.3	255	207	21.3	19.7	179	89.3
AAC-1	18.8		218		18.0		154	
AAD-1	18.0		361		19.9		221	
AAF-1	21.9		311		20.9		181	
AAG-1	20.2		265		20.3		139	
AAK-1	23.2		311		22.4		171	
AAM-1	20.5	21.6	263	193	20.0	19.9	169	83.1

[a]Contact time 0.5 ms.
[b]Contact time 1.0 ms.
[c]( ), repeat experiment.

IV. ALIPHATIC CARBON MOLECULAR MOTIONS AND RHEOLOGICAL PROPERTIES

The rheological properties and binder performance parameters of asphalts are based upon the properties of the whole asphalts, which include the chemical composition, molecular structure, and molecular association. For the eight SHRP core asphalts the rheological and performance properties vary extensively from one another. However, for the asphalts, the variation in the mass percent of carbon (80.6 to 86.7), the fraction of aliphatic carbons (0.672 to 0.766), and the relative percent of flexible aliphatic carbons (47–61) is small. It should be noted that any one of these chemical parameters alone does not correlate with any of the rheological properties of asphalts.

Because asphalts contain ~70% of aliphatic carbons [16], it can be assumed, to a very good first approximation, that initially (before aging and other effects) the rheological properties of "tank" asphalts may be dominated by the motion of these carbons. It is well known in polymer rheology that the rheological properties depend on the segmental and rotational motions of the aliphatic carbons [25]. Thus, a correlation should exist between the aliphatic flexible carbons of an asphalt and its rheological properties. Undoubtedly, aromatic carbons and heteroatom content in asphalts also influence the rheological properties through restricted motion due to the size of the aromatic cluster and molecular association based on the heteroatom functionality. In addition, an aged asphalt, because of changes in its chemical composition and molecular structure due to oxidation, will also have more restricted motion and, thus, affect the rheological properties of the asphalts.

To correlate the aliphatic flexible carbons for the different asphalts with their rheological properties, the mass percent of aliphatic flexible carbons in a whole asphalt must be known. This value is calculated from Equation 6.

$$C_{al}^{Flex} = Cf_{al}^{C} f_{al\ Flex}^{C} \qquad (6)$$

where C_{al}^{Flex} = mass percent of aliphatic flexible carbons

C = mass percent of carbon (from elemental analysis)

f_{al}^{C} = fraction of aliphatic carbons (from liquid-state NMR)

$f_{al\ Flex}^{C}$ = fraction of aliphatic flexible carbons (from CP/MAS/dd NMR)

Table 12 lists the mass percent of carbon, the fraction of aliphatic carbons, the fraction of flexible aliphatic carbons, and the mass percent of flexible aliphatic carbons for the SHRP core asphalts at 23°C. A temperature of 23°C was chosen because at very low temperatures most of the molecular motion is significantly slowed in all asphalts and at higher temperatures all asphalts have excessive molecular motion of the carbon types and, thus, the motions in the core asphalts are similar and cannot be resolved from one another at either temperature extreme.

The dipolar dephasing relaxation time measurements were repeated for asphalt AAA-1 at 23°C and at a contact time of 0.5 ms. The calculation of the aliphatic flexible carbon content (33.1 mass %) was found to be in close agreement with the first experiment (35.8 mass %). Despite possible small errors in the determination of the elemental carbon content and the aromaticity value, reproduc-

Table 12   Amount of Rigid and Flexible Aliphatic Carbons in SHRP Core Asphalts at 23°C Using CP/MAS/DD[a] NMR

Asphalt	Mass % of carbons[b]	Fraction of aliphatic carbons (f_{al}^{C})[c]	Fraction of flexible aliphatic carbons $(f_{al\ Flex}^{C})$	Mass % of flexible aliphatic carbons
AAA-1	84.2	0.721	0.59 (0.55)[d]	35.8 (33.1)
AAB-1	82.3	0.688	0.58	32.6
AAC-1	86.5	0.722	0.53	33.2
AAD-1	81.6	0.766	0.54	33.7
AAF-1	84.5	0.672	0.48	27.1
AAG-1	85.6	0.710	0.47	28.6
AAK-1	80.7	0.738	0.61	36.4
AAM-1	86.7	0.741	0.49	31.6

[a]Cross polarization/magic angle spinning/dipole dephasing (contact time = 0.5 ms).
[b,c]Ref. 16.
[d]( ), repeat experiment.

ibility of the aliphatic flexible carbon content from the solid-state NMR data is quite good. The absolute error is estimated to be about ± 1.4 mass % for the aliphatic flexible carbons.

A. Rheological Properties

The mass percent of aliphatic flexible carbons for the SHRP core asphalts was found to correlate with several rheological properties that are related to the binder (asphalt) performance characteristics, such as permanent deformation (rutting), fatigue, cracking, and low-temperature thermal cracking. In all correlations, as far as it can be ascertained, the rheological data used were obtained for neat (tank) asphalts and have not been pretreated according to recent SHRP recommended procedures.

1. Dynamic Viscosity

The correlation of the aliphatic flexible carbons with the dynamic viscosity of the asphalts at 25°C is shown in Figure 16. Separating the asphalts into two groups based upon their upper temperature grade (see Table 8) gives excellent correlations of the aliphatic flexible carbons with dynamic viscosity. It should be noted that the lower temperature gradation within each upper temperature grade decreases uniformly with increased dynamic viscosity. Separating the asphalts into two groups was also noted in the section on the correlation of the carbon-13 relaxation time ratio and the softening point of asphalts.

Because no correlations exist between the various single chemical components of asphalts such as saturates, polars, asphaltenes, and their measured rheo-

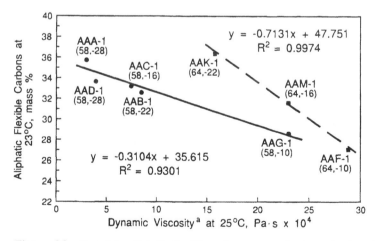

Figure 16 Plot of the aliphatic flexible carbons versus the dynamic viscosity for SHRP core asphalts: [a]From Ref. 26.

logical properties, the difference in the rheological behavior of the two groups may be due to the secondary effects of molecular structuring of the aliphatic chain and aromatic cluster size and to molecular association. For asphalts AAF-1, AAK-1, and AAM-1, the influence of these secondary effects on the correlation is presumably similar but differs from that for the other five asphalts. If asphalt AAK-1 is eliminated from the correlation because of the high paramagnetic content affecting the dipolar dephasing relaxation time, a correlation of flexible carbons of the remaining asphalts with their dynamic viscosity gives a reasonably good correlation coefficient of 0.844 (Table 13). This coefficient is considerably less than the correlation coefficients assuming two groups of asphalts.

2. Inverse of the Loss Compliance in Shear

The inverse of the loss compliance, in shear, $1/J''$, is related to rutting resistance and is given in Equation 7 [26,27].

$$1/J'' = G^*/\sin \delta \qquad (7)$$

where G^* = dynamic complex modulus in shear
        δ = phase angle

The value of G^* is a measure of the total resistance to deformation and the value of $\sin \delta$ reflects the relative nonelasticity of the asphalt [27]. To increase

Table 13 Correlation Coefficients for Rheological Properties and Binder Performance Parameters with Aliphatic Flexible Carbons for Asphalts

	Correlation coefficient (R^2)		
	Single set	Binary set	
Parameter	Asphalt A,B,C,D,F,G,M	Asphalt A,B,C,D,G	Asphalt F,K,M
Rheological			
Dynamic viscosity at 25°C	0.844	0.930	0.997
Inverse of the loss compliance in shear, $G^*/\sin \delta$ at 25°C (a rutting parameter)	0.901	0.901	0.962
Loss modulus in shear, $G^* \sin \delta$ at 25°C (a fatigue parameter)	0.894	0.889	0.914
Log S (0.1 s) at 20°C (a rutting parameter)	0.892	0.929	0.967
Binder performance			
Limiting fatigue temperature ($G^* \sin \delta$ = 3 MPa)	0.876	0.957	0.990
Fracture temperature	0.930	0.979	0.887
Limiting stiffness temperature [$S(t)$ = 200 MPa, 2 h]	0.802	0.873	1.000

the rutting resistance of an asphalt, $1/J''$ should increase. That is, the value of G^* should be large and sin δ small. The elasticity of the asphalt should be reflected in the aliphatic carbon chain length and the segmental motion of the aliphatic carbons. If the mass percent of segmental motion of the aliphatic carbons in asphalt is low, the rutting resistance of the asphalt should be high. Figure 17 shows a plot of the mass percent of the aliphatic flexible carbons as a function of the inverse of the loss compliance in shear. As shown in the plot for the two grades of asphalts, the correlations are quite good. An equally good correlation can be obtained if only seven of the asphalts are plotted. Asphalt AAK-1 was eliminated for the reason discussed previously.

3. Loss Modulus in Shear

The loss modulus in shear, G'', is given by Equation 8 and can be used to describe the resistance of an asphalt to fatigue cracking [26,27].

$$G'' = G^* \sin \delta \tag{8}$$

where G^* and δ have the same definitions as described for Equation 7. Fatigue cracking depends upon the ability of the asphalt to limit the dissipation of energy. A "soft" asphalt has a low value of G^*. That is, it can deform without developing large stresses. In addition, asphalts with low sin δ values are more elastic and recover to their original condition without dissipating energy. Asphalts with large amounts of aliphatic flexible carbons should be soft and more elastic than asphalts

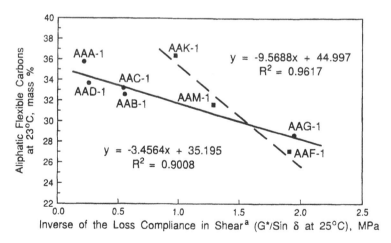

Figure 17 Plot of the aliphatic flexible carbons versus the inverse of the loss compliance in shear (a rutting parameter) for SHRP core asphalts. [a]From Ref. 28.

with less flexible aliphatic carbons. Figure 18 shows a plot of the mass percent of aliphatic flexible carbons as a function of the loss modulus in shear (28). Again reasonably good correlations are obtained for the two grades of asphalt. The correlation coefficient for a single data set is comparable to the coefficients for the two groups of asphalts (Table 13).

4. Stiffness

Thermal cracking associated with asphalts is a low-temperature phenomenon that depends on the stiffness of the asphalt. Stiffness, S, defined as the inverse of the extensional creep compliance, $1/D$, is given in Equation 9 [26].

$$S = 1/D \qquad (9)$$

However, with assumptions, the stiffness can be related approximately to the dynamic complex modulus in shear [26], G^*.

$$S = 3G^* \qquad (10)$$

Asphalts with low values of G^* are soft and have a low stiffness value. Thus, the more segmental and rotational motion an asphalt has at a given temperature, the lower its stiffness value. Figure 19 shows a plot of the mass percent of aliphatic flexible carbons in the SHRP core asphalts as a function of the logarithm of their stiffness at 20°C (29). As can be seen, a fairly good correlation exists for the two grades of asphalts. A single data set gives a correlation coefficient that is slightly less than the coefficients for the two groups of asphalts.

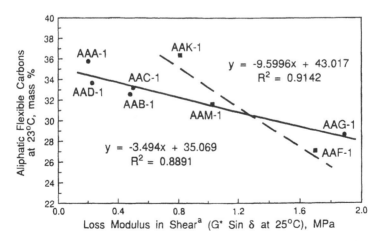

Figure 18 Plot of the aliphatic flexible carbons versus the loss modulus in shear (a fatigue parameter) for SHRP core asphalts. [a]From Ref. 28.

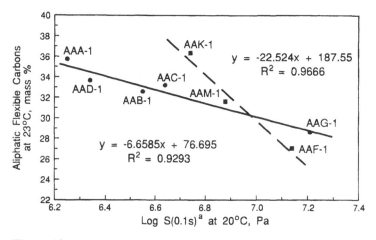

Figure 19 Plot of the aliphatic flexible carbons versus the log stiffness (a thermal cracking parameter) for SHRP core asphalts. ªFrom Ref. 29.

B. Binder Performance Parameters

The mass percent of aliphatic flexible carbons was found to correlate well with several binder performance parameters. These parameters include (a) the limiting fatigue temperature, (b) the fracture temperature, and (c) the limiting stiffness temperature. Even for these parameters, the best correlations are obtained if the two grades of asphalts are independently correlated with the flexible carbon content.

1. Limiting Fatigue Temperature

The limiting fatigue temperature is defined as that temperature at which the loss modulus in shear (G^* sin δ) equals 3 MPa. A plot of the flexible carbon content as a function of the limiting fatigue temperature for the various asphalts is shown in Figure 20. A high correlation exists for the two groups of asphalts. As shown, the higher the flexible carbon content, the lower the limiting fatigue temperature. A single data set gives a correlation coefficient that is considerably smaller than the correlation coefficients for the two groups of asphalts (see Table 13).

2. Fracture Temperature

Jung and Vinson [30] reported a thermal stress restrained specimen test to evaluate the thermal cracking resistance of asphalt concrete mixtures. These authors determined the fracture temperature of 14 SHRP asphalts and found that their ranking of the asphalt compared favorably with a ranking based on fundamental properties of the asphalt. A plot of the aliphatic flexible carbon content as a function of

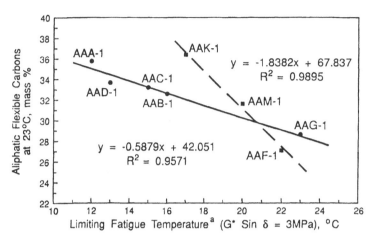

Figure 20 Plot of the aliphatic flexible carbons versus the limiting fatigue temperature for SHRP core asphalts. [a]From Ref. 24.

fracture temperature as determined by Jung and Vinson is shown in Figure 21. A near-excellent correlation exists for the five asphalts AAA-1, AAB-1, AAC-1, AAD-1, and AAG-1. A reasonable correlation exists for asphalts AAF-1, AAK-1, and AAM-1. These correlations parallel all of the other temperature correlations in that as the flexible carbon content in an asphalt increases, the

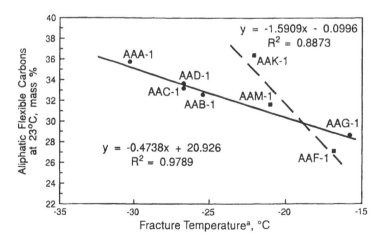

Figure 21 Plot of the aliphatic flexible carbons versus the fracture temperature (a thermal cracking parameter) for SHRP core asphalts. [a]From Ref. 30.

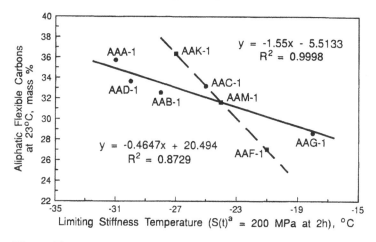

Figure 22 Plot of the aliphatic flexible carbons versus the limiting stiffness temperature (a thermal cracking parameter) for the SHRP core asphalts. [a]From Ref. 29.

fracture temperature decreases, indicating again the importance of molecular motion to road performance of an asphalt. A single data set for seven asphalts also gives a reasonably good correlation coefficient.

3. Limiting Stiffness Temperature

The limiting stiffness temperature is defined as the temperature at which the stiffness of the asphalt equals 200 MPa at a loading time of 2 h. The data in the report by Anderson et al. [29] were plotted against the aliphatic flexible carbon content as shown in Figure 22. An excellent correlation was found for asphalts AAF-1, AAK-1, and AAM-1, and a reasonably good correlation was obtained for the other five asphalts. However, a low correlation coefficient is obtained if only the seven asphalts are plotted as a single set.

V. SUMMARY

Various carbon and hydrogen relaxation times were determined for the SHRP core asphalts at 20°C. Attempts to correlate the relaxation times with any one chemical property (molecular weight, asphaltenes, polar, saturates, and waxes) of the asphalts were successful (as one would have expected) with the exception of the total metal concentration. It was observed that the aromatic carbon cross-polarization relaxation time, T_{CH}, depends on the total nickel, iron, and vanadium concentration. The small variation in the magnitude of the other relaxation times precluded any attempt at a correlation with the metal concentration. However, the effect of the metal content is evident on the various relaxation times measured for asphalt AAK-1. The ratio $T_{CH}/T_{1\rho}^{CS}$ for the main chain methylene carbons in asphalts was shown, with some reservation, to correlate with the softening point of the asphalts.

It was found that methylene and methyl carbons of straight and branched alkane hydrocarbons (attached to an aromatic cluster or paraffinic) have considerable segmental and rotational motions (in the 50-kHz range) at room temperature. However, almost all methylene segmental motions are stopped at $-45°C$. At this temperature extensive methyl rotation is still present. The aromatic quaternary carbons of the polycondensed aromatic cluster have little motion at room temperature. To explain the observed motions of tertiary aromatic carbons, rotation or twisting motion of "free" phenyl groups attached to the more rigid aromatic cluster must be invoked.

Many of the rheological properties and binder performance parameters associated with rutting and fatigue and thermal cracking depend on the flexibility of the aliphatic carbons, which are the dominant carbon type in asphalts (70%). Excellent correlations of the flexible carbons with rheological parameters were obtained if the asphalts were separated into two groups. However, because of the small number of asphalts in the two data sets, the correlations may be fortuitous. Assuming a single data set for seven of the SHRP core asphalts (AAK-1 was eliminated because of its high metal concentration effect on relaxation data), the correlations of the asphalts' rheological properties and binder performance parameters with flexible aliphatic carbons have correlation coefficients ranging from 0.80 to 0.93, which are only slightly lower than the correlation coefficients obtained when assuming the asphalts fall into two groups (0.87 to 1.0). Nevertheless, the data suggest that molecular motions of the aliphatic carbons in asphalts (the dominant carbon types) play a significant role in the rheological properties of asphalts. Thus, from a single ^{13}C dipolar dephasing NMR experiment at room temperature many rheological properties and binder performance parameters of an asphalt can be estimated. In addition, with a knowledge of the molecular mobility in asphalts it may be possible to modify an asphalt to achieve a desired binder performance.

ACKNOWLEDGMENTS

The authors express their appreciation for the financial support from the Federal Highway Administration under contract. DTFH61-92C-00170, to J. Greaser for typing the manuscript, and to Dr. James S. Frye for many helpful technical discussions.

DISCLAIMER

Mention of specific brand names does not imply endorsement by Western Research Institute or the Federal Highway Administration. The contents of this chapter reflect the views of the authors, who are responsible for the facts and accuracy of the data presented. The contents do not necessarily reflect the official views or policies of the Federal Highway Administration. This chapter does not constitute a standard, specification, or regulation.

REFERENCES

1. D. W. McCall, *Mol. Relax. Polym.*, **4**: 223 (1971).
2. W. P. Slichter, *Rubber Chem. Technol.*, **34**, 1574 (1961).
3. *High Resolution NMR Spectroscopy of Synthetic Polymers in Bulk* (R. A. Komoroski, ed.), VCH Publishers, Deerfield Beach, FL (1986).
4. *Solid-State NMR of Polymers* (L. J. Mathias, ed.), Plenum Press, New York (1991).
5. A. Dekmezian, D. E. Axelson, J. J. Dechter, B. Borah, and L. J. Mandelkern, *Polym. Sci. Polym. Phys. Ed.*, **23**: 367 (1985).
6. J. Schaefer, M. D. Sefcik, E. O. Stejskal, R. A. McKay, W. T. Dixon, and R. E. Cais, *Macromolecules*, **17**: 1107 (1984).
7. J. Schaefer, E. O. Stejskal, and R. Buchdahl, *Macromolecules*, **10**: 384 (1977).
8. T. Kohno, T. Yokono, Y. Sanada, and J. W. Harrell, Jr., *Fuel*, **64**: 1329 (1985).
9. F. P. Miknis and D. A. Netzel, in *Magnetic Resonance in Colloid and Interface Science*, ACS Symp. Ser. No. 34 (H. A. Resing and C. G. Wade, eds.), American Chemical Society, Washington, DC, p. 182 (1976).
10. D. L. VanderHart, *SHRP-A-335, Binder Characterization and Evaluation by Nuclear Magnetic Resonance Spectroscopy*, Strategic Highway Research Program, National Research Council, Washington, DC (1993).
11. D. A. Netzel, WRI/FHWA Annual Technical Report, November 1, 1994–May 15, 1995, Western Research Institute, Laramie, WY.
12. Z. Gülsün, *Fuel*, **66**: 1449 (1987).
13. D. L. VanderHart, W. F. Manders, and G. C. Campbell, *Preprint Am. Chem. Soc. Div. Fuel Chem.* **35**(3): 445 (1990).
14. D. L. VanderHart, SHRP A-002C Final Report, Strategic Highway Research Program, National Research Council, Washington, DC, February, p. 139 (1991).
15. E. D. Becker, *High Resolution NMR: Theory and Chemical Application*, Academic Press, New York (1980).
16. P. W. Jennings, in *SHRP-A-335, Binder Characterization and Evaluation by Nuclear Magnetic Resonance Spectroscopy*, Strategic Highway Research Program, National Research Council, Washington, DC (1993).
17. D. A. Netzel, D. R. McKay, R. A. Heppner, F. D. Guffey, S. D. Cooke, D. L. Varie, and D. E. Linn, *Fuel*, **60**: 307 (1981).
18. J. Schaefer, in *Molecular Basis of Transitions and Relaxations* (D. J. Meier, ed.), Gordon & Breach Science Publishers, New York, p. 108 (1978).
19. W. L. Earl, in *NMR of Humic Substances and Coal* (K. L. Warshaw and M. A. Mikita, eds.), Lewis Publishers, Chelsea, MI, Chapter 9 (1987).
20. K. J. Packer, *Prog. Nucl. Magn. Reson. Spectrosc.*, **2**: 87 (1967).
21. R. F. Boyer, *Polym. Eng. Sci.*, **8**(3): 161 (1968).
22. J. A. Sauer, *J. Polym. Sci. Part C*, **32**: 69 (1971).

23. P. I. Vincent, *Polymer*, **15**: 111 (1974).
24. Materials Reference Library Report on Asphalt Properties, Strategic Highway Research Program, National Research Council, Washington, DC (1992).
25. J. E. Mark, A. Eisenberg, W. W. Graessley, L. Mandelkern, E. T. Samulski, J. L. Koenig, and G. D. Wignall, *Physical Properties of Polymers*, 2nd Ed., American Chemical Society, Washington, DC, Chapter 3 (1993).
26. *SHRP-A-369, Binder Characterization and Evaluation*, Vol. 3: *Physical Characterization*, Strategic Highway Research Program, National Research Council, Washington, DC, p. 33 (1993).
27. H. U. Bahia and D. A. Anderson, Transportation Research Board Preprint, paper 950793 (1995).
28. S. S. Kim, 1995, private communication.
29. D. A. Anderson, D. W. Christensen, and M. G. Sharma, SHRP-A-002, WRI/SHRP Quarterly Report, pp. 113–114, June 1991, Western Research Institute, Laramie, WY.
30. D. Jung and T. S. Vinson, *Transport. Res. Rec.*, **1417**: 12 (1993).
31. D. A. Netzel and F. P. Miknis, *Appl. Spectros.*, **31**: 365 (1977).
32. F. P. Miknis, V. J. Bartuska, and G. E. Maciel, *Am. Lab.*, **11**:19 (1979).

3

DSC Studies of Asphalts and Asphalt Components

Thomas F. Turner and Jan F. Branthaver
Western Research Institute, Laramie, Wyoming

I. INTRODUCTION

The low-temperature properties of asphalt binders are of interest because low-temperature cracking is one of the primary asphalt pavement failure modes observed in cold-climate roads. Low-temperature cracking typically occurs during extreme low-temperature weather, usually within the first year or two after pavement construction. This failure mode can be recognized by transverse cracks in the pavement caused by the inability of the asphalt binder to deform to reduce stress. As the temperature of an asphalt binder decreases, its specific volume decreases. At the glass transition temperature (T_g) large-scale molecular motion is hindered by a "freezing in" of the system. This transforms the asphalt binder from a viscoelastic material to a brittle, amorphous solid. Because some reduction in specific volume occurs even below the glass transition temperature, internal stresses increase. In a roadbed, some shrinkage can occur laterally to relieve stress; however, in the longitudinal direction, the road length is effectively infinite, limiting shrinkage and stress relief. When the stresses exceed the asphalt binder's strength, the transverse cracks typical of low-temperature cracking appear. To reduce the frequency of low-temperature cracking failure, the glass transition temperature must be lowered or the binder strength increased. The problem for the asphalt producer is knowing how to blend asphalt components to control the low-temperature properties.

The goals of this work are to identify the asphalt components that most strongly affect the glass transition and to quantify the extent of the affect. This goal could be achieved by measuring the glass transition by dilatometry, calorimetry, or rheology or by methods that measure molecular motion such as nuclear magnetic resonance (NMR) and infrared (IR) spectroscopy. The methods discussed in this chapter are based on differential scanning calorimetry (DSC), specifically, modulated differential scanning calorimetry (MDSC). The advantages of DSC and MDSC are speed, small sample size, and the capability of monitoring crystallization/association and melting phenomena as well as the glass transition.

A. DSC and MDSC

Differential scanning calorimetry measures heat flow into and out of a sample (compared to a reference) as that sample is subjected to a temperature program [1]. The heat flow measured by a DSC is a function of the heating rate, sample heat capacity, and any endo- or exothermic events occurring in the sample. This is summarized by:

$$dQ/dt = C_p \, dT/dt + f(t, T) \tag{1}$$

where dQ/dt = total heat flow
C_p = heat capacity
dT/dt = heating rate
$f(t, T)$ = kinetic component

The first term on the right side of the equal sign in Equation 1 is the heat flow contribution due to heat capacity and heating the sample. Because there is a shift in heat capacity at the glass transition, this term contains the information about the glass transition. The second term on the right is the kinetic component, which comes from the time dependence of crystallization, melting, or chemical reactions. Difficulties arise with conventional DSC when searching for weak transitions and when analyzing systems with complex thermograms containing glass transitions and multiple overlapping exotherms and endotherms. The sensitivity of conventional DSC is proportional to the heating rate, as can be seen from Equation 1. Unfortunately, the resolution is inversely proportional to heating rate.

Modulated DSC is an adaptation in which the temperature (and, therefore, the heating rate) is modulated about an underlying heating or cooling program. The average temperature follows that of the conventional DSC; however, a periodic sine wave perturbation of the temperature is superimposed. This gives the advantage of providing a slow underlying heating rate for good resolution while still having short-duration high heating rates providing good sensitivity. Through signal processing of the resultant heat flow signal five thermograms are produced: modulated heat flow, total heat flow, heat capacity, reversing heat flow, and nonreversing heat flow. The modulated heat flow signal is the measured heat flow to the sample. The total heat flow is calculated as the average of the modulated heat flow signal and is equivalent to the signal from a conventional

DSC. The heat capacity signal is calculated from the ratio of the modulated heat flow amplitude to the modulated heating rate amplitude. It contains the sample heat capacity plus generally minor influences from changes in heat capacity caused by crystallization or other time-dependent phenomena. The reversing heat flow signal is the heat capacity component of the total heat flow calculated as the negative product of the heat capacity and the average heating rate. The nonreversing signal is the difference between the total heat flow and the reversing heat flow. This signal contains time-dependent phenomena and any baseline irregularities that might have been in the total heat flow signal. With MDSC there is less uncertainty in selecting baselines and end points than in conventional DSC when analyzing where a change in C_p occurs such as the glass transition.

The benefit of modulated DSC can be demonstrated by examining conventional and modulated DSC data for a high-wax asphalt. Figure 1 shows a total heat flow thermogram from an MDSC experiment indicating heat flow into the sample increasing in two stages separated by a plateau. Although not shown here, the size of the plateau is sensitive to the sample cooling rate prior to the measurement scan. This plateau and its dependence on cooling rate have led to what we believe are incorrect interpretations of the data in the literature. Figure 2 shows the reversing heat flow and nonreversing heat flow signals for the modulated experiment. The reversing signal is a relatively simple, clean signal

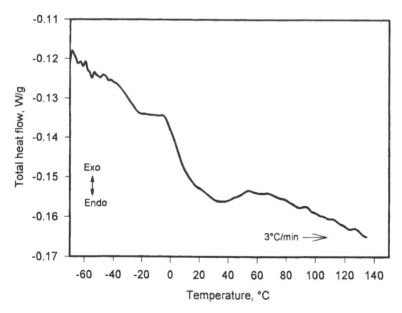

Figure 1 Total heat flow thermogram for a waxy asphalt.

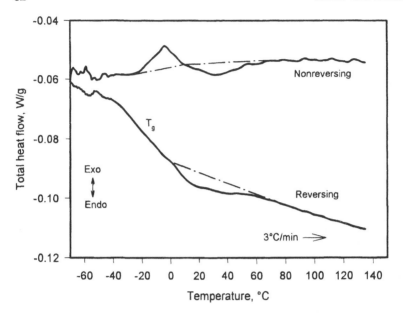

Figure 2   Reversing and nonreversing heat flow thermograms for a waxy asphalt.

with no plateau and one glass transition. The nonreversing signal contains an exotherm located at the same temperature as the plateau in the total heat flow signal.

B.   Glass Transition Properties

The glass transition is a fundamental property of amorphous materials, including asphalt binders. Below the glass transition temperature there is insufficient thermal energy in a material to allow large-amplitude molecular motion. Without this motion and the viscous flow it allows, the approach to a thermodynamic equilibrium is slow (this slow approach to equilibrium is called physical hardening). The transition is manifested by changes in slope of primary thermodynamic quantities such as specific volume and enthalpy and by discontinuities or jumps in secondary quantities such as thermal expansion coefficient and specific heat capacity.

The glass transition has been studied extensively in the polymer sciences because it marks a distinct change in polymer properties as a function of temperature and, thus, is important to tailoring product properties. Many excellent texts have been written containing large sections about the glass transition and its relation to physical and mechanical properties of polymeric materials [2–5]. Three main theoretical developments have been applied to the glass transition: the free-volume theory, the kinetic theory, and the thermodynamic theory. Sperling [5] discusses the development of these theories in detail. For the purposes of this discussion, it is important only to recognize that the glass transition depends on the thermal history of a sample and that comparisons of glass transition temperatures should be made with this in mind.

Some of the early research on polymers is relevant to the study of behavior near the glass transition in asphalts and asphalt components. For example, it is generally recognized that the glass transition temperature increases with increasing molecular weight. This was formalized in an expression from Fox and Flory [6] for weakly interacting molecules:

$$T_g = T_{g\infty} - K/M_n \tag{2}$$

where K is a constant, $T_{g\infty}$ is a limiting transition temperature at high molecular weight, and M_n is the number average molecular weight. The use of M_n rather than the mass average molecular weight shows the importance of small molecules in controlling low-temperature behavior. This can also be recognized as showing the importance of end-group concentration in low-temperature behavior.

It has also been recognized that the glass transition temperature of a homogeneous blend of two polymers can be described by the Fox [7] equation:

$$1/T_g = w_1/T_{g1} + w_2/T_{g2} \tag{3}$$

where T_g, T_{g1}, and T_{g2} are the glass transition temperatures of the whole blend, component 1, and component 2. and w_1 and w_2 are the mass fractions of the two components. Generalizing this relationship to multiple components gives

$$1/T_g = \sum (w_i/T_{gi}) \tag{4}$$

where, in this case, the summation is over all fractions, i [8]. With this relationship, one can calculate the relative influence of a single component on the T_g of a mixture. It may also allow us to identify fractions that do not blend homogeneously with others.

Asphalts are extremely complex materials containing literally millions of different molecular species. Components of the asphalt can be crystalline, amorphous solids, or liquids, depending on temperature and thermal history. In addition, under typical usage, the asphalt is in contact with mineral aggregate and is exposed to oxygen and moisture for long periods over a potentially wide temperature range. These conditions cause continuing changes in the asphalt binder chemical composition and mechanical performance. This complexity has led us to simplify our studies by examining chemically distinct asphalt components using techniques adapted from polymer and petroleum sciences.

C. DSC/MDSC Analysis Methods

A perceived weakness of using the glass transition to characterize an asphalt is the variation of the glass transition temperature (T_g) with thermal history, mainly cooling and heating rates. This variation is caused by the time-dependent processes (kinetic theory of the glass transition) occurring in the sample near the glass transition temperature. However, when similar test methods are used for samples, when similar thermal histories are imposed, and when samples are annealed at high temperature, the glass transition temperature is a good parameter for comparing the effects of asphalt additives and for characterizing asphalt components. Because the glass transition has a finite width somewhat characteristic of the material complexity (typically from 30 to 50°C wide in asphalts compared with 5 to 20°C in single-component systems), the width of the transition is also reported here.

Characterizing the glass transition region using DSC is not always straightforward. The asphalt designated AAA-1 is an example of a relatively simple case (Figure 3). This asphalt is one of the eight that was intensively studied during the Strategic Highway Research Program (SHRP). They are referred to as "core" asphalts and are coded with three capital letters to designate the asphalt source, followed by an Arabic numeral to specify viscosity grade. The specific heat capacity of asphalt AAA-1 changes smoothly through the glass transition

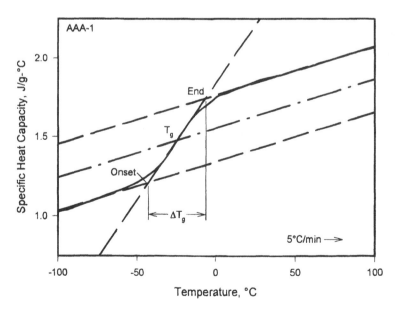

Figure 3 Specific heat capacity profile of the glass transition region of asphalt AAA-1.

region, making the measurement of T_g and of the transition width relatively easy. T_g is typically chosen as the half-vitrification temperature. The T_g is determined by first drawing tangents of the thermal signal above and below the glass transition region. Another line is then drawn halfway between the upper and lower tangents. The temperature at which this halfway line intersects the thermal signal in the glass transition region is T_g. This procedure is performed by computer in much less time than it takes to describe. The glass transition onset is defined as the temperature at which the tangent line below the glass transition intersects a tangent drawn through the steepest part of the glass transition. The glass transition end is similarly defined using the tangent above the glass transition region. The width of the transition, ΔT_g, is the temperature difference between the onset and end.

In some materials the selection of appropriate tangents above, below, or through the glass transition region is complicated by profile deviations caused by the crystallization or melting of asphalt components. This is evident in the heat capacity profile of asphalt AAF-1, where melting perturbs the profile above the glass transition region (Figure 4). To aid in the selection of tangents, we can fit an integrated Gaussian line shape to selected, unperturbed portions of thermograms. This is shown in Figure 4, where only portions of the signal are included in the curve fit. Because there is an underlying rise in heat capacity

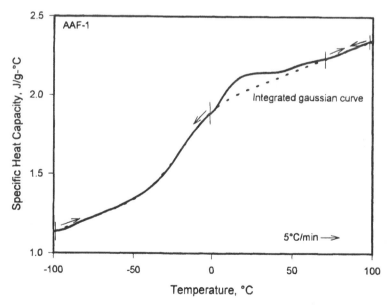

Figure 4 Specific heat capacity profile of the glass transition region of asphalt AAF-1 and integrated Gaussian shape.

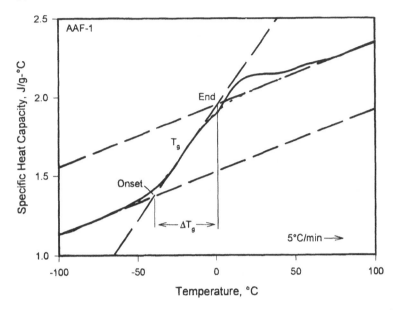

Figure 5 Specific heat capacity profile of the glass transition region of asphalt AAF-1 and selection of parameters using integrated Gaussian shape.

with increasing temperature, a sloped baseline must also be included in the fit. The slope is in units of J/(g °C^2). Figure 5 shows the tangents and parameters calculated from the fitted line shape. In this case the glass transition temperature is the midpoint of the Gaussian distribution.

II. EXAMINATION OF ASPHALTS AND ASPHALT COMPONENTS

A. Measurements on SHRP Core Asphalts

DSC/MDSC experiments were conducted on the SHRP core asphalts to provide a base for evaluating the effects of dopants and asphalt components. Glass transition parameters for the SHRP core asphalts were measured using the following standard method: anneal at 150°C for 15 min, cool at 10°C/min to −140°C, hold for 15 min, modulate ±0.75°C at 1 cycle/min, heat to 150°C at 5°C/min or 2°C/min. We have since modified our standard procedure to a 3°C/min heating rate with ±0.64°C modulation to accommodate larger samples, but we observe only minor changes in thermograms. Glass transition parameters were measured from the modulated heatup portion of each experiment. The sample (typically 5–6 mg) was spread into a standard aluminum pan with a crimped lid and then placed in the DSC cell. A pan and lid, the same weight as the sample pan and lid, were placed on the cell reference pedestal. All heating and cooling ramps and soaks were preprogrammed into the DSC controller.

The results of DSC/MDSC experiments on core asphalts at two heating rates are shown in Table 1. Parameters were calculated both with the standard instrument software (Tg, Onset, End) and with the Gaussian-fit method described above [Tg(G), Onset(G), End(G), Slope(G)]. Comparisons of glass transition temperatures measured by both methods are shown in Figure 6. There is a small difference between the two with the Gaussian method tending to somewhat lower values for the glass transition. For asphalts with low wax contents, such as AAA-1 and AAG-1, there is less than a degree difference between the methods. For asphalts with higher wax contents the scatter is greater. The most apparent outlier is the 5°C/min point for asphalt AAM-1, a relatively high-wax complex thermogram. Because wax crystallization and melting are just the phenomena used to justify the need for the integrated-Gaussian method, its not surprising that higher wax asphalts show larger deviations. More tests are needed to determine which method is more consistent.

Figure 7 compares the glass transition temperatures at the two heating rates. In the tests at the average 2°C/min heating rate, the modulation amplitude was large enough that the instantaneous heating rate was actually negative during part of each cycle. This is a good procedure for measuring reversible properties

Table 1 Glass Transition Characteristics of SHRP Asphalts

Asphalt	Htg. rate (°C/min)	T_g (G) (°C)	Onset (G) (°C)	End (G) (°C)	ΔT_g (G) (°C)	T_g (°C)	Onset (°C)	End (°C)	ΔT_g (°C)
AAA-1	5	−24.6	−42.8	−6.3	36.5	−23.9	−39.3	−8.3	30.9
AAB-1	5	−20.9	−45.6	3.8	49.4	−20.2	−41.6	1.5	43.1
AAC-1	5	−19.4	−41.7	2.9	44.5	−19.1	−38.6	0.6	39.2
AAD-1	5	−24.0	−43.9	−4.1	39.8	−24.9	−40.5	−9.2	31.3
AAF-1	5	−19.2	−39.7	1.2	41.0	−19.3	−37.3	−1.3	36.0
AAG-1	5	−7.8	−24.7	9.2	33.9	−7.6	−24.1	9.0	33.1
AAK-1	5	−20.6	−39.9	−1.3	38.6	−20.0	−36.9	−2.8	34.1
AAM-1	5	−20.8	−48.9	7.3	56.2	−18.0	−37.5	1.8	39.2
AAA-1	2	−23.4	−44.1	−2.7	41.5	−22.4	−39.1	−5.5	33.7
AAB-1	2	−21.7	−44.0	0.6	44.5	−20.9	−42.1	0.6	42.7
AAC-1	2	−18.1	−44.5	8.2	52.7	−18.6	−38.2	0.9	39.1
AAD-1	2	−23.4	−41.2	−5.6	35.6	−24.6	−40.9	−8.2	32.7
AAF-1	2	−16.2	−40.4	8.0	48.5	−17.4	−37.7	2.9	40.6
AAG-1	2	−7.2	−25.0	10.6	35.6	−7.0	−23.9	9.8	33.7
AAK-1	2	−20.1	−40.0	−0.2	39.8	−19.9	−37.0	−2.6	34.4
AAM-1	2	−20.0	−45.1	5.2	50.3	−19.9	−39.1	−0.3	38.8

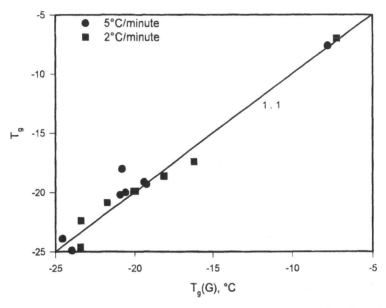

Figure 6 Comparison of glass transition temperatures calculated from half-vitrification and integrated gaussian fit.

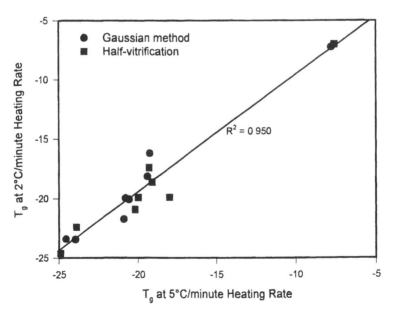

Figure 7 Comparison of glass transition temperatures measured at two heating rates.

like the heat capacity but is not recommended for kinetic phenomena like melting and crystallization. In the 5°C/min tests, the heating rate was always positive. As shown in Table 1 and Figure 7, the maximum differences in glass transition temperature measurement are about 3°C for the two heating rates with the 2°C/min values tending slightly to higher temperature.

B. Studies of IEC Neutral Fractions

Asphalts may be separated into their polar and nonpolar components by ion exchange chromatography (IEC). This separation produces defined chemical fractions (neutral, acidic, basic, amphoteric), not solubility classes. The IEC neutral fractions comprise all nonpolar asphalt components, which make up 50–60% of neat asphalts. The IEC neutral fractions used in this study were prepared in the SHRP program using the methods described in the *Binder Characterization and Evaluation* final report, Volume 4. These materials are relatively volatile, losing weight at high annealing temperatures. The DSC/MDSC methods were modified to limit the high temperature to 100°C to minimize sample weight loss.

DSC/MDSC results show narrow glass transitions located 10°C or more below the transitions for the whole asphalts (Table 2). Individual thermograms were complex with substantial exo- and endothermic excursions caused by crystallization and melting. The glass transition temperatures are plotted as a function of reciprocal molecular weight in Figure 8. A linear least-squares fit to the data produces an expression in the form of equation 4, where $T_{g\infty} = -17.5°C$ and $K = 12728°C$-daltons. Presumably, this relationship fits the data because the IEC neutrals contain similar, relatively noninteracting, molecules at all molecular weights. Because glass transitions occur only in amorphous structure and substan-

Table 2 DSC/MDSC Results for IEC Neutral Fractions

Asphalt	Heating rate (°C/min)	Cooling rate (°C/min)	T_g (°C)	Onset (°C)	End (°C)	ΔT_g	M_n
AAA-1	2	10	−38.4	−44.3	−32.3	12.1	590
AAA-1	5	10	−39.3	−47.5	−32.1	15.4	590
AAB-1	2	10	−36.8	−45.4	−30.7	14.7	660
AAC-1	2	10	−33.8	−42.2	−25.3	16.9	750
AAD-1	2	10	−42.9	−52.2	−33.8	18.4	510
AAD-1	5	10	−43.4	−50.9	−36.2	14.7	510
AAF-1	2	10	−31.1	−40.3	−22.1	18.2	700
AAK-1	2	10	−37.7	−45.8	−29.6	16.2	590
AAM-1	2	10	−28.6	−41.0	−15.9	25.1	1140
AAM-1	5	10	−29.1	−40.2	−18.3	21.9	1140

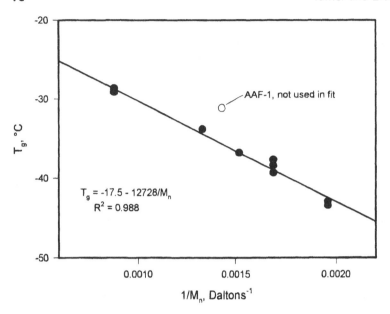

Figure 8 Comparison of IEC neutral fraction molecular masses with glass transition temperatures.

tial crystallization occurred during cooldown in these tests, the glass transition temperatures measured are properties of the portion of the IEC neutrals that did not associate.

C. Studies of Wax Content

Many asphalts are known to contain significant amounts of waxes, whereas others are virtually wax free. Most waxes found in asphalts are long-chain *n*-alkanes that are crystalline solids at ambient temperatures. It is not definitely known whether waxes in asphalts form separate phases during pavement service lifetimes. If they do, low-temperature properties must surely be affected.

Many procedures can be used to separate waxes from petroleum and petroleum products, and each produces a wax unique to that procedure. In some crude oils, waxes are concentrated using a solvent precipitation method. When applied to asphalts, this simple, one-step method results in the precipitation of much polar material along with waxes. Thus waxes concentrated from asphalts by this method are highly impure. A purer wax fraction may be isolated from asphalts by a two-step procedure. Polar and nonpolar components are first separated by IEC. The nonpolar components (neutrals) are then dissolved in toluene, and then methyl ethyl ketone is added to the chilled solution to precipitate waxes. Some of the asphalts used in this study contain very small amounts of waxes, but in others wax contents are substantial [9].

Wax samples were examined using the DSC/MDSC techniques described earlier with minor modifications. Crystallization exotherms were integrated in both the unmodulated cooldown and the modulated heatup signals. The total exotherm areas in joules per gram of asphalt were converted to crystallized fractions using a value of 200 joules per gram of crystallized material [10], which assumes the crystallized material is predominantly long-chain, normal alkanes.

Asphalts AAB-1, AAF-1, and AAG-1 were dewaxed in one step by solvent precipitation. This procedure yields a dewaxed asphalt and a wax concentrate, as explained above. The wax concentrate was not used in subsequent doping experiments. DSC/MDSC results for these whole asphalts, dewaxed asphalts, and wax concentrates are listed in Table 3. The three wax concentrate samples showed little evidence of any glass transition but had large crystallization exotherms during cooling. The dewaxed asphalts had smooth, well-defined glass transition profiles and only minor crystallization exotherms. Crystallized fractions of the whole asphalts ranged from 0.2 wt % for AAG-1 to 2.5 wt % for AAB-1. The ratio of crystallized fraction to wax content for the AAG-1 wax indicates that only 3 wt % of the AAG-1 wax contributes to the crystallization exotherms compared with 8 wt % and 13 wt % for AAB-1 and AAF-1 waxes. In general, the same can be said for the contributions of the waxes in the whole asphalts. The crystallized fraction/wax content ratios for the whole asphalts are similar to those for the waxes.

Waxes isolated from asphalts using the two-step procedure do not contain polar impurities. These waxes were added in small amounts as "dopants" to various asphalts. In some experiments, asphalts were doped with their indigenous waxes. In others, asphalts were doped with waxes from different asphalts. The effect of wax content on glass transition temperature was quantified using a dopant efficiency term defined as the magnitude of the shift in glass transition in °C divided by the amount of dopant added in wt %. A similar expression is used in the polymer industry to quantify plasticizer efficiency [11]. To reinforce the concept that a shift of the glass transition to lower temperature improves pavement performance, the efficiency term in this work is defined as negative when the glass transition temperature decreases with increasing dopant concentration. The last column in Table 3 lists the efficiencies of the waxes for changing the glass transition temperatures. None of the three asphalts show any improvement with increasing wax content, although asphalt AAF-1 shows no sensitivity to wax content at all. In asphalt AAG-1, each percent of wax added increases the glass transition temperature by 1°C.

The results of wax-doping experiments are summarized in Table 4 through Table 16. The first four tables (4, 5, 6, and 7) describe sample types, pretreatment conditions, DSC heating and cooling rates, glass transition parameters, and crys-

Table 3 DSC/MDSC Results for Asphalts AAB-1, AAF-1, and AAG-1

Asphalt	Type of material	Wax content (mass %)	Sample no.	Heating rate (°C/min)	Cooling rate (°C/min)	T_g (°C)	Onset (°C)	End (°C)	ΔT_g (°C)	Total exotherm (J/g)	% CF assuming 200 J/g	% CF/ % wax ratio	Dopant efficiency (°C/mass %)
AAB-1	Asphalt	20.47		5	10	−20.2	−40.9	0.5	41.4	4.9	2.5	0.12	0.45
AAB-1	Waxes	100	843-11-W1	5	10	—				16.9	8.4	0.08	
AAB-1	Dewaxed	0	843-11-M1	5	10	−29.5	−43.0	−15.8	27.2	0.8	0.4		
AAF-1	Asphalt	14.89		5	10	−18.0	−36.3	0.3	36.7	4.9	2.4	0.16	0.03
AAF-1	Waxes	100	843-13-W1	5	10	—				26.1	13.0	0.13	
AAF-1	Dewaxed	0	843-13-M1	5	10	−18.4	−38.3	1.6	39.9	2.4	1.2		
AAG-1	Asphalt	4.07		5	10	−7.9	−24.7	8.9	33.6	0.4	0.2	0.05	1.03
AAG-1	Waxes	100	843-12-W1	5	10	—				6.9	3.4	0.03	
AAG-1	Dewaxed	0	843-12-M1	5	10	−12.1	−31.0	7.2	38.3	0.0	0.0		

Table 4 Asphalt AAA-1 Doping Experiments: DSC Results

Amount added (mass %)	Source of material added	Type of material added	Sample pretreatment conditions	Heating rate (°C/min)	Cooling rate (°C/min)	T_g (°C)	Onset (°C)	End (°C)	ΔT_g (°C)	Total exotherm (J/g)
		Whole	a	2	10	−23.9	−39.3	−7.5	31.8	0.9
				5	10	−24.4	−40.6	−7.6	33.0	0.2
		Whole	b	2	10	−24.6	−39.3	−10.0	29.3	1.0
		Neutrals		5	10	−39.5	−47.6	−31.7	15.9	2.9
		Waxes		5	10	−19.1	−43.5	5.5	48.9	58.0
		Waxes		2	Quench	−18.2	−36.0	−1.0	35.0	
4.25	AAA-1	Waxes	c	5	10	−26.3	−42.1	−9.9	32.1	4.5
4.84	AAD-1	Waxes	c	2	10	−25.6	−40.3	−10.6	29.7	6.4
				5	10	−25.6	−42.7	−8.2	34.6	5.8
5.08	AAM-1	Waxes	c	5	10	−26.1	−43.6	−7.7	35.9	2.0

a Anneal at 150°C for 30 min, no solvent treatment.
b Anneal at 150°C for 30 min after solvent treatment.
c Anneal at 150°C for 30 min after solvent removal.

Table 5 Asphalt AAC-1 Doping Experiments: DSC Results

Amount added (mass %)	Source of material added	Type of material added	Sample pretreatment conditions	Heating rate (°C/min)	Cooling rate (°C/min)	T_g (°C)	Onset (°C)	End (°C)	ΔT_g (°C)	Total exotherm (J/g)
		Whole	a	5	10	−19.1	−38.6	0.6	39.2	5.9
		Whole	b	5	10	−19.6	−39.4	1.5	40.8	3.7
		Dewaxed Waxes	c	5	10	−21.9	−38.7	−4.9	33.8	34.6
4.18	AAC-1	Waxes	c	5	10	−20.3	−38.6	−1.7	36.9	6.7
4.42	AAC-1	Waxes	c	5	10	−18.6	−38.4	2.2	40.7	4.6
5.73	AAM-1	Waxes	c	5	10	−20.4	−38.7	−1.1	37.6	10.2

[a] Anneal at 150°C for 30 min. no solvent treatment.
[b] Anneal at 150°C for 30 min after solvent treatment.
[c] Anneal at 150°C for 30 min after solvent removal.

Table 6 Asphalt AAD-1 Doping Experiments: DSC Results

Amount added (mass %)	Source of material added	Type of material added	Sample pretreatment conditions	Heating rate (°C/min)	Cooling rate (°C/min)	T_g (°C)	Onset (°C)	End (°C)	ΔT_g (°C)	Total exotherm (J/g)
		Whole	a	2	10	−23.8	−40.4	−6.6	33.8	2.0
		Whole		5	10	−24.7	−41.1	−7.6	33.5	2.0
		Whole	b	2	10	−23.4	−42.0	−4.2	37.8	2.4
		Neutrals		5	10	−43.7	−50.7	−36.7	14.0	5.8
		Waxes		5	10	−5.0	−40.7	30.7	71.4	120.3
5.34	AAA-1	Waxes	c	5	10	−25.3	−45.5	−5.4	40.1	4.8
5.21	AAD-1	Waxes	c	2	10	−23.0	−43.4	−1.5	41.9	7.8
				5	10	−24.0	−44.9	−1.5	43.4	7.9
4.46	AAM-1	Waxes	c	5	10	−25.2	−42.7	−7.5	35.2	2.8
4.96	AAM-1	Waxes	c	5	10	−24.5	−42.9	−5.7	37.2	2.6

a Anneal at 150°C for 30 min. no solvent treatment.
b Anneal at 150°C for 30 min after solvent treatment.
c Anneal at 150°C for 30 min after solvent removal.

Table 7 Asphalt AAM-1 Doping Experiments: DSC Results

Amount added (mass %)	Source of material added	Type of material added	Sample pretreatment conditions	Heating rate (°C/min)	Cooling rate (°C/min)	T_g (°C)	Onset (°C)	End (°C)	ΔT_g (°C)	Total exotherm (J/g)
		Whole	a	5	10	−20.2	−39.3	0.5	39.8	7.5
		Neutrals		5	10	−29.0	−39.0	−19.1	19.9	14.4
		Waxes		5	10	−36.1	−46.8	−25.4	21.5	33.6
		Waxes		2	Quench	−36.8	−51.3	−20.9	30.4	
4.11	AAA-1	Waxes	c	5	10	−19.3	−39.4	2.3	41.7	9.8
6.09	AAD-1	Waxes	c	5	10	−12.8	−40.3	15.7	56.0	9.3
5.91	AAD-1	Waxes	c	5	10	−19.5	−38.3	0.1	38.4	10.1
4.84	AAM-1	Waxes		5	10	−26.3	−44.0	−8.0	36.0	3.1

[a] Anneal at 150°C for 30 min. no solvent treatment.
[c] Anneal at 150°C for 30 min after solvent removal.

tallization exotherms for asphalts AAA-1, AAC-1, AAD-1, and AAM-1. The IEC neutral fractions of these asphalts always have lower glass transition temperatures than the whole asphalts and the widths of the glass transition regions of the IEC neutral fractions are about half the widths of the whole asphalts and are more typical of simple miscible systems. The wax from asphalt AAM-1 has an anomalously low glass transition temperature, substantially lower even than the glass transition temperature of the AAM-1 IEC neutral fraction. Waxes from asphalts AAA-1 and AAD-1 have glass transition temperatures above those of the whole asphalts and the IEC neutral fractions.

Tables 8 through 11 list the calculated crystallized fractions and total wax contents for the same four asphalts. The wax contents for asphalts AAA-1, AAD-1, and AAM-1 were calculated from the wax contents of the IEC neutral fractions and the weight percent IEC neutrals in the whole asphalts. For the doped asphalts, the total wax content was assumed to be the sum of the doped wax and the native wax contents. The relationships between crystallized fraction and wax content for the seven asphalts (now including the limited data for AAB-1, AAF-1, and AAG-1) are shown in Figure 9. The plot for each asphalt used in the doping study contains the data for the whole asphalt, the IEC neutrals (except AAC-1), the wax, and the whole asphalt doped with its own wax. The plots for asphalts AAB-1, AAF-1, and AAG-1 use the data listed in Table 3. Although the evidence is not conclusive, it appears that the waxes from different asphalts have characteristic crystallized fractions that are not dependent on the concentration of the wax in the sample. The slopes of the lines in Figure 9 provide the characteristic ratios

Table 8 Asphalt AAA-1 Doping Experiments: Crystallized Fraction and Total Wax

Amount added (mass %)	Source of material added	Type of material added	Sample pretreatment conditions	Heating rate (°C/min)	% CF assuming 200 J/g	Estimated total wax (mass %)	% CF/ % wax ratio
		Whole	*a*	2	0.5	1.2	0.36
		Whole		5	0.1	1.2	0.07
		Whole	*b*	2	0.5	1.2	0.39
		Neutrals		5	1.4	2.1	0.69
		Waxes		5	29.0	100.0	0.29
		Waxes		2		100.0	
4.25	AAA-1	Waxes	*c*	5	2.2	5.4	0.41
4.84	AAD-1	Waxes	*c*	2	3.2	6.0	0.53
		Waxes		5	2.9	6.0	0.48
5.08	AAM-1	Waxes	*c*	5	1.0	6.3	0.16

[a] Anneal at 150°C for 30 min, no solvent treatment.
[b] Anneal at 150°C for 30 min after solvent treatment.
[c] Anneal at 150°C for 30 min after solvent removal.

Table 9 Asphalt AAC-1 Doping Experiments: Crystallized Fraction and Total Wax

Amount added (mass %)	Source of material added	Type of material added	Sample pretreatment conditions	Heating rate (°C/min)	% CF assuming 200 J/g	Estimated total wax (mass %)	% CF/ % wax ratio
		Whole	a	5		11.9	
		Whole	b	5	2.9	11.9	0.25
		Dewaxed	c	5	1.9	0.0	
		Waxes		5	17.3	100.0	0.17
4.18	AAC-1	Waxes	c	5	3.4	15.6	0.21
4.42	AAC-1	Waxes	c	5	2.3	15.8	0.14
5.73	AAM-1	Waxes	c	5	5.1	17.0	0.30

[a]Anneal at 150°C for 30 min, no solvent treatment.
[b]Anneal at 150°C for 30 min after solvent treatment.
[c]Anneal at 150°C for 30 min after solvent removal.

Table 10 Asphalt AAD-1 Doping Experiments: Crystallized Fraction and Total Wax

Amount added (mass %)	Source of material added	Type of material added	Sample pretreatment conditions	Heating rate (°C/min)	% CF assuming 200 J/g	Estimated total wax (mass %)	% CF/ % wax ratio
		Whole	a	2	1.0	1.7	0.59
				5	1.0	1.7	0.59
		Whole	b	2	1.2	1.7	0.71
		Neutrals		5	2.9	3.3	0.89
		Waxes		5	60.1	100.0	0.60
		Waxes		2			
5.34	AAA-1	Waxes	c	5	2.4	6.9	0.35
5.21	AAD-1	Waxes	c	2	3.9	6.8	0.57
				5	3.9	6.8	0.58
4.46	AAM-1	Waxes	c	5	1.4	6.1	0.23
4.96	AAM-1	Waxes	c	5	1.3	6.6	0.20

[a]Anneal at 150°C for 30 min, no solvent treatment.
[b]Anneal at 150°C for 30 min after solvent treatment.
[c]Anneal at 150°C for 30 min after solvent removal.

Table 11 Asphalt AAM-1 Doping Experiments: Crystallized Fraction and Total Wax

Amount added (mass %)	Source of material added	Type of material added	Sample pretreatment conditions	Heating rate (°C/min)	% CF assuming 200 J/g	Estimated total wax (mass %)	% CF/ % wax ratio
		Whole	a	5	3.7	27.5	0.14
		Neutrals		5	7.2	51.5	0.14
		Waxes		5	16.8	100.0	0.17
		Waxes		2			
4.11	AAA-1	Waxes	b	5	4.9	30.5	0.16
6.09	AAD-1	Waxes	b	5	4.7	31.9	0.15
5.91	AAD-1	Waxes	b	5	5.1	31.8	0.16
4.84	AAM-1	Waxes	b	5	1.6	31.0	0.05

[a] Anneal at 150°C for 30 min, no solvent treatment.
[b] Anneal at 150°C for 30 min after solvent removal.

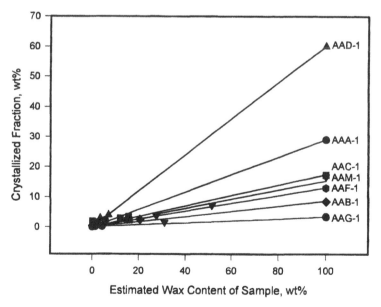

Figure 9 Crystallized fractions of waxes, wax-doped asphalts, and asphalts compared to sample wax contents.

is not conclusive, it appears that the waxes from different asphalts have characteristic crystallized fractions that are not dependent on the concentration of the wax in the sample. The slopes of the lines in Figure 9 provide the characteristic ratios for the asphalts (Table 12). Individual sample ratios of crystallized fraction to wax content are listed in the last columns of Tables 8 through 11.

If the wax from each asphalt has a characteristic crystallizable fraction, as was suggested above, it should be possible to predict the crystallized fractions of all the cross-doped asphalt mixtures in this study. This is accomplished for each mixture by multiplying the weight percent of each wax type in the mixture by its characteristic crystallized fraction and then adding the products for both asphalts in the mixture. This simple procedure assumes the crystallizable fractions will add and that no interactions occur between the different waxes. The results of this exercise are shown in Figure 10, where the calculated and measured crystallized fractions are compared. Outlying points above the 1:1 line in this figure are combinations that result in less measured crystallization than expected. The points below the line are combinations resulting in more crystallization than expected. All combinations that deviate substantially from the 1:1 line have asphalt AAM-1 involved either as the host asphalt or as the dopant wax source. Asphalt AAM-1 wax even appears to inhibit crystallization in asphalt AAM-1.

The dopant efficiency was calculated for all the wax-doped asphalts and is listed in Tables 13 through 16 for host asphalts AAA-1, AAC-1, AAD-1, and AAM-1, respectively. AAA-1 wax is beneficial in hosts AAA-1 and AAD-1 but is detrimental in AAM-1. AAD-1 wax is beneficial only in host AAA-1 and is strongly detrimental in host AAM-1. AAM-1 wax is beneficial in all hosts, but the effect is greatest in host AAM-1.

D. Studies of SEC Fractions

In addition to being separated into defined chemical fractions by IEC, asphalts were separated into fractions of different molecular sizes by means of size

Table 12 Characteristic Crystallized Fraction/Wax Content Ratios

Asphalt	% CF/% wax ratio
AAA-1	0.29
AAB-1	0.086
AAC-1	0.17
AAD-1	0.60
AAF-1	0.13
AAG-1	0.034
AAM-1	0.15

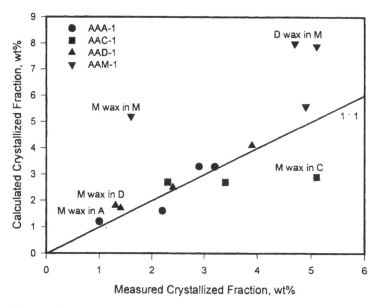

Figure 10 Demonstration of additive behavior of waxes in asphalts.

Table 13 Asphalt AAA-1 Doping Experiments: Dopant Efficiency

Asphalt	Amount added (mass %)	Source of material added	Type of material added	Dopant efficiency (°C/mass %)
AAA-1			Whole	0.00
AAA-1	4.25	AAA-1	Waxes	−0.56
AAA-1	4.84	AAD-1	Waxes	−0.35
AAA-1	5.08	AAM-1	Waxes	−0.42

Table 14 Asphalt AAC-1 Doping Experiments: Dopant Efficiency

Asphalt	Amount added (mass %)	Source of material added	Type of material added	Dopant efficiency (°C/mass %)
AAC-1			Whole	0.00
AAC-1	4.18	AAC-1	Waxes	−0.17
AAC-1	4.42	AAC-1	Waxes	0.23
AAC-1	5.73	AAM-1	Waxes	−0.13

Table 15 Asphalt AAD-1 Doping Experiments: Dopant Efficiency

Asphalt	Amount added (mass %)	Source of material added	Type of material added	Dopant efficiency (°C/mass %)
AAD-1			Whole	0.00
AAD-1	5.34	AAA-1	Waxes	−0.28
AAD-1	5.21	AAD-1	Waxes	0.15
AAD-1	4.46	AAM-1	Waxes	−0.32
AAD-1	4.96	AAM-1	Waxes	−0.14

Table 16 Asphalt AAM-1 Doping Experiments: Dopant Efficiency

Asphalt	Amount added (mass %)	Source of material added	Type of material added	Dopant efficiency (°C/mass %)
AAM-1			Whole	0.00
AAM-1	4.11	AAA-1	Waxes	0.22
AAM-1	6.09	AAD-1	Waxes	1.21
AAM-1	5.91	AAD-1	Waxes	0.12
AAM-1	4.84	AAM-1	Waxes	−1.27

exclusion chromatography (SEC). In this technique, solutions of an asphalt are injected into a column containing specially prepared gels swelled in a solvent. The gel beads contain numerous pores of varying sizes. Larger components of asphalts cannot enter the gel pores, so they pass through the column between gel particles along with the solvent front. Smaller molecules reside in the pores for times that are inversely related to their molecular size. Thus, in an SEC separation of asphalts, the largest entities elute first, followed by those of moderate size, and the smallest molecules elute last. SEC fractions were isolated using a preparative size exclusion chromatography method described in detail elsewhere [12]. The SEC I (or 1) fraction contains the highest molecular size materials. Fractions 2–9 have decreasing molecular sizes. Fractions 8 and 9 are combined with fraction 7 to provide enough material for analyses. It is important for this discussion to recognize also that these separations are performed on solutions of asphalts in toluene, not in the condensed phase. DSC/MDSC experiments were conducted as described earlier except that the maximum temperature was limited to 100°C to reduce weight loss during testing.

Thermal analysis results for the SEC fractions of asphalts AAA-1, AAC-1, AAD-1, and AAF-1 are listed in Tables 17 through 20 along with the mass

Table 17 DSC of SEC Fractions of Asphalt AAA-1

SEC fraction	Mass fraction	M_n (daltons)	T_g (°C)	Onset (°C)	End (°C)	ΔT_g (°C)
1	0.216	11000	>100	—	—	—
2	0.104	2200	−4.2	−25.1	17.1	42.1
3	0.140	1200	−27.4	−42.7	−11.8	30.9
4	0.213	730	−44.6	−53.1	−36.5	16.8
5	0.211	540	−45.6	−55.6	−34.8	20.8
6	0.093	390	−11.1	−18.1	−4.4	13.8
7	0.021	NA	−3.2	−9.4	2.6	12.0
II (2–7)	0.782		−29.5	−43.6	−15.6	28.1
Total	0.998		−23.9	(whole asphalt)		
T_g calculated from 2–7			−32.4			

Table 18 DSC of SEC Fractions of Asphalt AAC-1

SEC fraction	Mass fraction	M_n (daltons)	T_g (°C)	Onset (°C)	End (°C)	ΔT_g (°C)
1	0.136	7380	>100	—	—	—
2	0.12	1610	−29.1	−44.3	−14.4	29.8
3	0.256	1000	−38.3	−49.4	−27.3	22.1
4	0.291	780	−26.1	−40.5	−11.3	29.2
5	0.138	610	16.9	6.0	27.7	21.7
6	0.047	490	35.7	25.9	45.6	19.7
7	0.012	NA	−9.0	−18.7	0.1	18.8
II (2–7)	0.864		−18.8	−40.1	4.1	44.2
Total	1.000		−19.1	(whole asphalt)		
T_g calculated from 2–7			−21.5			

fractions and number average molecular weights for each fraction taken from Volume 3, Chapter 2 of the SHRP *Binder Characterization and Evaluation* final report [13]. Also included in the tables are the glass transition parameters for tests on SEC II fractions, which are equivalent to the sum of fractions 2 through 7. The glass transition temperatures for the fractions range from above the maximum test temperature for SEC fraction 1 to 15 or more degrees below the whole asphalt glass transition temperature. Applying Equation 4 to the data in Tables 17 to 20 gives a calculated estimate of the glass transition temperature for the SEC II fraction for each asphalt. This calculated value is listed in the tables. For asphalts AAA-1, AAC-1, and AAF-1 the difference from the measured

Table 19 DSC of SEC Fractions of Asphalt AAD-1

SEC fraction	Mass fraction	M_n (daltons)	T_g (°C)	Onset (°C)	End (°C)	ΔT_g (°C)
1	0.234	7000	>100	—	—	—
2	0.094	2200	18.8	−2.9	40.3	43.2
3	0.112	1200	−10.4	−27.6	6.6	34.2
4	0.176	700	−38.3	−52.0	−24.3	27.8
5	0.223	470	−34.1	−42.6	−25.7	16.9
6	0.123	360	−29.6	−36.5	−22.9	13.7
7	0.030	NA	−14.3	−20.2	−8.8	11.4
II (2–7)	0.758		−31.9	−43.5	−20.0	23.5
Total	0.992		−24.9	(whole asphalt)		
T_g calculated from 2–7			−24.7			

Table 20 DSC of SEC Fractions of Asphalt AAF-1

SEC fraction	Mass fraction	M_n (daltons)	T_g (°C)	Onset (°C)	End (°C)	ΔT_g (°C)
1	0.139	8690	>100	—	—	—
2	0.118	1970	−9.0	−40.4	24.3	64.7
3	0.199	1110	−30.4	−45.0	−16.1	28.9
4	0.270	770	−34.4	−46.8	−21.9	24.9
5	0.182	610	−10.4	−21.5	0.7	22.2
6	0.076	450	7.2	−2.7	17.2	19.9
7	0.020	NA	−0.1	−9.5	9.2	18.7
II (2–7)	0.865		−22.1	−39.2	−4.7	34.6
Total	1.004		−19.3	(whole asphalt)		
T_g calculated from 2–7			−21.3			

SEC II T_g is less than 5°C. The exception is AAD-1, for which the difference is 10°C. This indicates that the AAD-1 SEC fractions may be less compatible with each other than in the other three asphalts.

Because measurements were made on the whole asphalts and on the SEC II fractions, it is possible to approximate the effect of the SEC I fraction on the glass transition temperature. This is accomplished by calculating efficiencies using the mass percent of SEC fraction 1 and the difference between the whole asphalt and SEC II glass transition temperatures. The efficiencies are 0.26, −0.03, 0.28, and 0.20 for asphalts AAA-1, AAC-1, AAD-1, and AAF-1, respectively (Table 21). The addition of fraction 1 raises the glass transition temperature in all asphalts except AAC-1, where it has little effect.

The effects of the other SEC fractions for the asphalts can be estimated because Equation 3 predicts the SEC II glass transition temperature reasonably well for all except AAD-1. Efficiencies were calculated for the addition of a fraction to a mixture of all other fractions. The results are shown in Table 21. Based on these estimates, it might be more advantageous to remove high or low fractions than to add more of one of the middle fractions. This cannot be confirmed without testing specially reconstituted asphalts containing and deficient in the critical fractions.

Glass transition temperatures for the SEC fractions are plotted against reciprocal molecular weights in Figure 11. Fraction 1 is on the left side of each plot with higher fractions following in order. Obviously, the relationship expressed in Equation 2 cannot describe the SEC data. At molecular masses above 1000 daltons the glass transitions follow the trend of increasing temperature with increasing molecular weight. However, below 1000 daltons all the asphalts examined show minima in their glass transition temperatures followed by substantial increases as the molecular mass decreases further. This behavior may be caused by the presence of low-molecular-weight polar materials in the asphalts. These molecules would be separated in the toluene solution passing through the SEC column but would interact significantly in the condensed phase examined with the DSC. Evidence in favor of this idea can be seen in the increased nitrogen, oxygen, and sulfur contents of SEC fraction 6 in asphalts AAB-1 and AAM-1. Similar results for other asphalts have been reported by Huynh et al. [14].

DSC/MDSC experiments were conducted on mixtures of SEC I and SEC II fractions to evaluate the relative influence of the fractions on the glass transition of the mixtures. The SEC II fraction of a core asphalt was mixed with SEC I fractions from all eight core asphalts at the natural abundance level of the SEC II asphalt. At this time seven of the eight sets of samples have been examined using MDSC.

Table 21   SEC Fraction Efficiencies

	SEC fraction						
Asphalt	1[a]	2[b]	3[b]	4[b]	5[b]	6[b]	7[b]
AAA-1	0.26	0.29	0.06	−0.18	−0.20	0.22	0.27
AAC-1	−0.03	−0.09	−0.26	−0.07	0.39	0.49	0.12
AAD-1	0.30	0.41	0.16	−0.19	−0.14	−0.06	0.10
AAF-1	0.20	0.13	−0.12	−0.21	0.13	0.28	0.20

[a] Calculated for whole asphalt.
[b] Calculated for SEC-II.

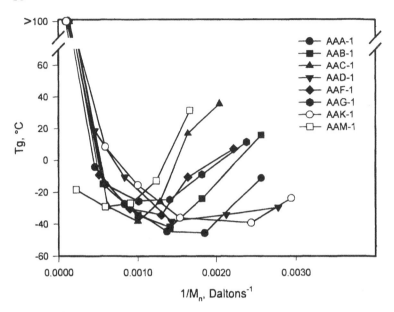

Figure 11   Number average molecular mass effect on the glass transition temperatures of SEC fractions.

Glass transition parameters were measured from the modulated heatup portion of each experiment. The sample (typically ~15 mg) was spread into an aluminum pan with a hermetically sealed lid and then placed in the DSC cell. A pan and lid, the same weight as the sample pan and lid, were placed on the cell reference pedestal. All heating and cooling ramps and soaks were preprogrammed into the DSC controller.

The results of DSC/MDSC experiments on the core asphalts and their SEC II fractions are shown in Table 22 for reference. The glass transition temperatures listed are averages of five determinations of separate samples (same storage can). The differences between the asphalt T_g values and the SEC II T_g values range up to 8°C, with the SEC II fractions having the lower T_g values, except for AAC-1, where the values are very close. The differences between asphalt and SEC II fraction T_g values must be caused by the SEC I fractions in the asphalts.

DSC results for blends with asphalt AAA-1 SEC II fraction show that the SEC I source asphalt has little influence on the glass transition except for SEC I from asphalt AAM-1 (Table 23). Because asphalts AAM-1 and AAF-1 were solvent deasphaltened in the refining process and might be expected to behave somewhat differently than the other core asphalts, Table 23 has been organized to show average values and standard deviations with and without these asphalts.

Table 22 Summary of SEC-I Source Asphalt Properties

Asphalt	$T_g{}^a$ (°C)	SEC-II T_g (°C)
AAA-1	−24.8	−29.5
AAB-1	−20.9	−26.5
AAC-1	−20.6	−18.8
AAD-1	−25.3	−31.9
AAF-1	−14.1	−22.1
AAG-1	−8.4	−10.6
AAK-1	−20.8	−29.6
AAM-1	−19.0	−20.4

a Average of five samples.

The average T_g for the six blends is the same as the T_g of whole asphalt AAA-1 with a standard deviation of less than 1°C. The blend with the SEC I fraction from asphalt AAM-1 has a lower T_g than the AAA-1 SEC II fraction. This indicates that the SEC I fraction of AAM-1 must have properties such that it is more closely related to the solvent component of AAM-1 than are SEC I fractions of other asphalts.

Table 23 DSC Results for Mixtures of SEC Fraction II of AAA-1 (78.3%) with SEC Fraction I (21.7%) of Eight Asphalts

SEC-I source asphalt	Mixture properties					
	T_g (°C)	ΔC_p (J/g-K)	ΔT_g (°C)	Onset (°C)	End (°C)	CF^a (mass %)
AAA-1	−24.7	0.25	31.8	−40.6	−8.9	0.5
AAB-1	−24.3	0.27	30.9	−39.6	−8.8	0.4
AAC-1	−23.5	0.26	34.2	−40.6	−6.3	0.4
AAD-1	−22.3	0.27	34.3	−39.4	−5.1	0.2
AAF-1	−25.0	0.27	32.2	−40.9	−8.7	0.6
AAG-1	−24.5	0.24	28.6	−38.7	−10.1	0.5
AAK-1	−24.3	0.25	31.4	−39.9	−8.5	0.2
AAM-1	−30.6	0.31	25.0	−43.1	−18.1	0.9
Average	−24.9	0.26	31.0	−40.3	−9.3	0.5
Std. dev.	2.5	0.02	3.1	1.3	3.9	0.2
Excluding AAM-1 and AAF-1						
Average	−23.9	0.26	31.9	−39.8	−7.9	0.4
Std. dev.	0.9	0.01	2.1	0.7	1.9	0.1

a Assumes 200 J/g for CF heat of crystallization.

Similar results are observed with the other SEC blends, which are listed in Tables 24 through 28 for SEC II fractions of asphalts AAB-1, AAC-1, AAD-1, AAF-1, and AAG-1, respectively. The SEC II fraction determines the base glass transition temperature. Addition of SEC I fractions shifts the transition temperature. With one exception, the shift in glass transition temperature is independent of the source of the fraction of SEC I. For all SEC II fractions blended with SEC I from asphalt AAM-1, the blend has a lower T_g than the whole SEC II source asphalt.

E. Measurements on Asphaltene-Maltene Mixtures

Asphalts are complex mixtures of many different kinds of organic compounds. These compounds may differ from one another greatly in both polarity and molecular size or may closely resemble each other in these properties. The more polar compounds interact with one another by hydrogen bonding, π-π interactions, or other weak forces to form molecular associations (self-assemblies) of varying strengths and sizes.

One model of asphalt structure considers asphalts to be dispersions of polar, aromatic molecules in a less polar, less aromatic solvent component. The properties of an asphalt will be determined by the effectiveness with which the dispersed component is solvated. In some asphalts, the dispersed component is

Table 24 DSC Results for Mixtures of SEC Fraction II of AAB-1 (79.2%) with SEC Fraction I (20.8%) of Eight Asphalts

SEC-I source asphalt	Mixture properties					
	T_g (°C)	ΔC_p (J/g-K)	ΔT_g (°C)	Onset (°C)	End (°C)	CF[a] (mass %)
AAA-1	−21.2	0.32	41.8	−41.7	0.1	3.0
AAB-1	−15.4	0.32	43.5	−38.1	5.4	2.4
AAC-1	−19.3	0.31	39.0	−39.4	−0.4	2.5
AAD-1	−19.3	0.32	49.9	−43.8	6.1	2.3
AAF-1	−22.7	0.31	42.9	−43.5	−0.6	3.3
AAG-1	−22.3	0.30	39.8	−41.9	−2.1	2.8
AAK-1	−21.0	0.31	45.8	−43.2	2.6	3.1
AAM-1	−25.2	0.35	34.4	−42.1	−7.7	2.0
Average	−20.8	0.32	42.1	−41.7	0.4	2.7
Std. dev.	2.9	0.01	4.6	2.0	4.4	0.4
Excluding AAM-1 and AAF-1						
Average	−19.8	0.31	43.3	−41.3	2.0	2.7
Std. dev.	2.4	0.01	4.1	2.2	3.3	0.3

[a] Assumes 200 J/g for CF heat of crystallization.

Table 25   DSC Results for Mixtures of SEC Fraction II of AAC-1 (86.4%) with SEC Fraction I (13.6%) of Eight Asphalts

SEC-I source asphalt	Mixture properties					
	T_g (°C)	ΔC_p (J/g-K)	ΔT_g (°C)	Onset (°C)	End (°C)	CF^a (mass %)
AAA-1	−18.0	0.40	46.7	−40.8	5.9	2.8
AAB-1	−18.9	0.39	42.6	−39.6	3.1	3.1
AAC-1	−19.1	0.38	42.9	−40.0	2.9	2.7
AAD-1	−20.0	0.41	50.4	−44.9	5.4	2.7
AAF-1	−18.8	0.40	42.6	−39.6	3.0	3.4
AAG-1	−17.5	0.39	43.9	−39.1	4.8	2.8
AAK-1	−18.4	0.41	42.8	−39.5	3.3	3.1
AAM-1	−21.0	0.37	37.9	−39.7	−1.8	3.6
Average	−19.0	0.39	43.7	−40.4	3.3	3.0
Std. dev.	1.1	0.01	3.6	1.9	2.4	0.3
Excluding AAM-1 and AAF-1						
Average	−18.6	0.40	44.9	−40.6	4.3	2.9
Std. dev.	0.9	0.01	3.1	2.2	1.3	0.2

[a] Assumes 200 J/g for CF heat of crystallization.

Table 26   DSC Results for Mixtures of SEC Fraction II of AAD-1 (78.8%) with SEC Fraction I (21.2%) of Eight Asphalts

SEC-I source asphalt	Mixture properties					
	T_g (°C)	ΔC_p (J/g-K)	ΔT_g (°C)	Onset (°C)	End (°C)	CF^a (mass %)
AAA-1	−23.6	0.36	37.5	−42.0	−4.5	1.5
AAB-1	−25.7	0.34	32.1	−41.7	−9.6	1.5
AAC-1	−26.1	0.35	32.1	−42.1	−10.0	1.7
AAD-1	−24.5	0.32	34.5	−41.6	−7.0	1.5
AAF-1	−25.5	0.33	32.9	−42.0	−9.1	1.4
AAG-1	−26.2	0.33	31.0	−41.6	−10.6	1.7
AAK-1	−25.5	0.35	36.1	−43.1	−7.0	1.8
AAM-1	−28.4	0.35	32.7	−44.6	−11.9	1.7
Average	−25.7	0.34	33.6	−42.3	−8.7	1.6
Std. dev.	1.4	0.02	2.2	1.0	2.4	0.1
Excluding AAM-1 and AAF-1						
Average	−25.3	0.34	33.9	−42.0	−8.1	1.6
Std. dev.	1.0	0.02	2.6	0.6	2.3	0.1

[a] Assumes 200 J/g for CF heat of crystallization.

Table 27   DSC Results for Mixtures of SEC Fraction II of AAF-1 (86.7%) with
SEC Fraction I (13.3%) of Eight Asphalts

SEC-I source asphalt	Mixture properties					
	T_g (°C)	ΔC_p (J/g-K)	ΔT_g (°C)	Onset (°C)	End (°C)	CF[a] (mass %)
AAA-1	−14.1	0.39	51.6	−39.4	12.2	2.7
AAB-1	−15.5	0.40	45.8	−38.1	7.6	2.8
AAC-1	−14.3	0.39	50.3	−38.9	11.4	2.5
AAD-1	−12.7	0.37	52.1	−38.5	13.6	2.9
AAF-1	−14.9	0.38	48.1	−38.5	9.6	2.0
AAG-1	−12.7	0.37	53.5	−39.1	14.4	2.0
AAK-1	−13.9	0.30	49.9	−38.5	11.4	3.0
AAM-1	−17.1	0.41	43.9	−38.8	5.0	2.7
Average	−14.4	0.38	49.4	−38.7	10.7	2.6
Std. dev.	1.5	0.03	3.3	0.4	3.1	0.4
Excluding AAM-1 and AAF-1						
Average	−13.9	0.37	50.5	−38.7	11.8	2.6
Std. dev.	1.1	0.04	2.7	0.5	2.4	0.4

[a] Assumes 200 J/g for CF heat of crystallization.

Table 28   DSC Results for Mixtures of SEC Fraction II of AAG-1 (88.8%) with
SEC Fraction I (11.2%) of Eight Asphalts

SEC-I source asphalt	Mixture properties					
	T_g (°C)	ΔC_p (J/g-K)	ΔT_g (°C)	Onset (°C)	End (°C)	CF[a] (mass %)
AAA-1	−7.3	0.37	37.1	−25.9	11.2	0.0
AAB-1	−6.8	0.36	38.1	−25.9	12.2	0.0
AAC-1	−8.0	0.35	35.2	−25.7	9.5	0.0
AAD-1	−7.3	0.36	38.5	−26.7	11.8	0.0
AAF-1	−6.8	0.38	36.5	−25.0	11.6	0.0
AAG-1	−6.8	0.38	39.5	−26.5	13.0	0.0
AAK-1	−6.4	0.35	36.4	−24.5	11.9	0.0
AAM-1	NM[b]	NM	NM	NM	NM	NM
Average	−7.0	0.36	37.3	−25.7	11.6	0.0
Std. dev.	0.5	0.01	1.4	0.8	1.1	0.0
Excluding AAM-1 and AAF-1						
Average	−7.1	0.36	37.5	−25.9	11.6	0.0
Std. dev.	0.6	0.01	1.5	0.8	1.2	0.0

[a] Assumes 200 J/g for CF heat of crystallization.
[b] Not measured.

large, poorly solvated, and forms extensive molecular associations. In others, the dispersed component is small, well solvated, and molecular associations are minimized. If this model is correct, the dispersed component of any given asphalt should be of much larger molecular size than the solvent component. The two components should be amenable to separation by size exclusion chromatography. The SEC technique separates complex mixtures into fractions according to molecular size. Thus, if components of a mixture are all of approximately the same size, an SEC fractionation is not possible. Asphalts can be separated by SEC into two fractions that differ greatly in properties. The properties are what would be predicted by the model for the dispersed and solvent components. The components are designated SEC fraction I (dispersed) and SEC fraction II (solvent). The former is much more aromatic than the latter and has more of the heteroatom-containing constituents than the latter. The SEC fraction I materials are also much higher in molecular weight than the SEC fraction II materials. In the subsequent discussion asphalt dispersed components are treated as equivalent to SEC fraction I and the solvent component considered to be SEC fraction II.

In the previous section it was shown that the source of the dispersed component had little effect on T_g. The difference between SEC II T_g values and the neat asphalt T_g values, however, does indicate that the amount of SEC I influences low-temperature properties. The effect of varying the concentration of the dispersed component is being examined in asphaltene-maltene mixtures. Heptane asphaltenes, maltenes, and asphaltene-maltene mixtures of varying concentrations were prepared during SHRP. Descriptions of the preparation and properties of these samples are contained in the SHRP final report [15]. DSC work on the samples is continuing, but six asphalt sets have been examined so far: AAA-1, AAD-1, AAF-1, two sets of AAG-1, and AAK-1. The DSC procedure for examining the SHRP asphaltene-maltene samples was the same as for the SEC mixtures described above.

DSC results for the five asphalts and mixtures of their maltenes and asphaltenes are listed in Tables 29 through 33 for asphalts AAA-1, AAD-1, AAF-1, AAG-1, and AAK-1, respectively. Plots of the data are shown in Figure 12 for easier comparison. The addition of asphaltenes to the maltenes has little effect on the onset temperature of the glass transition. There is no significant variation of onset temperatures except at very high asphaltene contents as seen at the 40% level for asphalt AAK-1. The transition end temperature, however, is sensitive to asphaltene content. Increasing asphaltene concentration results in higher transition end temperatures and, as a result, wider glass transitions. Glass transition temperatures, which are usually near the middle of the transition, follow the same trend. Transition end temperatures for all asphalts show rapid increases at asphaltene concentrations between 5 and 15 mass %, indicating some change in structure or decrease in system compatibility. It is in this concentration range that Storm et al. [16] suggest the occurrence of the critical concentration for asphaltene

Table 29 Thermal Analyses of Asphalt AAA-1 Maltene-Asphaltene Mixtures

Mixture		Mixture properties			
Asphaltenes (mass %)	Maltenes (mass %)	T_g (°C)	ΔT_g (°C)	Onset (°C)	End (°C)
0.00	100.00	−28.7	25.4	−41.3	−16.0
1.00	99.00	−24.9	29.7	−39.6	−9.9
4.20	95.80	−27.2	27.5	−40.9	−13.5
8.40	91.60	−26.3	28.1	−40.4	−12.3
12.60	87.40	−24.4	33.7	−41.1	−7.4
16.80[a]	83.20	−23.2	33.8	−39.9	−6.2
21.00	79.00	−21.4	37.8	−40.1	−2.4
33.60	66.40	−15.3	46.9	−38.4	8.5
	Average			−40.2	
	Std. dev.			0.9	

[a] Solvent-treated neat asphalt.

association. Further work is necessary to pursue this hypothesis. Results for asphalt AAG-1 show a similar variation with asphaltene content. However, asphalt AAG-1 is the only asphalt examined with a natural asphaltene abundance below the level of the rapid increase. This may indicate that asphalt AAG-1 is a single-phase system. The overall increase in T_g for asphalt AAG-1 is 0.41°C/mass %, compared with 0.34, 0.47, 0.54, and 0.43°C/mass % for asphalts AAA-1, AAD-

Table 30 Thermal Analyses of Asphalt AAD-1 Maltene-Asphaltene Mixtures

Mixture		Mixture properties			
Asphaltenes (mass %)	Maltenes (mass %)	T_g (°C)	ΔT_g (°C)	Onset (°C)	End (°C)
0.00	100.00	−32.6	23.2	−44.2	−21.0
1.10	98.90	−30.8	25.1	−43.1	−18.1
4.30	95.70	−30.9	24.1	−42.9	−18.8
10.80	89.20	−26.6	30.1	−41.5	−11.4
16.10	83.90	−26.4	32.2	−42.3	−10.2
21.50	78.50	−24.1	35.4	−41.6	−6.3
21.50[a]	78.50	−18.2	39.6	−39.4	0.1
26.90	73.10	−20.2	41.9	−40.9	1.0
	Average			−42.0	
	Std. dev.			1.5	

[a] Solvent-treated neat asphalt.

Table 31 Thermal Analyses of Asphalt AAF-1 Maltene-Asphaltene Mixtures

Mixture		Mixture properties			
Asphaltenes (mass %)	Maltenes (mass %)	T_g (°C)	ΔT_g (°C)	Onset (°C)	End (°C)
0.0	100.0	−20.6	38.7	−39.7	−1.0
0.7	99.3	−19.3	42.9	−40.5	2.5
2.6	97.4	−18.6	42.9	−39.7	3.2
6.5	93.5	−17.1	45.3	−39.5	5.9
9.8	90.2	−13.4	53.8	−40.2	13.6
13.0	87.0	−13.0	53.5	−39.4	14.1
13.0[a]	87.0	−13.5	53.8	−40.1	13.7
16.3	83.7	−11.7	55.2	−39.1	16.1
	Average			−39.8	
	Std. dev.			0.5	

[a] Neat asphalt.

Table 32 Thermal Analyses of Asphalt AAG-1 Maltene-Asphaltene Mixtures

Mixture		Mixture properties			
Asphaltenes (mass %)	Maltenes (mass %)	T_g (°C)	ΔT_g (°C)	Onset (°C)	End (°C)
0.00	100.00	−8.2	36.9	−26.6	10.3
0.00	100.00	−7.9	34.8	−25.3	9.5
0.50	99.50	−7.7	37.2	−26.3	10.9
1.25	98.75	−7.8	36.1	−25.8	10.3
2.50	97.50	−7.1	38.7	−26.5	12.2
3.75	96.25	−6.4	38.2	−25.6	12.7
5.00	95.00	−6.0	38.5	−25.2	13.3
6.25	93.75	−5.6	41.6	−26.2	15.4
	Average			−25.9	
	Std. dev.			0.5	
0.00	100.00	−6.7	37.6	−25.3	12.3
5.00	95.00	−5.6	39.6	−25.5	14.1
10.00	90.00	−2.3	43.8	−24.2	19.7
15.00	85.00	−1.6	49.3	−26.1	23.2
20.00	80.00	1.2	51.0	−24.4	26.7
25.00	75.00	3.5	51.7	−22.1	29.5
	Average			−24.6	
	Std. dev.			1.4	

Table 33 Thermal Analyses of Asphalt AAK-1 Maltene-Asphaltene Mixtures

Mixture		Mixture properties			
Asphaltenes (mass %)	Maltenes (mass %)	T_g (°C)	ΔT_g (°C)	Onset (°C)	End (°C)
0.00	100.00	−26.1	26.8	−39.6	−12.8
5.05	94.95	−24.6	28.9	−39.2	−10.3
10.10	89.90	−21.4	36.2	−39.5	−3.3
15.15	84.85	−19.3	38.3	−38.3	0.0
20.20	79.80	−17.9	40.6	−37.8	2.8
25.25	74.75	−16.4	44.6	−38.3	6.3
40.40	59.60	−8.4	51.8	−33.9	17.9
	Average			−38.1	
	Std. dev.			2.0	

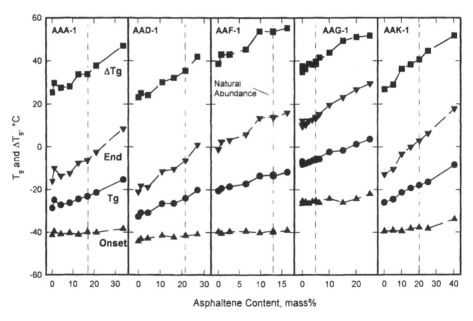

Figure 12 Glass transition changes with asphaltene concentration for five SHRP asphalts.

1, AAF-1, and AAK-1, respectively. The idea that AAG-1 might be a single-phase material is supported by the data shown in Figure 13 (AAG-1 components are circled). The AAG-1 point for the neat asphalt is near the trend line for solvent components (IEC neutrals, SEC II fraction, and maltenes). AAG-1 is the only whole asphalt near this line. The peptizing power of the maltenes (p_0) and the state of peptization (P) of asphalt AAG-1 have been shown to be exceptionally high in Heithaus tests [17], which measure overall compatibility of maltene-asphaltene mixtures. The lower peptizing power of the maltenes in the other asphalts may explain the distinct differences between their neat asphalt viscosities and solvent component viscosities shown in Figure 13.

F. Studies of Supercritical Fluid Extracts

Preliminary tests were performed to evaluate supercritical fluid extraction (SFE) as a potential method for generating fractions of the neutral components from asphalt. Although supercritical carbon dioxide may not have sufficient solvent strength to extract all of the asphalt, the solvent properties can be readily changed by temperature and pressure. This can provide a solvent that can be easily tuned by the use of temperature and pressure. It permits one to fine-tune the solvent properties, rather than make quantum leaps from one solvent to another.

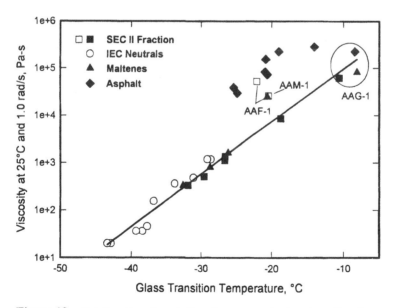

Figure 13 Relationship of asphalt and component glass transition temperatures to viscosity.

Samples of the eight SHRP core asphalts were fractionated using supercritical fluid extraction techniques. The fractions were extracted from a 6 mass % mixture of asphalt on aggregate (SHRP aggregate RJ). The aggregate was used as a convenient support for the asphalt as it permits the flow of supercritical fluid. Earlier work [18] suggested that the type of aggregate had no effect on the supercritical extraction of the asphalt. A limestone and a siliceous aggregate as well as glass helices were tested. Statistically, no affect on the extraction could be noted by either of the aggregates or the glass helices.

The asphalt-aggregate mixtures (6 mass % asphalt) used for SFE fractionation were prepared from asphalt and aggregate RJ. The asphalt and aggregate were heated to 150°C for approximately 20 min, and the aggregate was added to the asphalt and stirred until cool. Mix samples were placed in a 50-mL extraction vessel and successively extracted at five increasingly strong extraction conditions. The conditions selected were (a) 200 atm CO_2, (b) 300 atm CO_2, (c) 400 atm CO_2, (d) 500 atm CO_2, and (e) 500 atm CO_2 modified with 20 vol % THF. All of the extractions were performed at a temperature of 115°C. The SFE-extracted material was collected in a tared vial with a vented septum cap. The vent for the collection vial was introduced into a secondary collection vial containing methylene chloride to trap any entrained material. The material collected in the second vial was not used for any analysis but was used only for material balance purposes. The amounts of fractions generated using these conditions have been reported elsewhere [19].

DSC/MDSC experiments were run on the extracts from the supercritical fluid extractions and the results are listed in Table 34. Also included are the results of analyses of extracts from AAD-1 obtained earlier. Not all analyses are complete, but some trends are apparent. Increasing the extraction pressure results in the extraction of materials with higher glass transition temperatures (Figure 14). With only two exceptions, the slopes of the glass transition temperatures as functions of extraction pressure are nearly identical and the relationship is linear. The exceptions are the AAM-1 extracts, which have a much lower slope but still show a linear relationship with extraction pressure, and the most recent AAD-1 extracts, which have a higher slope than all the other asphalt extracts. The extracts from AAG-1 are shifted to higher glass transition temperatures compared with the other asphalt extracts.

G. Studies of Miscellaneous Dopants

The effects of model compound dopants on the glass transitions of asphalts AAF-1, AAG-1, and AAM-1 were measured. Dopants were selected on the basis of molecular size, functional groups, and conformation (linear or branched). Linear hydrocarbon dopants included octadecane (n-$C_{18}H_{38}$), tetracosane (n-$C_{24}H_{50}$), and hexacontane (n-$C_{60}H_{122}$). Branched hydrocarbon dopants included pristane

Table 34 Glass Transition Temperature of Supercritical Fluid Extracts

					T_g (°C)			
			Carbon dioxide pressure (atm)					Extrac-
	Heating	Cooling					500	tion
	rate	rate					w/THF	residue
Asphalt	(°C/min)	(°C/min)	200	300	400	500		
AAA-1	5	10	−67.9	−55.1	−40.8	−25.4	−18.1	>100
AAB-1	5	10	−70.8	−55.5	−45.7	−28.7	−31.0	>100
AAC-1	5	10	—	−51.7	−44.5	−33.7	−10.4	81.6
		Quench	−72.0					
AAD-1	5	10	−68.4	−51.4	−32.6	−8.4	−9.7	>100
(prev.)	2	10	−65.3	−52.0	−37.2	−26.4	14.3	>100
AAF-1	5	10	−72.8	−56.7	−41.6	−30.5	−6.1	>100
AAG-1	5	10	−50.9	−38.4	−26.7	−14.1	7.1	>100
AAK-1	5	10	−71.8	−52.8	−38.3	−22.0	−18.0	
AAM-1	5	10		−42.6	−36.4	−30.4	−27.7	−12.8

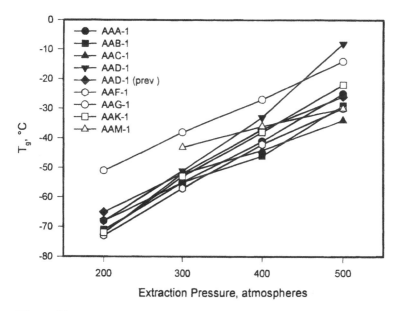

Figure 14 Variation of glass transition temperatures of SFE fractions with extraction pressure.

Table 35 Dopant Effects on Glass Transition Temperatures

Asphalt	Dopant	Efficiency (°C/mass %)	Range (mass %)
AAF-1	1,3,5,-Triethylbenzene	−2.6	0–2.5
AAF-1	Pristane ($C_{19}H_{40}$)	−2.5	0–5.0
AAF-1	Squalane ($C_{30}H_{62}$)	−1.9	0–5.1
AAG-1	1,3,5,-Triethylbenzene	−3.9	0–3.0
AAG-1	$nC_{18}H_{38}$	−3.5	0–3.0
AAG-1	$nC_{24}H_{50}$	−3.8	0–3.0
AAG-1	$nC_{60}H_{122}$	−0.4	0–3.0
AAG-1	12-Aminododecanoic acid (C_{12})	−0.3	0–3.0
AAG-1	12-Aminododecanoic acid (C_{12})	0.1	3.0–5.0
AAG-1	1,12-Diaminododecane (C_{12})	−1.1	0–2.0
AAM-1	1,3,5-Triethylbenzene	−1.6	0–10.0

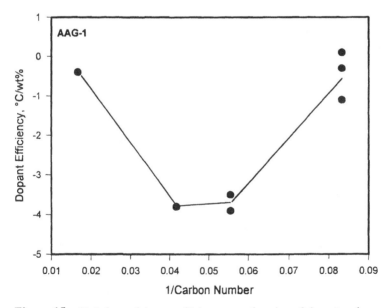

Figure 15 Variations of dopant efficiency as a function of dopant carbon number for asphalt AAG-1.

(2,6,10,14-tetramethylpentadecane) and squalane (2,6,10,15,19,23-hexamethylte-tracosane). 1,3,5-Triethylbenzene was an aromatic but, sterically hindered dopant. Dipolar function was represented by 1,12-diaminododecane and amphoteric function was represented by 12-aminododecanoic acid.

Dopant efficiencies and the weight ranges over which the efficiencies were calculated are listed in Table 35 for the three asphalts. The most effective dopants for lowering the glass transition temperature are triethylbenzene and the C_{18} and C_{24} normal hydrocarbons. Dopants with functional groups were the least effective, although, in this small study, none of the dopants actually increased the glass transition temperature over that of the neat asphalt. Enough tests were performed with asphalt AAG-1 to plot the resulting efficiencies against the reciprocal carbon number of the dopants as shown in Figure 15. The similarity between this graph and Figure 11 should be obvious and lends credence to the idea that low-molecular-weight polars are responsible for increasing glass transition temperatures at low molecular weights in the SEC fractions.

III. CONCLUSIONS

The introduction of modulated DSC has aided the interpretation of complex asphalt thermograms, especially in cases in which cold crystallization of asphalt components overlaps the broad glass transition region.

IEC neutral materials show thermal behavior consistent with the Fox and Flory [6] description of weakly interacting molecules. The glass transition temperature of these materials is controlled by their number average molecular weights.

Asphalt waxes crystallize in fractions dependent on the parent asphalt. The maximum crystallized fraction observed was approximately 60% for the wax from asphalt AAD-1. Most crystallized fractions were much lower.

The waxes from most of the SHRP core asphalts have little effect on the glass transition temperature. Wax from asphalt AAM-1 is the exception, causing over a 1°C drop in the transition temperature of asphalt AAM-1 for every 1% by mass added.

Cross-doping studies with the waxes show that the asphalt environment has little effect on the crystallization.

Application of a modified Fox relationship (Equation 4) is possible with SEC fractions 2–7 with reasonable agreement between calculated glass transition temperatures and those measured for the SEC II fraction. This is supportive evidence for fractions 2–7 being a solvent phase. SEC I (or 1) fractions have high glass transition temperatures or no transitions.

SEC fractions deviate from Fox-Flory behavior at low molecular weights. This is probably due to interactions among weakly polar materials. These weakly polar materials would be dispersed in solvent during the SEC separation but would interact in the neat samples used for DSC measurements. As discussed above, the IEC neutrals, which do not contain weakly polar materials, follow Fox-Flory behavior.

Cross-blending SEC I and SEC II fractions from different asphalts shows that although the amount of SEC I fraction in the blend influences the glass transition temperature of the blend, the source of SEC I has little affect. This shows that SEC I fractions are similar in character even though differing in chemical composition. The formation of molecular associations would be consistent with this observation.

DSC measurement on asphaltene-maltene mixtures of various concentrations shows that the addition of asphaltenes increases the upper end point temperature of the glass transition but has little affect on the onset temperature. The spreading of a glass transition profile is generally indicative of lessening compatibility.

Supercritical fluid extraction shows promise as a tunable method for removing waxlike asphalt components. The glass transition temperatures of the extracts are linearly related to the pressure of the extraction. Components extracted at pressures of 400 atm of carbon dioxide and below have glass transition temperatures below those of the whole asphalts.

Model compound dopant effects on the glass transition temperatures of asphalt are generally as might be expected. Freely mobile molecular groups tend to reduce the glass transition temperature. Polar molecules, especially amphoterics, provide some structure or association that has little effect on the glass transition for the concentrations and dopants used.

ACKNOWLEDGMENTS

The DSC results described in this chapter were obtained under contract with the Federal Highway Administration (FHWA) from 1992 through 1995. Asphalt component isolation procedures were developed at Western Research Institute (WRI) under FHWA contract and in the earlier Strategic Highway Research Program (SHRP) Binder Characterization and Evaluation research. WRI's project manager for the FHWA project is R. E. Robertson and for the SHRP project was J. C. Petersen. Asphalt and component samples described in this chapter were prepared under both contracts at WRI by teams directed by: J. F. McKay for waxes; J. J. Duvall for SEC, asphaltenes, and maltenes; and F. A. Barbour for SFE. Significant suggestions and guidance for the DSC studies came from S.-S. Kim. DSC analyses were conducted by G. E. Forney, Jr. The authors are grateful to Mr. Leonard C. Thomas, V.P. Marketing, TA Instruments, for clarifications and comments on modulated DSC.

REFERENCES

1. B. Wunderlich, *Thermal Analysis*, Academic Press, San Diego, p. 222 (1990).
2. D. W. Van Krevelen, *Properties of Polymers*, 3rd Edition, Elsevier, Amsterdam (1990).
3. B. Wunderlich, *Thermal Analysis*, Academic Press, San Diego, (1990).
4. *Thermal Characterization of Polymeric Materials* (E. A. Turi, ed.), Academic Press, New York (1981).
5. L. H. Sperling, *Introduction to Physical Polymer Science*, 2nd Edition, Wiley, New York (1992).
6. T. G. Fox and P. J. Flory, *J. Appl. Phys.*, **21**: 581 (1950).
7. T. G. Fox, *Bull. Am. Phys. Soc.*, [2] **1**: 123 (1956).
8. M. K. Gupta, *J. Coatings Technol.*, **67** (846): 53 (1995).
9. J. F. McKay, J. F. Branthaver, and R. E. Robertson, Isolation of Waxes from Asphalts and the Influence of Waxes on Asphalt Rheological Properties, *Prep. Am. Chem. Soc. Div. Petr. Chem.*, **40**(4): 794 (1995).
10. P. Claudy, J. M. Letoffe, G. N. King, J. P. Planche, and B. Brule., *Fuel Sci. Technol. Int.*, **9**: 71 (1991).
11. H. E. Bair, in *Thermal Characterization of Polymeric Materials* (E. A. Turi, ed.), Academic Press, New York, p. 882 (1981).
12. SHRP-A-370, *Binder Characterization and Evaluation*, Vol. 4: *Test Methods*, Strategic Highway Research Program, National Research Council, Washington, DC (1994).
13. SHRP-A-369, *Binder Characterization and Evaluation*, Vol. 3: *Physical Characterization*, Strategic Highway Research Program, National Research Council, Washington, DC, Chapter 9 (1994).
14. H. K. Huynh, T. D. Khong, S. L. Malhotra, and L-P. Blanchard, *Anal. Chem.*, **50**(7): 976 (1978).
15. SHRP-A-368, *Binder Characterization and Evaluation*, Vol. 2: *Chemistry*, Strategic Highway Research Program, National Research Council, Washington, DC (1993).
16. D. A. Storm, R. J. Barresi, and E. Y. Sheu, Evidence for the Micellization of Asphaltenic Molecules in Vacuum Residue, *Prep. Am. Chem. Soc. Div. Petr. Chem.*, **40**(4): 776 (1995).
17. FHWA Annual Technical Report, Contract DTFH61-92-C-00170, p. 251, November–May (1995).
18. F. A. Barbour, Preliminary Evaluation of Supercritical Fluid Extraction of Asphalt Binders from Asphalt-Aggregate Mixes, FHWA topical report (1994).
19. FHWA Annual Technical Report, Contract DTFH61-92-C-00170, November–May (1995).

4

Detection of Strongly Acidic Compounds in Extensively Aged Asphalts

Sang-Soo Kim, Gerald W. Gardner,
John F. McKay, Raymond E. Robertson,
and Jan F. Branthaver
Western Research Institute, Laramie, Wyoming

I. INTRODUCTION

Oxidative aging of asphalt pavements causes irreversible increases in asphalt viscosity. In time, the viscosity increase in the asphaltic component of the pavement will be large enough that the pavement will become excessively brittle. Failures of pavements due to embrittlement caused by oxidative aging are more rapid in warmer climates than in colder climates.

Petersen [1] discussed the types of compounds formed when oxygen reacts with asphalts. Based on infrared (IR) spectrometric data, the main products are ketones and sulfoxides. In extensively aged asphalts, small amounts of carboxylic acids and carboxylic acid anhydrides are observed by IR. All these compound types are observed in laboratory-aged asphalts and materials recovered from pavements by solvent extraction. Potentiometric titrations of oxidatively aged asphalts show that the concentration of weak bases increases markedly with aging [2]. This is mainly due to buildup of sulfoxides. It was also observed by these workers that concentrations of strong bases, which in asphalts are mostly pyridine-type compounds, decrease as a result of oxidative aging. Because pyridine-type compounds are not easily oxidized under the aging conditions pavements experience, the latter result was somewhat surprising.

103

Inverse gas-liquid chromatography (IGLC) has long been used to study chemical properties of asphalts and chemical changes that occur during aging [3]. Because of the involatility of asphalts, they cannot be analyzed by conventional gas-liquid chromatographic methods. In IGLC, asphalts are coated on an inert support, and the coated support serves as the stationary phase. Test compounds are injected into columns filled with the coated support, and the retention behavior of the test compounds is measured. The retention behavior is a function of the nature of the test compound and the asphalt. As asphalts become oxidatively aged, they become more polar and their interactions with polar test compounds change. The changes brought about by aging determined by IGLC were correlated with pavement performance in the Zaca-Wigmore study [4]. In the laboratory, oxidative aging can be performed in IGLC columns, or asphalts aged by other methods can be coated on an inert support and used as the stationary phase in IGLC columns.

In the study described in the following, asphalts have been investigated using IGLC with the objective of developing the technique so that it can be used to predict pavement durability.

II. EXPERIMENTAL

A. IGLC Procedure

A Shimadzu GC-14A gas chromatograph with thermal conductivity detector was used for analysis of asphalt columns, and an HP 5700A gas chromatograph was used for aging and conditioning the columns. Development of the IGLC technique and experimental procedure is discussed in detail by Davis et al. [3] and by Barbour et al. [5]. In brief, the experimental procedure used in this study is as follows. A sample of asphalt (about 5 g) is dissolved in 40 mL of benzene. About 50 g of an inert support (Teflon, 40/60 mesh size), 10 times the mass of the asphalt, is placed in a mixing bowl and the dissolved asphalt solution is added and thoroughly mixed, followed by evaporation of benzene in a good fume hood. The asphalt-coated support is packed into aluminum tubing, 6.35 mm (1/4 inch) inner diameter and 3.81 m (12.5 feet) length. The column is then coiled so as to fit into the GC oven, with care taken not to bend the column. Prior to the column packing, the inside of the column should be thoroughly rinsed with acetone to remove oil residue from column manufacturing. The packed column is placed in the GC apparatus, and helium is passed through the column at a flow rate of about 30 mL/min while the instrument operating temperature is raised to 130°C (266°F). The column is maintained under these conditions for 6 h. Following column conditioning, a standard solution and a series of test compounds are injected individually in the amount of 1.0 μL plus 1.0 μL of air as an internal reference. Because phenol is not in liquid state at ambient temperature, it is dissolved in an equal amount (mass/volume) of toluene and the solution is injected in the amount of 1.2 μL. The standard solution is a mixture of equal proportions of five normal alkanes: pentane, hexane, heptane, octane, and nonane. Corrected retention times (CRTs) of each of the normal alkanes and test compounds are obtained by subtracting air retention time from the recorded retention time of each peak. When the logarithms of CRTs of the normal alkanes are plotted against their molecular weights, a linear relationship is obtained as shown in Figure 1. Then, based on the plot for the standard solution of alkanes, the IGLC interaction coefficient (I_p) of each test compound is calculated as follows:

$$I_p = 100(\log \text{CRT}_{\text{test}} - \log \text{CRT}_{\text{hyp}})$$

where CRT_{test} = CRT of a test compound

CRT_{hyp} = hypothetical CRT of a hypothetical normal alkane having the same molecular weight as the test compound. It is determined from the linear plot of log CRT and molecular weight of the normal alkane standard solution (Figure 1)

The method of obtaining the interaction coefficient for phenol on an IGLC column containing unaged AAD-1 asphalt is illustrated in Figure 1. Phenol has a molecular weight (MW) of 94. The logarithm of its measured CRT on a column packed with asphalt AAD-1 is observed to be approximately 1.74. The logarithm of the hypothetical CRT of a hypothetical *n*-alkane of MW = 94 is 0.43. Subtraction of 0.43 from 1.74 and multiplying the difference by 100 yields an I_p for phenol of 131. Since CRT_{hyp} approximates the retention behavior of the test

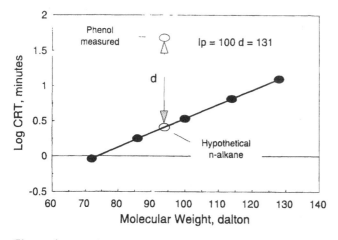

Figure 1   Relationship between corrected retention time and molecular weight of normal alkane and calculation of interaction coefficient (I_p) for phenol on AAD-1 asphalt.

compound with the column due to nonpolar forces as in normal alkanes, I_p is a measure of the retention behavior due to polar interactions between the test compound and the column.

After injecting all test compounds used, the column is aged by passing air through at 130°C (266°F) with a 30 mL/min flow rate for 6 h. An attachment at the exit end of the IGLC column was assembled to control back pressure and to obtain more uniform aging. Air inlet and outlet pressures are kept constant at 6.3 and 2.3 psig, respectively. Then the interaction coefficients are again determined as described above.

Test compounds used in this IGLC analysis are propionic acid, phenol, toluene, and 1-methylpyrrolidine (1-MPD). The test compounds were selected on the basis of the different kinds of chemical interactions that can occur between the test compounds and molecules in asphalts.

Precautions must be taken to vent all vapors of benzene and other organic materials into a fume hood.

B. Potentiometric Titrations

The equipment used is a Metrohm 665 Dosimat connected to a Metrohm 670 Titroprocessor. A glass pH electrode and a silver/silver chloride reference electrode are used for making measurements. The reference electrode is filled with a saturated solution of lithium chloride in ethanol. This constitutes a complete potentiometric titration apparatus.

Titration for acids is carried out using 0.1 N tetrabutyl ammonium hydroxide (Aldrich) in toluene (Aldrich) as the titrant. A mixture of ethanol (Baker) and chlorobenzene (Aldrich) (1:9) is used as the solvent for asphalts and model compounds. All titrations were performed by submerging the tip of the microburette into the asphalt solution to avoid localized concentration effects. Cut points for strong and weak acids are determined by titrating 5 mL of 0.01 N solutions of reference compounds, e.g., palmitic acid (Aldrich) or naphthol (Baker), in the ethanol-chlorobenzene solvent. For the mixtures of reference compounds, 5 mL of each solution was mixed and titrated. Asphalt solutions were prepared by weighing 500 mg of asphalt into a tared beaker and then adding 30 mL of solvent, followed by stirring.

Titrations are started at a rate of 200 μL/min. The rate is automatically reduced as the end points are approached. A plot is produced on the screen of the Titroprocessor as the titration progresses. The instrument locates the half-neutralization points using preestablished sensitivity parameters and carries out the necessary calculations. Results are printed out with the titration curve. Corresponding values for mL of titrant used and millivolt reading for each half-neutralization point are reported.

III. DISCUSSION

Four asphalts studied during the Strategic Highway Research Program (SHRP) were selected for IGLC and potentiometric titration studies. One asphalt, coded AAD-1, contains over 5% sulfur and is highly susceptible to oxidative aging [6]. This means that, when aged under mild conditions, the viscosity of the aged asphalt at a given temperature is much greater than that of the original asphalt. A second asphalt, coded AAG-1, is low in sulfur and is not susceptible to oxidative aging. The viscosity increase of this asphalt after mild aging is not very large. Asphalt AAG-1 was lime treated during the asphalt manufacturing process. Studies also were performed on asphalt ABD, which is derived from the same crude source as AAG-1 but was not lime treated. The fourth asphalt, coded AAM-1, is low in sulfur but is also susceptible to oxidative aging.

Tables 1–4 list IGLC interaction coefficients for the four test compounds toluene, phenol, propionic acid, and 1-methylpyrrolidine with asphalts AAD-1, AAM-1, AAG-1, and ABD. Interaction coefficients of toluene for all four neat asphalts are essentially identical. Toluene can interact only weakly with asphalts because it is a simple aromatic compound. Phenol, on the other hand, can interact with basic compounds native to the asphalts much more strongly than by simple aromatic interactions. Phenol can also hydrogen bond to almost any functional group containing nitrogen, oxygen, or sulfur. There is some variation among the four phenol interaction coefficients of the neat asphalts because the heteroatom contents (N, O, S) of the asphalts vary; those of AAD-1, AAG-1, and ABD greatly exceed that of AAM-1, and the difference is reflected in the phenol

Table 1 Interaction Coefficients of Unaged and Aged AAD-1

Sample	IGLC aging time (h)	Test compounds and interaction coefficients			
		Toluene	Phenol	Propionic acid	1-Methyl pyrrolidine
AAD-1	0	52;53	127;134	89;96	65;68
	6	54	158	108	83
	12	54;55	170;171	114;116	166;206
	24	55;55	180;181	119;121	NE[a]
	48	55;55	186;186	122;123	NE[a]
RTFOT-treated AAD-1	0	53;53	137;138	95;95	67;69
	24	55;55	176;182	116;122	NE[a]
TFO-PAV (80°C, 144 h) treated AAD-1	0	54;54	175;178	120;122	146;181
	24	55;56	186;191	123;126	NE[a]

[a]NE, Not eluted.

Table 2 Interaction Coefficients of Unaged and Aged AAM-1

Sample	IGLC aging time (h)	Test compounds and interaction coefficients			
		Toluene	Phenol	Propionic acid	1-Methyl pyrrolidine
AAM-1	0	50;50	113;113	69;69	59;60
	6	50;51	127;126	75;75	131;117
	12	50;50	128;127	75;76	120
	24	50;51	129;130	77;77	—
	48	51;51	132;132	79;79	185
RTFOT-treated	0	50;50	118;117	74;70	67;66
AAM-1	24	51;50	132;130	83;77	NE[a]
TFO-PAV	0	50;51	127;127	76;76	133;153
(80°C, 144 h)	24	51;51	132;132	81;80	NE[a]
treated AAM-1					

[a]NE, Not eluted.

Table 3 Interaction Coefficients of Unaged and Aged AAG-1

Sample	IGLC aging time (h)	Test compounds and interaction coefficients			
		Toluene	Phenol	Propionic acid	1-Methyl pyrrolidine
AAG-1	0	53;52	127;129	NE[a]	60;61
	6	53	143	188	65
	12	53;53	147;148	148;150	73;47
	24	54;54	150;149	122;130	87;87
	48	54;54	152;152	113;116	128;116
RTFOT-treated	0	53;52	132;131	NE[a]	62
AAG-1	24	55;54	151;151	126;122	106;95
TFO-PAV	0	53;54	150;151	159;161	70;70
80°C, 144 h)	24	55;54	154;152	110;116	153;109
treated AAG-1					

[a]NE, Not eluted.

Table 4 Interaction Coefficients of Unaged and Aged ABD

Sample	IGLC aging time (h)	Test compounds and interaction coefficients			
		Toluene	Phenol	Propionic acid	1-Methyl pyrrolidine
ABD	0	53	133	96	45
	6	55	155	106	158
	12	56	161	112	—
	24	55	164	110	177
	48	55	168	112	NE[a]

[a]NE, Not eluted.

interaction coefficients. Propionic acid can interact with asphalts only through its carboxylic acid function. It will be attracted to basic molecules, amphoteric molecules, and possibly other acidic species. Propionic acid is characterized by larger interaction coefficients than toluene but lower interaction coefficients than phenol in three of the asphalts. In lime-treated AAG-1, propionic acid is nonemergent. It forms the calcium salt and is retained on the column. It may be fortuitous that for three of the asphalts, the sum of the toluene and propionic acid interaction coefficients nearly equals the phenol interaction coefficient. The basic compound 1-methylpyrrolidine does not interact very strongly with any of the neat unaged asphalts. It is possible that the acidic compounds in the asphalts are engaged in sufficiently strong associations that they do not interact with this test compound.

All four asphalts were column aged at 130°C (266°F) for varying lengths of time, up to 48 h. The interaction coefficients were determined during the course of the column oxidations at regular intervals. These results are also listed in Tables 1–4. The changes in interaction coefficients during the oxidations are represented graphically in Figures 2–5. Interaction coefficients of toluene remain virtually unchanged regardless of the degree of oxidation. No great increases in aromatic compounds are observed as a result of oxidation. Phenol interaction coefficients increase substantially with increasing column oxidation time, reflecting incorporation of oxygen into the asphalts.

For asphalts ABD and AAM-1, propionic acid interaction coefficients increase moderately with aging. For asphalt AAD-1, the interaction coefficient increase is large. For aged AAG-1, an interaction coefficient for propionic acid is measurable, unlike the situation for the neat asphalt, and is smaller than that of phenol. This result indicates that the strongly basic sites in neat AAG-1 have been neutralized during the aging process. Immediately coming to mind as a cause of the neutralization is the generation of acidic species greater in acidity than indigenous carboxylic acids.

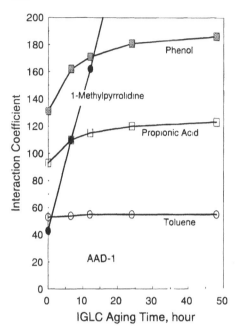

Figure 2   Interaction coefficients versus aging time for AAD-1 asphalt.

The changes in interaction coefficients of 1-methylpyrrolidine with aging of the asphalts are very large. This compound does not strongly interact with any of the neat asphalts. For 48-h-aged AAD-1 and ABD. this compound is nonemergent. For the other two asphalts aged 48 h. interaction coefficients of 1-methylpyrrolidine are observed to have doubled. Again, the results imply the buildup of a strongly acidic material during the aging process.

Tables 1–3 also list interaction coefficient data for asphalts aged by other methods and then coated on Teflon and used as the stationary phase in IGLC. In the rolling thin-film oven test (RTFOT), samples of asphalts are placed in a bottle. The bottle is put in a carousel in an oven. The bottle is heated to 163°C (325.4°F) for 85 min while the carousel rotates (AASHTO Procedure T240). It is evident that, as a result of this process, there are not major changes in interaction coefficients of any of the test compounds. When the RTFOT-aged asphalts are subsequently coated on Teflon and subjected to 24 h of column oxidation at 130°C (266°F), significant changes in interaction coefficients are observed (Tables 1–3).

The SHRP investigated oxidative aging of many asphalts at relatively low temperatures in a pressure-aging vessel (PAV). The purpose of these experiments was to simulate pavement aging. It is known that the course of asphalt aging at high temperatures (>80°C; 176°F) is different from low-temperature aging, which

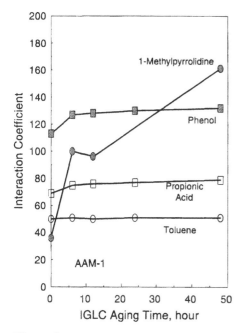

Figure 3   Interaction coefficients versus aging time for AAM-1 asphalt.

is what takes place during pavement service. Aging tests must accelerate the process of oxidation but should be performed at as low a temperature as possible. In the PAV aging test, pressure is used to accelerate reaction with oxygen. A preliminary thin-film oven (TFO) test was performed in these experiments. This test involves heating asphalt in a pan in an oven at 163°C (325.4°F) for 5 h. Tables 1–3 list interaction coefficients of four test compounds in columns filled with Teflon coated with three asphalts aged by the TFO-PAV method for 144 h at 80°C (176°F). The interaction coefficients are, for the most part, considerably larger than for the RTFO-treated samples. Further aging of the TFO-PAV–treated asphalts (after coating them on Teflon) for 24 h in the columns causes greater increases in the interaction coefficients. From these data, it appears that the column oxidation is a severe oxidation compared with the TFO-PAV procedure. About 6–12 h of IGLC column aging yields results similar to those for the 144-h TFO-PAV aging.

The foregoing data show that interaction coefficients are sensitive and consistent indicators of aging levels. They also suggest that, as aging proceeds, there is buildup of a strongly acidic species more acidic than most organic carboxylic acids. If so, such species should be observable by nonaqueous potentiometric titration (NAPT).

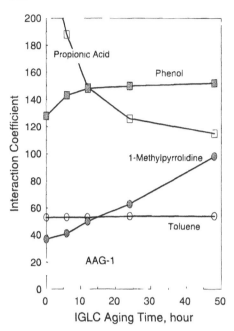

Figure 4   Interaction coefficients versus aging time for AAG-1 asphalt.

The NAPT of asphalt solutions was performed as described in the experimental section. Preliminary results showed that asphalts contain many acids of widely varying strengths, so that titration curves with sharp inflections are not observed. In solvents that have dielectric constants much lower than that of water, acids of different strengths interact with one another in mixtures such that end points are shifted relative to end points of each acid titrated alone. Table 5 lists half-neutralization points in millivolts of various model compounds in chlorobenzene-ethanol titrated with tetrabutyl ammonium hydroxide. The millivolt data show that benzoic acid is slightly stronger than palmitic acid and that these acids are much stronger than the phenolic compounds 1-naphthol and 4-*t*-butylphenol. In mixtures of carboxylic acids and phenols, half-neutralization points of the carboxylic acids are shifted slightly. Similar shifts are observed when model compounds are mixed with neat asphalts (Table 6). When asphalt AAD-1 is titrated by itself, an inflection in the titration curve is observed at about −225 to −250 mV. This corresponds to the titration of carboxylic acids naturally present in AAD-1. Distinct inflection points corresponding to phenols were not observed in AAD-1. Titration of asphalt ABD (Table 7) also shows that it contains considerable amounts of carboxylic acids, based on the inflection in the titration curve at −200 to −225 mV. Asphalt AAM-1 (Table 7) contains small amounts of carbox-

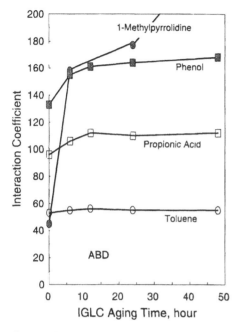

Figure 5 Interaction coefficients versus aging time for ABD asphalt.

Table 5 Nonaqueous Potentiometric Titration of Model
Compounds in Chlorobenzene-Ethanol

| Model compound | Run | Half-neutralization points | |
		mL titrant	Millivolts
Benzoic acid	1	0.45	−225
	2	0.45	−240
Palmitic acid	1	0.45	−255
	2	0.47	−255
1-Naphthol	1	0.52	−335
	2	0.55	−340
4-*tert*-Butylphenol	1	0.70	−350
	2	0.74	−350

Table 6 Nonaqueous Potentiometric Titration of Model Compounds and Mixtures of Model Compounds with AAD-1 in Chlorobenzene-Ethanol

		Half-neutralization points	
Components of mixtures	Run	mL titrant	mV
Palmitic acid + 1-naphthol	1	0.43	−150
		0.88	−325
	2	0.43	−190
		0.87	−340
Benzoic acid + 4-*tert*-butylphenol	1	0.40	−170
		0.98	−340
	2	0.41	−175
		1.00	−340
Benzoic acid + AAD-1	1	0.56	−150
		0.68	−275
	2	0.55	−155
		0.68	−275
Palmitic acid + AAD-1	1	0.56	−175
		0.67	−260
	2	0.58	−175
		0.67	−260
1-Naphthol + AAD-1	1	0.17	−175
		0.79	−320
	2	0.17	−200
		0.76	−325
AAD-1	1	0.2	−250
	2	0.2	−250

ylic acids. These results are confirmed by infrared functional group analyses (IR-FGAs) [6].

Samples of asphalts AAD-1, AAG-1, and AAM-1 were TFO-PAV aged for 400 h at 60°C (140°F). These conditions, which differ somewhat from the 80°C (176°F), 144-h PAV discussed earlier, should oxidize asphalts to a degree that corresponds to oxidation resulting from several years of pavement aging. The aged asphalts were titrated for acids by NAPT. Results in Table 7 show that small amounts of material characterized by a half-neutralization point of −10 to −70 mV are detected, corresponding to a strongly acidic species. Moreover, substantial amounts of what appear to be carboxylic acids are observed in the aged AAG-1 sample, which is a lime-treated asphalt. These carboxylic acids could be observed only if their calcium salts were neutralized by a stronger acid.

Table 7 Nonaqueous Potentiometric Titration of Unaged and Aged (60°C, 400 h) Asphalts in Chlorobenzene-Ethanol

Sample	Run	Half-neutralization points	
		mL titrant	mV
ABD, unaged	1	0.19	−210
	2	0.21	−225
AAM-1 unaged	1	0.065	−180
	2	0.070	−220
AAM-1, aged	1	0.050	−25
		0.155	−275
AAD-1, unaged	1	0.20	−225
	2	0.20	−225
AAD-1, aged	1	0.050	−10
		0.31	−300
	2	0.048	−25
		0.32	−300
	3	0.070	−40
		0.36	−315
AAG-1, unaged	1	0.08	−220
	2	0.10	−235
AAG-1, aged	1	0.05	−60
		0.24	−230
	2	0.05	−70
		0.24	−235

To further verify the existence of the strongly acidic species, samples of unaged AAD-1 and aged AAD-1 [400 h, 60°C (140°F) TFO-PAV] were dissolved in toluene (1.0 g in 20 mL) and shaken in a separatory funnel with water. After shaking, the water that had been in contact with the unaged AAD-1 was observed to be clear, and the pH of the water was measured and found to be 4.9. The mixture of water with the toluene solution of the aged AAD-1 formed an emulsion that took several hours to break. Afterward, the water layer was observed to be colored, and the pH of the water was measured and found to be 3.5.

The buildup of such a highly acidic species could have a significant influence on pavement durability. The highly acidic material could be involved in moisture damage (leading to stripping) by attacking asphalt-aggregate interfaces in pavements containing acidic aggregates.

The nature of the highly acidic species is speculative. Sulfoxides are known to be major products of asphalt oxidation. These compounds may cleave at the carbon-sulfur bond and oxidize further to sulfinic or sulfonic acids. In fuel oils, sulfonic acids are believed to form as a result of reaction of thiophenols with oxygen [7]. The same process could occur in paving asphalts. Sulfonic or sulfinic acids would be much stronger than native carboxylic acids and also would be surface active. Alternatively, polybasic aromatic carboxylic acids may form by some unknown pathway. Some of these acids are among the strongest organic acids. We plan to characterize the acidic species in future work.

IV. CONCLUSIONS

The IGLC method is responsive to chemical changes that occur in paving asphalts during oxidative aging. For the polar compounds phenol and 1-methylpyrrolidine, interaction coefficients become larger as a function of aging, particularly for the latter compound. This behavior parallels oxygen incorporation in asphalts to form more polar molecules, both acidic and basic. Interaction coefficients of toluene do not change with oxidation, suggesting that aromatic centers are not generated. Interaction coefficients of propionic acid rise somewhat with oxidative aging of asphalts that are not lime treated. For one lime-treated asphalt, the interaction coefficient of propionic acid decreases with aging. These results suggest generation of a strongly acidic species as a result of oxidation. The buildup of a strong acid, possibly a sulfonic or sulfinic acid, is supported by NAPT measurements on aged paving asphalts and extraction of the acid into water. Strong acids in extensively aged asphalts could affect the course of moisture damage in pavements consisting of binders and acidic aggregates.

The acidic and basic functional groups generated during oxidative aging of asphalts do not neutralize one another, hence they do interact with acidic and basic test compounds under the conditions of the IGLC experiments. During binder oxidation, acidic sites may be generated in basic molecules, basic sites may be generated in acidic molecules, and either or both functionalities may be generated in neutral molecules. Previous work has shown that concentrations of amphoteric molecules increase as a result of binder oxidation [6].

ACKNOWLEDGMENTS

The authors thank the Federal Highway Administration for financial support of this work. IGLC experiments were performed by C. Brommer. RTFOT and TFO-PAV oxidations were performed by J. Wolf. The manuscript was typed by J. Greaser.

DISCLAIMER

Mention of specific brand names does not imply endorsement by Western Research Institute or the Federal Highway Administration. The contents of this chapter reflect the views of the authors, who are responsible for the facts and accuracy of the data presented. The contents do not necessarily reflect the official views or policies of the Federal Highway Administration. This chapter does not constitute a standard, specification, or regulation.

REFERENCES

1. J. C. Petersen, *Transport. Res. Rec.*, **999**: 13, (1984).
2. R. E. Robertson, J. F. Schabron, A. A. Gwin, and J. F. Branthaver, *Prepr. Div. Petrol. Chem.*, Am. Chem. Soc. **37**(3): 913, (1992).
3. T. C. Davis, J. C. Petersen, and W. E. Haines, *Anal. Chem.*, **38**: 241, (1966).
4. T. C. Davis and J. C. Petersen, *Proc. Assoc. Asphalt Paving Technologists*, **36**, 1, (1967).
5. F. A. Barbour, S. M. Dorrence, and J. C. Petersen, *Anal. Chem.*, **42**: 668, (1970).
6. SHRP A-368, *Binder Characterization and Evaluation*, Vol. 2: *Chemistry*. Strategic Highway Research Program, National Research Council, Washington, DC (1993).
7. J. F. Pedley, R. W. Hiley, and R. A. Hancock, *Fuel*, **66**: 1646, (1987).

5

Oxidation of Asphalt Fractions

Jinmo Huang
The College of New Jersey, Trenton, New Jersey

I. INTRODUCTION

The oxidation of asphalt is believed to be one of the major factors contributing to the failure of asphalt pavement. The hardening and embrittlement of asphalt are due to the oxidation of certain asphalt molecules resulting in the formation of highly polar and strongly interacting functional groups containing oxygen [1]. Ultraviolet (UV) light on pavement surface [2] and oxygen diffusion through pavement air voids [3] are believed to be the principal causes of asphalt oxidation. Previous studies have shown that the major functional groups produced during oxidation of asphalt are carbonyl groups [4], sulfoxides [5], and sulfones [6].

Many instrumental techniques including electron spin resonance (ESR) [7], high-performance liquid chromatography (HPLC) [8], infrared (IR) spectrometry [9–11], mass spectrometry (MS) [12], nuclear magnetic resonance (NMR) [13,14], and x-ray diffraction [15] have been used to characterize asphaltic materials. Among these techniques, IR is the fastest and the most sensitive technique for the detection of functional groups in asphalt and has proved to be a useful tool for the analysis of tar sand fractions [16]. Infrared spectrometry was first used by Hadzi [17] to study bituminous substances qualitatively in 1953. Brandes [18] first used IR to quantify aromatic, naphthenic, and paraffinic carbons of gas oils in 1956. Other researchers [9,10,11,19] have used IR to study asphaltic materials.

Accurate molecular weight distributions are essential to the structural analysis of asphalt. Various methods including vapor pressure osmometry [20], ultracentrifugation [21], x-ray diffraction and scattering [15], mass spectrometry [21], electron microscopy [21], and gel permeation chromatography [22] have been used to study the molecular weights of asphaltic materials. Among these, vapor pressure osmometry (VPO) was reported as a convenient method for the molecular weight determinations [23]. However, divergent results have been generated by VPO at different sample concentrations in different solvents [20]. Also, VPO provides only number average molecular weight and no information on molecular size profile.

Gel permeation chromatography (GPC), with separation based on the size and shape of molecules, gives the number average molecular weight (M_n), the weight average molecular weight (M_w), the dispersity (M_w/M_n), and the profile of molecular size distribution of samples in one run. Because of its capability, simplicity, and rapidness, GPC has been used in the characterization of petroleum products [23–27] since the early 1960s. In the determination of molecular weights of asphaltic materials by GPC, several factors including eluting solvent [28], compatible calibration standard [29], concentration of sample [30], and interaction of sample molecules to form molecular aggregates [27] have been reported to affect the measurement.

Most of the past studies of oxidized asphalt were done prior to separation, thus preventing full understanding of the oxidation process. Because asphalt is composed of a large number of species, total separation has not been achieved [19]. However, there have been effective separations by functionality. One of the most used procedures was developed by Corbett [31].

The Corbett technique utilizes the difference in the polarity of chemical species in asphalt as the basis of separation. In this research, the asphalt sample was separated into saturates, naphthalene aromatics, polar aromatics, and asphaltenes by the Corbett method [32]. These Corbett fractions were then placed under the irradiation of ultraviolet light to enhance oxidation. Fourier transform infrared spectrometry (FTIR) and GPC were used to study the structural changes and the molecular weight distributions of Corbett asphalt fractions during photooxidation.

II. EXPERIMENTAL SECTION

A. Corbett Separation

Three grams of AC-20 asphalt sample (prepared by Citgo Asphalt Refining Co. and provided by the New Jersey Department of Transportation) were separated into saturates, naphthalene aromatics, polar aromatics, and asphaltenes using the Corbett separation technique [32]. The asphaltenes were separated as the insoluble phase when 300 mL of n-heptane (J.T. Baker HPLC solvent) was added to the asphalt sample, stirred for 2 h, and allowed to settle overnight. The insoluble asphaltenes were obtained by filtering the asphalt solution through filtering paper (Schleicher & Schuell no. 560 prepleated filters). The solid asphaltenes were then washed with warm *n*-heptane and dried in a 50°C oven for 2 h. The *n*-heptane solubles, known as maltenes, were concentrated under vacuum at room temperature. Then the concentrated maltenes were loaded on a basic aluminum oxide (Aldrich standard grade, activated, 150 mesh) column. The saturates were eluted out of the column by *n*-heptane. The naphthalene aromatics were eluted from the column by toluene (Sigma-Aldrich HPLC grade). The polar aromatics were eluted from the column by a mixture of methanol (Sigma-Aldrich HPLC grade) and toluene (50:50), then followed by methylene chloride (EM Science GR grade). The solvents were removed under vacuum at room temperature, and the residues were used for the study.

B. Photooxidation and FTIR Analysis of Corbett Asphalt Fractions

The Corbett fractions were dissolved in methylene chloride, and the fraction solutions were cast into thin film on potassium bromide (KBr) plates (International Crystal 25 × 4 mm disk). The sample plates were analyzed by a Perkin-Elmer 1720X model FTIR with a resolution of 4 cm^{-1}. The sample plates were then placed under the irradiation of an ultraviolet light (Supelco Spectroline E-series UV lamp, short wavelength used) for 4 weeks. The UV-irradiated plates were again analyzed by FTIR.

C. GPC Analysis of Corbett Asphalt Fractions

Sample solutions of preoxidized and oxidized Corbett fractions were prepared by adding 1.0 mL of toluene to each 5 mg of Corbett fractions. The standard polystyrene solution was obtained by dissolving 5 mg of each of the standards 580, 1680, 3250, and 28,500 in 1.0 mL of toluene. Thus, the concentrations of samples and calibration standards are all approximately 5 mg/mL.

The chromatographic system consisted of a Perkin-Elmer series 200 HPLC with a refractive index detector. The column was a 300 × 7.5 mm PL-Gel 3-μm Mixed-E column (Polymer Laboratories), with a working molecular weight range of 100–40,000. Fresh tetrahydrofuran (Fisher HPLC grade) was used as the eluent with a flow rate of 0.7 mL/min. The column temperature was maintained at 30°C by a Perkin-Elmer series 101 LC oven. The injection volume was 20 μL. Data were acquired and analyzed by Perkin-Elmer Turbochrom TC4-SEC software.

III. RESULTS AND DISCUSSION

A. FTIR Analysis of Corbett Asphalt Fractions

The FTIR spectra of preoxidized Corbett fractions are shown in Figure 1. The spectrum of the saturates has features expected in heavy alkanes. There are three

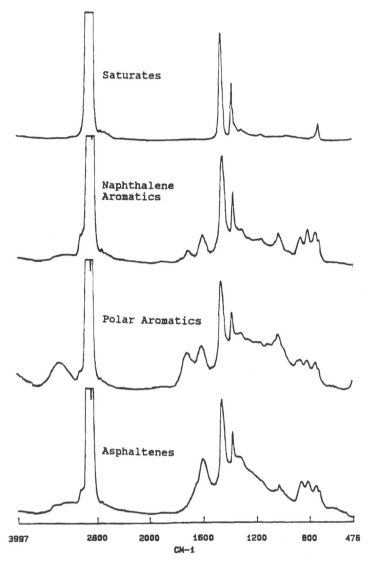

Figure 1   FTIR spectra of asphalt fractions [40].

peaks (2950, 2925, and 2855 cm^{-1}) featuring saturated CH stretching vibrations, two peaks (1460 and 1375 cm^{-1}) due to CH bending vibrations, a weak band around 1155 cm^{-1} characteristic of CH$_2$ next to an unoxidized sulfur [33], and a peak at 720 cm^{-1} assigned to a long chain of CH$_2$ [33]. No OH and NH stretching vibrations are observed in the region 3500–3100 cm^{-1} and no carbonyl (C=O) peak around 1700 cm^{-1}.

Naphthalene aromatics and polar aromatics have similar spectral features. Both have a small band at 3060 cm^{-1} due to aromatic hydrogen stretching vibrations, carbonyl peak at around 1710 cm^{-1}, C=C stretching vibrations at 1600 cm^{-1}, and sulfoxides (S=O) at 1030 cm^{-1} [34]. Both also have peaks around 865, 815, and 760 cm^{-1} characteristic of aromatic out-of-plane frequencies [10]. The difference between naphthalene aromatics and polar aromatics is that the peaks at around 3400 and 1710 cm^{-1} are more intense in the spectrum of polar aromatics. This difference explains why the polar aromatics can be eluted out of the column only by increasing the polarity of eluent by adding methanol to toluene to a 50:50 ratio.

Asphaltenes show the characteristics of aromatic structure by the peaks at 3060, 1600, 865, 815, and 760 cm^{-1}. No carbonyl peak around 1700 cm^{-1} is apparently found, and a relatively small and broad band in the region 3500–3100 cm^{-1} is observed. A peak at 1030 cm^{-1} characteristic of sulfoxides is also found.

The spectra of naphthalene aromatics, polar aromatics, and asphaltenes all show relatively small bands in the region 1320–1060 cm^{-1}. These bands indicate the presence of compounds that have sulfur-containing functional groups [35]. Also, these three spectra show saturated CH, stretching vibrations represented by the peaks at 2950, 2925, and 2850 cm^{-1}. This indicates that there are saturated aliphatic substituents in the aromatic structures of these three fractions [36].

B. FTIR Analysis of Oxidized Corbett Asphalt Fractions

FTIR spectra of the oxidized Corbett fractions are shown in Figure 2. The spectrum of the oxidized saturates shows the emergence of a weak band around 3420 cm^{-1} characteristic of aliphatic OH stretching vibrations [19], an intense peak around 1730 cm^{-1} due to carbonyl stretching vibrations, and a relatively intense band in the region 1320–1000 cm^{-1} that includes a sharp peak at 1180 cm^{-1} and two shoulders around 1230 and 1030 cm^{-1}. This indicates the presence of different sulfur-containing compounds including sulfoxides, sulfones [35], and sulfates due to oxidation of sulfur-containing molecules in asphalt.

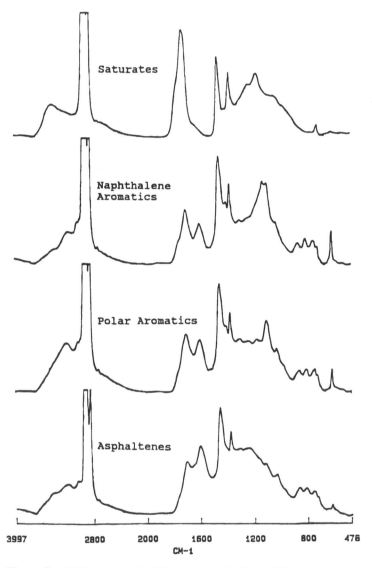

Figure 2   FTIR spectra of oxidized asphalt fractions [40].

The spectra of the oxidized naphthalene aromatics and the oxidized polar aromatics are again similar in features, with the exception of a peak found at 1180 cm^{-1} in the oxidized naphthalene aromatics, which is apparently related to the aliphatic sulfur-containing groups [33]. Both spectra have shown significant increases in intensity around 3240, 1710, and 1135 cm^{-1}. The broad peak at around 3240 cm^{-1} in the spectrum of the oxidized naphthalene aromatics results from OH or NH stretching vibrations. The shoulder around 3430 cm^{-1} may correspond to aliphatic OH stretching vibrations. The broad band at 3400 cm^{-1} in the spectrum of the preoxidized polar aromatics has shifted to 3240 cm^{-1} during photooxidation. This results from hydrogen bonding due to increasing amounts of these polar functional groups [19].

The spectrum of the oxidized asphaltenes has shown increases in the intensity of OH or NH bands in the region 3500–3100 cm^{-1} and the emergence of a carbonyl peak at 1705 cm^{-1}. The sulfur-containing groups also increase significantly in the region 1320–1000 cm^{-1}.

Like the spectrum of the oxidized saturates, the spectra of the other three oxidized fractions have shown significant increases of carbonyl groups. However, the absorption wave numbers (around 1715–1705 cm^{-1}) are relatively lower than that (1730 cm^{-1}) of the oxidized saturates. The wavelength difference among carbonyl groups in various Corbett fractions indicates that a variety of carbonyl compounds exist in the oxidized asphalt [19]. Also, an upshift baseline in the region of 1320–1000 cm^{-1} in these three spectra can be explained by the formation of sulfoxides, sulfones, and sulfates in each Corbett fraction during oxidation. The small difference in wave number of corresponding peaks among these three spectra is due to the different chemical environments surrounding the sulfur-containing functional groups.

The emergence of a peak at 620 cm^{-1} in all Corbett fractions, although relatively small in the spectra of the oxidized saturates and the oxidized asphaltenes, is a clear evidence of sulfates [33]. To further confirm the existence of sulfates in the oxidized fractions, the sample plates of the oxidized naphthalene aromatics and the oxidized polar aromatics were rinsed with methylene chloride. The residues, which are methylene chloride insoluble, on the plates were then analyzed by FTIR. The spectra were compared with that of pure potassium sulfate (EM Science GR grade, run as a KBr pellet) in Figure 3. Three characteristic peaks of potassium sulfate at 1120, 980, and 620 cm^{-1} are clearly observed in the spectra of the oxidized naphthalene aromatics and the oxidized polar aromatics. The association of metallic sulfates with water migration in asphalt pavement has been reported in an evaluation of sulfur-extended asphalt pavement [37]. The oxidation-produced sulfates may provide an important key to explaining deterioration of asphalt pavement through interaction of sulfates and water.

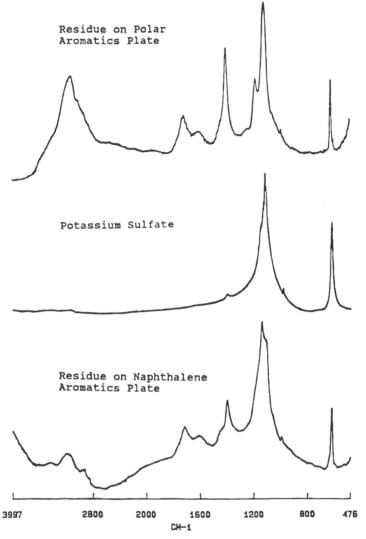

Figure 3 FTIR spectra of potassium sulfate and residues on plates of oxidized naphthalene and oxidized polar aromatics [40].

C. Calibration of GPC Column by Nondispersive Polystyrene Standards

The GPC chromatogram and peak assignment [38] of polystyrene standards are shown in Figure 4. The calibration curve for the logarithm of molecular weight versus retention time is illustrated in Figure 5. This calibration curve is quite linear throughout the entire calibration range. Although the structural compatibility of polystyrene standards with asphaltic materials has been questioned by several researchers [26], polystyrenes are still by far the most common standards used in the calibration of GPC columns for asphaltic materials. Several chemists [39] have synthesized octylated asphaltenes as standards to calibrate GPC columns used for determinations of the molecular weight of asphaltenes. They found that the differences in the structure between polystyrenes and asphaltenes are not a major consideration as far as the GPC molecular weight values are concerned.

D. Comparison of Molecular Weight Distributions of Corbett Asphalt Fractions

The four GPC chromatograms of preoxidized Corbett fractions are overlaid in Figure 6. A comparison of M_w, M_n, peak weight, and dispersity for the four Corbett fractions is summarized in Table 1. Naphthalene aromatics have the lowest M_n, and saturates have the lowest M_w and dispersity. The M_w, M_n, and dispersity of polar aromatics are all higher than those of naphthalene aromatics.

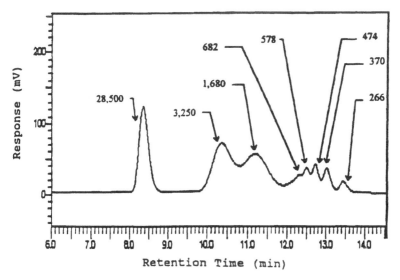

Figure 4 GPC chromatogram and peak assignment of polystyrene standards [41].

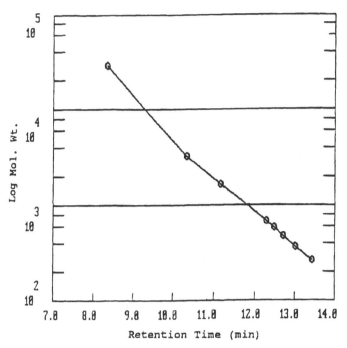

Figure 5   Calibration curve of molecular weight versus retention time [41].

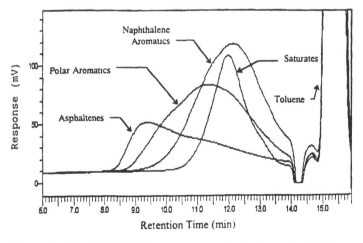

Figure 6   Overlaid GPC chromatograms of asphalt fractions [41].

Table 1 Comparison of Molecular Weight Distributions of Asphalt Fractions During Photooxidation [41]

Fraction	M_w	M_n	Peak weight	Dispersity
Saturates	874	732	852	1.19
Naphthalene aromatics	1083	630	754	1.72
Polar aromatics	2055	976	1463	2.10
Asphaltenes	4327	1615	6605	2.68

Asphaltenes have the highest M_w, M_n, and dispersity. This illustrates that asphaltenes have the most complicated molecular structure among all Corbett fractions. Also, the relatively high peak weight, 6605, of asphaltenes indicates that asphaltenes, contain associated molecules.

The M_n of the total asphalt, obtained by averaging the M_n values of the four Corbett fractions, is around 988. This number is lower than results reported by other researchers [20,21,29,39]. The reason might be that the degree of molecular self-association is decreased by Corbett fractionation run at a lower sample concentration (5 mg/mL). Thus, average molecular weights reported in this research are considered more reasonable.

E. Effect of Photooxidation on Molecular Weight Distributions of Corbett Asphalt Fractions

The overlaid GPC chromatograms of the preoxidized and oxidized Corbett fractions are shown in Figures 7, 8, 9, and 10. A quantitative comparison of molecular weights, peak weights, and dispersities for the preoxidized and oxidized Corbett fractions is summarized in Table 2. It is a common trend among all Corbett fractions that the molecular weight shifts to a higher value during photooxidation. Compared with the data for the preoxidized fractions, the M_n values of the oxidized Corbett fractions increase at least 130 in mass, and their M_w values increase more than 300 in mass. It is apparent that M_w is a more sensitive indicator for monitoring the oxidation of Corbett fractions by GPC. The oxidized saturates display the highest M_w, peak weight, and dispersity among all oxidized Corbett fractions. Observing a peak emerging at a value of 15,811 in the oxidized saturates, it is evident that polymerization reactions occur during the photooxidation of saturates. Also, the molecular size profile of the oxidized saturates illustrates the fact that at least two different groups of molecules coexist in the oxidized saturates.

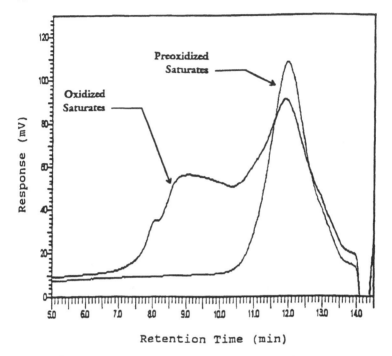

Figure 7   Overlaid GPC chromatograms of saturates [42].

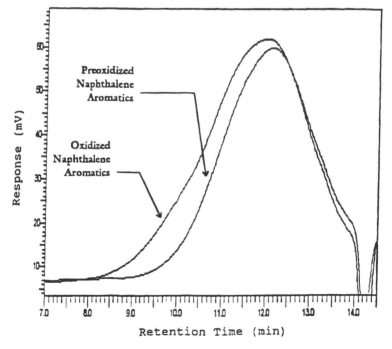

Figure 8   Overlaid GPC chromatograms of naphthalene aromatics [42].

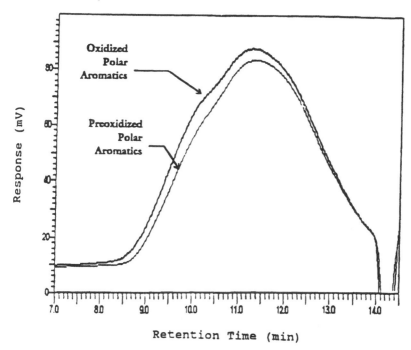

Figure 9   Overlaid GPC chromatograms of polar aromatics [42].

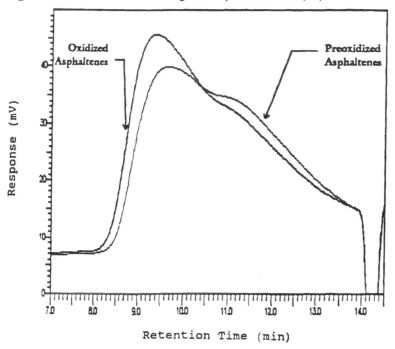

Figure 10   Overlaid GPC chromatograms of asphaltenes [42].

Table 2 Comparison of Molecular Weight Distributions of Asphalt Fractions
During Photooxidation [42]

Fraction	M_w	M_n	Peak weight	Dispersity
Saturates				
Preoxidized	874	732	852	1.19
Oxidized	6555	1316	15,811; 911	4.98
Naphthalene aromatics				
Preoxidized	1083	630	754	1.72
Oxidized	1453	783	829	1.86
Polar aromatics				
Preoxidized	2055	976	1463	2.10
Oxidized	2342	1092	1504	2.14
Asphaltenes				
Preoxidized	4327	1615	6605	2.68
Oxidized	5434	1810	8677	3.00

The M_n of the total oxidized asphalt, obtained by averaging the M_n values
of the four oxidized Corbett fractions, is around 1250. This number is about 260
higher than the M_n (988) of the total asphalt sample prior to photooxidation.
Several researchers [20,21,29,39] have reported different results for the number
average molecular weights of asphalts. This might be due to the various degrees
of oxidation in different asphalt samples used for molecular weight determination.

IV. CONCLUSION

The FTIR analysis of preoxidized and oxidized Corbett fractions does provide
more information than analysis of whole asphalt. However, the similarity between
naphthalene aromatics and polar aromatics suggests that a better separation technique needs to be developed to separate the aromatics into fractions with clear
distinctions. The formation of sulfates during photooxidation leads to a reasonable
conclusion that different kinds of sulfur-oxygen compounds coexist in the oxidized asphalt. The formation of sulfates during asphalt oxidation may also explain
deterioration of asphalt pavement.

The GPC analyses of Corbett fractions give relatively low average molecular weights of asphaltic fractions. This result supports the conclusion that molecules in asphalt might not be as big but rather partially associated. The GPC
comparison of preoxidized and oxidized Corbett fractions proves that the molecular weight of asphalt increases during photooxidation. The high molecular weight
due to photooxidation is certainly related to the hardening and embrittlement
of asphalts.

ACKNOWLEDGMENTS

The support of this work by The College of New Jersey through the FIRSL grant and the provision of the asphalt samples by the New Jersey Department of Transportation are appreciated. I would like to express my gratitude for the participation of G. Romeo, R. Yuro. K. Q. Mac, D. S. Bertholf, and A. Shah in this project through their independent study.

REFERENCES

1. J. C. Petersen, *Fuel Sci. Technol. Int.*, **110**: 57 (1993).
2. Y. Tropsha and A. Anthony, Photodegradation of Asphalt and Its Sensitivity to the Different Wavelengths of the Sunlight Spectrum, *Proc. ACS Div. Polym. Mater.*, **66**, 305 (1992).
3. T. Mill, D. S. Tse, B. Loo, C. C. D. Yao, and E. Canavesi, Oxidation Pathways for Asphalt, *Prepr. ACS Div. Fuel Chem.*, **37**(3): 1367 (1992).
4. S. M. Dorrence. F. A. Barbour, and J. C. Petersen, *Anal. Chem.*, **46**: 2242 (1974).
5. J. C. Petersen, S. M. Dorrence, M. Nazir, H. Plancher, and F. A. Barbour, Oxidation of Sulfur Compounds in Petroleum Residues: Reactivity-Structural relationships, *Prepr. Div. Petrol. Chem. ACS*, **26**(4): 898 (1981).
6. O. P. Strausz and E. M. Lown, *Fuel Sci. Technol. Int.*, **9**(3): 269 (1991).
7. T. F. Yen. J. G. Erdman, and A. J. Saraceno, *Anal. Chem.*, **34**: 694 (1962).
8. P. W. Jennings, J. A. S. Pribanic, W. Cambell, K. R. Dawson, and R. B. Taylor, High Pressure Liquid Chromatography as a Method of Measuring Asphalt Composition. *Gov. Rep. Announce. Index (U.S.)*, **81**(4): 686 (1981).
9. J. W. Bunger. K. P. Thomas. and S. M. Dorrence. *Fuel*, **58**: 181 (1979).
10. T. F. Yen. W. H. Wu. and G. V. Chilingar. *Energy Sources*. **7**(3): 203 (1984).
11. P. W. Yang. H. H. Mantsch. L. S. Kotlyar. and J. R. Woods. *Energy Fuels*. **2**: 26 (1988).
12. P. J. Gale. and B. L. Bentz. *Fuel Sci. Technol. Int.*. **10**(4–6): 1059 (1992).
13. T. F. Yen and J. G. Erdman. *Am. Chem. Soc. Div. Petrol. Chem.*, **7**(3): 99 (1962).
14. P. W. Jennings, M. A. Desando. M. F. Raub. R. Moats. T. M. Mendez, F. F. Stewart, J. O. Hoberg, J. A. S. Pribanic, and J. A. Smith. *Fuel Sci. Technol. Int.*, **10**(4–6): 887 (1992).
15. T. F. Yen, J. G. Erdman, and S. S. Pollack, *Anal. Chem.*, **33**: 1587 (1960).
16. J. W. Bunger, K. P. Thomas, and S. M. Dorrence, *Fuel*, **58**: 181 (1979).
17. D. Hadzi, *Fuel*, **32**: 112 (1953).
18. G. Brandes. *Brennstaff Chem.*, **37**: 263 (1956).
19. J. C. Peterson. Quantitative Functional Group Analysis of Asphalts Using Differential Infrared Spectrometry and Selective Chemical Relations—Theory and Application, *Trans. Res. Rec.*, **1096**: 1 (1986).
20. M. M. F. Al-Jarrah, and R. L. Apikian. *J. Chem. Technol. Biotechnol.*, **39**: 231 (1987).
21. J. P. Dickie and T. F. Yen, *Anal. Chem.*, **39**: 1847 (1967).
22. L. R. Snyder, *Anal. Chem.*, **39**: 1223 (1967).
23. G. A. Haley, *Anal. Chem.*, **43**: 371 (1971).
24. H. H. Kiet, L. P. Blanchard, and S. L. Malhotra, *Sep. Sci.*, **12**: 607 (1977).

25. W. A. Dark and R. R. McGough, *J. Chromatogr. Sci.*, **16**: 610 (1978).
26. M. L. Selucky, S. S. Kim, F. Skinner, and O. P. Strausz, *Chemistry of Asphaltenes*, American Chemical Society, Washington, DC, Chapter 6, pp. 83–118 (1981).
27. P. W. Jennings, J. A. S. Pribanic, T. M. Mendes, and J. A. Smith, *Fuel Sci. Technol. Int.*, **10**: 809 (1992).
28. G. R. Donaldson, M. W. Hlavinka, J. A. Bullin, and C. J. Glover, *J. Liquid Chromatogr.*, **11**(3): 749 (1988).
29. H. Reerink and J. Lijenga, *Anal. Chem.*, **47**: 2160 (1975).
30. S. I. Anderson, *J. Liquid Chromatogr.*, **17**: 4065 (1994).
31. L. W. Corbett, *Anal. Chem.*, **41**: 576 (1969).
32. ASTM Method, Standard Test Method for Separation of Asphalt into Four Fractions, D412, p. 683 (1982).
33. N. B. Colthup, *Introduction to Infrared and Raman Spectroscopy*, Academic Press, New York (1964).
34. J. B. Green, S. K.-T. Yu, C. D. Pearson, and J. W. Reynolds, *Energy Fuels*, **7**: 119 (1993).
35. M. Komatsu and M. Nakamizo, *Kyushu Kogyo Gijutsu Shikensho Hokoku*, **31**: 2031 (1983).
36. O. P. Strausz, T. W. Mojelsky, and E. M. Lown, *Fuel*, **71**: 1355 (1992).
37. W. A. Liddle and J. P. Mahoney, Sulfur Extended Asphalt Pavement Evaluation for the Baker River Highway, Washington, Proceedings of Sulfur Dev. Inst., Calgary, Alberta, Int. Conference, pp. 592–603 (1985).
38. *Polymer Laboratories Application Note*, No. 235 (1995).
39. S. Acevedo, G. Escobar, L. B. Gutierrez, and J. D'Aquino, *Fuel*, **71**: 1077 (1992).
40. J. Huang, R. Yuro, and G. A. Romeo, Jr., *Fuel Sci. Technol. Int.*, **13**: 1121 (1995).
41. J. Huang and D. Berthoff, *Fuel Sci. Technol. Int.*, **14**: 1037 (1996).
42. J. Huang and D. Berthoff, *Fuel Sci. Technol. Int.*, in press (1997).

6

Reduction of Odor from Hot Asphalt

Peter Kalinger
Canadian Roofing Contractors' Association,
Ottawa, Ontario, Canada

Robert J. Booth
Hansed-Booth, Dalkeith, Ontario, Canada

Ralph M. Paroli
National Research Council of Canada,
Ottawa, Ontario, Canada

I. INTRODUCTION

Roofing contractors, engaged in hot process built-up roofing, are often forced to contend with any number of challenges in the course of their on-site operations. Not the least of these is the negative perception of the public about the nature of their activities. Hot kettles and asphalt conveyance equipment with billowing clouds of white smoke and steam often evoke images reminiscent of the factory smokestacks of the early industrial revolution. Not only are the sight and noise disconcerting to some individuals, but the unmistakable odor of hot asphalt, which can be carried over great distances, may result in a variety of complaints ranging from slight discomfort to mild nausea.

Roofing asphalt is normally transported to the construction site in the form of solid blocks or kegs, where it is heated in a gas fire kettle, or in bulk form as a preheated liquid in specially designed tankers. From these heating devices, the hot liquid asphalt is carried to various locations on the roof through hollow tubing, welded drums, or buckets for application. The fumes emitted tend to be concentrated at the locations where the asphalt is being heated, discharged into the carriers, and applied.

Unfortunately, current technology for heating roofing bitumens, relying on external heating sources and exhausting fumes into the air, makes its aroma an unavoidable externality that must be taken into consideration each time a hot built-up roof is installed. Complaints from individuals regarding the noxiousness of the fumes, particularly on such sensitive projects as schools and health care facilities, have, in some instances, led to costly interruptions in operations, severe restrictions to work schedules, and in extreme cases the complete cessation of all roofing activity.

In response to these reported complaints and the high costs imposed by them on roofing contractors, a research project was designed to investigate potential ways of reducing or mitigating the adverse effects with respect to the odors produced while heating asphalt. The project consisted of laboratory testing and field trails of commercial fragrant compounds to determine their effectiveness in masking the odors associated with heated asphalt. The results of this research indicated that these materials can be successfully employed by the roofing contractor during the application of built-up roofs to screen the aroma of hot bitumen and reduce the discomfort of individuals who find it offensive. Further research, including a comprehensive analysis of health, safety, and performance consequences of the use of the cover odor materials, is required. In addition, the most efficient method of introducing the fragrance masks into the roofing operations has not been determined and requires further study.

Although this preliminary research indicates that commercial cover odors can be effective in veiling the fragrance of hot asphalt, this report should not be construed as an endorsement of their use. Further research is required to determine their full effects, and until such time as this research is complete, the use of these masking agents is at the contractor's own risk.

II. BACKGROUND

The health consequences of exposure to asphalt fumes have been the subject of much debate and numerous research studies have been undertaken to determine their toxicity and long-term health effects [1–22]. Most of this research involves dermal studies in animals exposed to large concentration of condensed asphalt fumes and limited epidemiological studies of mortality rates among roofing workers. At best, this research has been inconclusive, and further study is required before the effects of asphalt fumes on humans can be determined with any degree of confidence. In any event, there is no evidence that the limited exposure that a healthy member of the public would be subjected to in the reroofing of a school, church, or office building will have any detrimental effect on his or her well-being. Nevertheless, numerous individuals have reported feeling ill upon smelling heated asphalt or have objected to the noxiousness of the odor associated with asphalt fumes.

The sense of smell is probably the most subjective and complex of all our sensory faculties. The variance among individuals as to the effect of an odor is astounding. What may be pleasing to one person, to another may be quite repugnant. Crude attempts to categorize aromas under various headings ranging from aromatic to fetid have failed and the reasons for the differences in human reactions have continued to elude scientists and researchers.

However, certain aromas tend, for the most part, to evoke similar reactions among the majority of the population. Rotting eggs and skunk spray are examples of odors that elicit an almost universally negative response. Laboratory analysis has revealed that the offending substances in skunk excretion are mercaptans, or thiols. The potency of these mercaptans (thiols) is truly remarkable. Research has shown that the average person's olfactory sense has the ability to detect them in concentration as low as 1 part in 30 billion (1:30,000,000,000).

The particular offending mercaptan in skunk spray has been identified as butanethiol, a derivative of hydrogen sulfide. Crude petroleum, from which asphalt originates, contains small percentages of such sulfur-containing compounds and thiols. Although most of them are removed in the oil refining process, a small residual amount can be found in roofing asphalt. Given the potency of these substances, a very small concentration can cause a strong odor that is easily detected when the asphalt is heated.

Roofing asphalt, when heated, results in a dispersion of these small amounts of thiols into the atmosphere. The pungency of the odors varies with a number of factors, including the source of the asphalt (composition), temperature of the asphalt, ambient air temperature, prevailing winds, and the type and condition of the melting equipment. The response to the odor by people near the construction site will differ from individual to individual. In recognition of the fact that some members of the public will find the odor of the heated bitumen displeasing and discomforting, a research program was designed to examine methods of mitigating the objectionable odor of asphalt fumes and reduce in both the number and severity of the complaints received from the public near and around hot asphalt roofing sites.

In order to meet the project objectives, the following criteria were considered to be absolute requirements.

1. The method of altering or suppressing the asphalt odor must be effective in reducing the discomfort and annoyance of individuals in the proximity of roofing operations.
2. The method of suppression must be safe and produce no adverse health, safety, or environmental consequences.

3. The method of suppression must be compatible with current technologies employed in roofing operations.
4. The method of suppression must not adversely affect the physical properties of the heated bitumen and the subsequent performance of the roofing system.

In addition to these requirements, the method of suppression should satisfy conditions of economic efficiency to the extent that it would not add incrementally to current roofing costs so as to be prohibitive.

III. SELECTION OF MASKING AGENTS

The initial stage of the project involved a review of potential alternative methods of odor suppression or reduction. To satisfy the previously established criteria and, specifically, to meet the requirement of compatibility with existing technologies (criterion 3), the introduction of a masking agent during the asphalt melting process was identified as the most viable alternative.

A wide variety of available products and compounds, ranging from household bleach to commercially manufactured fragrance masks, specifically designed for the purpose of eliminating obnoxious nontoxic odors have been extensively used at various industrial locations, such as mills and foundries. Refineries and chemical and asphalt plants have successfully used these commercial cover odors to reduce the in-plant and emission odors at their operating plants.

Although household bleach is the least costly product capable of suppressing or reducing the pungency of asphalt fumes, it is a strong oxidizing agent. When added to asphalt in concentrated quantities, it may have a deleterious effect on the properties of the bitumen. Anecdotal evidence of independent on-site experiments in which bleach was added to the heated asphalt supported this supposition. As a result, attention was focused on commercially available fragrance masking agents.

At the suggestion of a major oil refinery and roofing bitumen producer, a supplier of concentrated cover odors was contacted. Four fragrance masking agents were identified as potentially suitable for the purpose of eliminating the odor of heated asphalt. Each substance is a yellowish oil-soluble liquid. One sample of each masking agent, consisting of 114 mL each, was procured for testing purposes.

A. Laboratory Evaluation of the Masking Agents

The experiment consisted of heating approximately 340 g of CSA A123.4 type 2 asphalt to a temperature of 200°C and recording the initial odor. A known weight (less than 0.1% of the asphalt weight) of the fragrance mask, an oily liquid, was added to the asphalt using a medicine dropper. The odor of the asphalt-mask mixture was recorded. Increasing amounts of the masking agent were added to the asphalt and the degree to which the masking occurred was recorded at every stage of the experiment. The amount of masking agent required to completely veil the odor of asphalt was determined. The results are summarized in Table 1. Based on the results of this qualitative experiment, the masking agents were ranked in increasing order of effectiveness. Fragrance A was deemed to be the most effective and was selected for further testing.

B. Evaluation of Properties of Asphalt When Mixed with Fragrance Mask

The introduction of any foreign substance into the heated asphalt may adversely affect the physical properties of the bitumen and consequently its performance as an adhesive layer or waterproof coating in built-up roofs. In order to assess the effects of the masking agent on these properties, laboratory testing was carried out on the asphalt-mask mixtures.

1. Flash Point

The flash point temperature of the masking agent was of particular concern. The published Material Safety Data Sheet reports the flash point temperature of fragrance A to be above 93°C. As the supplier could not provide a more precise flash point temperature of the cover odor, a series of tests was conducted to

Table 1   Results on Effectiveness of Four Hot Asphalt Odor Masking Agents

Weight % of fragrance mask	Fragrance A	Fragrance B	Fragrance C	Fragrance D
0.06	Change in odor detected	Change in odor detected	Change in odor detected	Change in odor barely detected
0.15	Change in odor easily detected	Change in odor easily detected	Change in odor easily detected	Change in odor detected
0.19	Change in odor very easily detected	Significant change in odor	Significant change in odor	Change in odor easily detected
0.27	Asphalt odor masked completely	Asphalt odor masked	Asphalt odor masked	Very significant change in odor
0.30	Very strong fragrance odor	Strong fragrance odor	Mild fragrance odor	Asphalt odor masked

determine the flash point of the neat fragrance, the neat asphalt, and mixtures of masking agent and asphalt in a 50:50 and a 0.3:99.7 ratio.

The flash points were determined in accordance with ASTM D92, Standard Test Method for Flash and Fire Points by Cleveland Open Cup. The testing revealed the flash points of the substances to be:

Sample	Flash point temperature (°C)
Neat fragrance mask	110
50:50 mixture of fragrance mask with CSA A123.4 type 2 asphalt	120
0.3:99.7 mixture of fragrance mask with CSA A123.4 type 2 asphalt	270

These results indicate that the fragrance mask has a minimal effect on the flash point of the asphalt, but caution is needed when adding the agent to the asphalt to avoid the ignition of vapors.

2. Asphalt Compatibility

During the preliminary testing, a brown exudate on the surface of the asphalt-mask mixtures was visible after cooling. The presence of the exudate indicates a potential for incompatibility between the fragrance mask and the asphalt. As a result, contact compatibility, softening point, and slippage tests were performed on the 0.3:99.7 mixture of fragrance mask and asphalt. The results of the slippage and softening point tests were then compared with the properties of neat asphalt. The test procedures used were ASTM D1370, Standard Test Method for Contact Compatibility Between Asphaltic Materials (Oliensis Test), and ASTM D36, Standard Test Method for Softening Point of Bitumen (Ring and Ball Apparatus). The slippage test consisted of pouring the mixtures on aluminum plates and exposing them to a 1:16 slope at a temperature of 50°C.

The softening point of the fragrance mask-asphalt mixture was measured to be 73.8°C. The softening point of neat CSA A123.4 type 2 asphalt was 75.4°C. The slight drop in the softening point indicates that the masking agent may have a small solvent effect on the asphalt. The slippage test, conducted at 50°C, showed no difference in slippage between the neat type 2 asphalt and the 0.3:99.7 mask-asphalt mixture. It was decided to continue further slippage tests at higher temperatures. The results of those tests are shown in Table 2. It was observed that at higher temperatures there was a noticeable increase for the mask-asphalt mixture when compared with neat type 2 asphalt. The 24°C decrease in the point of slippage makes the type 2 asphalt behave similarly to a type I.

The change in the point of slippage and the slight fallback in softening point temperature in the asphalt when the masking agent has been added must

Table 2 Slippage (mm) of 1 kg/m^2 Material Bonded to Steel Plate on 1:16 Slope

Temperature (°C)	Neat type 2 asphalt (softening point = 75–83°C)	Mixture 0.3:99.7%
50	0	0
60	0	0
70	0	0
78	0	5 mm
84	0	10 mm
90	0	32 mm
102	5 mm	55 mm

be taken into account when the cover odor is introduced directly into the bitumen. Further study and testing as to the consequences of these changes are necessary to ensure that no adverse performance occurs as a result of these changes in the asphalt's physical characteristics.

IV. FIELD TRIALS

A series of field test trials were organized in order to determine the effectiveness of the masking agent under real job site conditions. The minimum order quantity of fragrance mask A was purchased on two separate occasions and distributed to roofing contractors who had volunteered to carry out the field tests and evaluations. A table of asphalt quantity (in pounds) to masking agent in milliliters was provided to the volunteers to use as a guide for the on-site mixing of masking agent with asphalt (see Table 3). In light of the cost of the fragrance mask, at $77.00/kg ($0.016/kg in a 1% fragrance-asphalt mixture), one of the objectives of the site testing would be to ascertain the most efficient means of introducing the masking agent into the roofing operations.

A. Field Trial No. 1

The first field trial was carried out at 9:30 a.m. on September 24, 1993 at a reroofing site located in Ontario. The roof being constructed covered a warehouse near the Canada-U.S. border. The ambient air temperature was approximately 18°C on a cloudless day with a 15 km/h wind from the west-northwest.

A 2250-L heating kettle, with a circulating pump containing approximately 2025 L of type 3 asphalt, was slowly heated until the bitumen reached a temperature of 292°C. Two liters of the fragrance mask, the amount required for a 0.1% mixture, was added to the asphalt in 0.5-L amounts by sprinkling the fragrance

Table 3 Ratio of Asphalt to Masking Agent A

Asphalt quantity kg	Masking agent quantity (mL)		
	At 0.1% concn.	At 0.2% concn.	At 0.3% concn.
100	44	88	132
200	88	176	263
300	132	263	395
400	176	351	527
500	220	439	659
700	307	615	922
1000	439	878	1317
1500	659	1317	1976
2000	878	1756	2634
2500	1098	2195	3293
3000	1317	2634	3951
3500	1537	3073	4610
4000	1756	3512	5268
4500	1976	3951	5927
5000	2195	4390	6585

liquid over the surface of the asphalt in the kettle. An immediate change in the odor of the fume emissions was detected both during and after the addition of the masking agent. The odor from the asphalt had been completely overwhelmed by the perfume fragrance of the masking agent.

The 0.1% mixture effectively masked the odor of the asphalt at all locations near the site. The roofing workers, the building occupant, and the two individuals conducting the test reported that the smell of asphalt remained undetectable for a period of approximately 1 h after the addition of the fragrance mask. The only detectable odor was the slightly perfume fragrance of the masking agent. This condition was reported at a distance of 100 m from the kettle location. In addition, there was no noticeable asphalt odor of asphalt pumped to the roof top level and applied onto the insulation. However, the fragrance of the masking agent was easily detected.

In the light of the potency of the fragrance mask when added to the asphalt, another test was conducted to determine its potential effectiveness when introduced and vaporized in the vicinity of the melting equipment but not mixed directly into the asphalt. Approximately 100 mL of masking agent was poured into a small deflection on the top of the kettle. The perfume odor was perceptible for the duration of the experiment. It was found that heat from the kettle evaporated sufficient fragrance to mask the asphalt odor at the kettle location. However, it was not possible to distinguish between the effects of the masking agent warmed on the surface of the kettle and the residual effects of the fragrance-asphalt mixture.

B. Field Trial No. 2

The second field trial was conducted during the month of January in Calgary, Alberta. This trial was carried out in two phases, the first consisting of experimentation in the contractor's yard and the second at a single-story office-warehouse complex that was the site of a built-up reroofing project.

At the storage yard depot, the contractor compared several alternative methods of introducing the masking agent into the asphalt heating operation in order to determine the most effective method of masking the fragrance of hot asphalt. The weather was clear and sunny with an approximate outside temperature of −10°C.

To evaluate the flashing properties of the cover odor, 100 mL of the liquid fragrance was heated in a metal can to a temperature well above its reported flash point. Although an abundance of smoke was emitted, no ignition occurred during the heating process. Subsequent to the heating of the masking agent, another 250 mL was poured directly into the exhaust flues of a fired asphalt heating kettle. A powerful perfume odor was reported that lasted for a duration of approximately 1 h. These preliminary tests indicated that although the material could be heated to the elevated temperatures of liquid roofing asphalt, the fragrance volatiles would be depleted rapidly at these temperatures, rendering the cover odor ineffective in a relatively short period of time.

The contractor then placed 250 mL of the agent in a 2-L tin can and suspended the can inside the kettle, above the heated asphalt. No masking of the asphalt fumes was discernible. It was assumed that this method did not heat the cover odor sufficiently to activate the perfume to the extent that the asphalt fragrance would be masked.

At the site of the reroofing project, 500 mL of the masking agent was poured into the 2-L can. However, in this instance, the can was suspended in an enclosed 13,500-kg asphalt tanker. This method proved to be very effective, masking the fragrance of the hot asphalt both in the vicinity of the tanker and at the roof level where it was being applied. To provide a means of heating the fragrance mask to the required temperature and avoid introducing the cover odor directly into the hot asphalt, the contractor constructed a container consisting of a 1.5-m-long steel pipe, 75 mm in diameter. One end of the pipe was capped. Five hundred milliliters of the masking agent was poured into the pipe, which was in turn suspended in the inside of the asphalt tanker. The steel tube was attached so that the closed end was submerged into the liquid asphalt and the open end was above the top surface of the liquid asphalt.

The fragrance of the hot asphalt was completely masked almost immediately at both ground and roof level. An inspection of the building interior did not reveal any detectable asphalt odor during the entire operation. The asphalt odors were effectively suppressed for a duration of approximately 8 h. At the end of that period, the metal pipe was removed from the tanker. It was observed that approximately 100 mL of the masking agent remained in the pipe, indicating that this method of introducing the masking agent was efficient, maximizing the effectiveness over an extended period of time.

C. Field Trial No. 3

The third series of field trials was conducted by a contractor in Edmonton, Alberta. As with trial 2, the testing was carried out both on the contractor's facilities and on a project site. The masking agent was introduced to the asphalt heating operation both by adding the cover odor directly into the hot liquid asphalt and by heating the substance in a shallow metal pan placed on top of a kettle and tanker.

At the roofing project, approximately 25 mL of the masking agent was added to a kettle with a capacity of 1600 kg. The asphalt and fragrance mask were mixed and heated to a temperature of 205°C. It was observed that the asphalt odor was indiscernible at both ground and roof level. The aroma of the fragrance mask was clearly prevalent over that of asphalt for a period of 1 h after the addition of the agent, after which the asphalt aroma once again became noticeable. Within a period of 2.5 h, the asphalt odor became increasingly conspic-uous, until it completely overpowered the perfume odor of the masking agent.

To ascertain the efficacy of the masking agent when mixed with hot asphalt at the point of application, the masking agent was added to a hot tank (lugger) container with a capacity of 160 kg. The temperature of the asphalt in the container was recorded as 190°C. On three separate occasions, approximately 15 mL was added to the liquid asphalt prior to its being removed for application. The asphalt odor was generally, but not totally, masked as the asphalt was progressively drawn from the hot tank.

The final test of adding the masking agent directly to the hot asphalt consisted of pouring 150 mL of the agent into a liquid asphalt tanker containing roofing bitumen at a temperature of 180°C. The asphalt was then allowed to cool for 48 h until its temperature had fallen to 150°C, after which it was reheated to 200°C. It was observed that during the reheating process, the asphalt fragrance was effectively veiled until the bitumen reached a temperature of 170°C. At that point, the asphalt odor became clearly discernible.

The second battery of tests involved the extrinsic usage of the masking agent by heating the agent near the liquid bitumen but with no direct contact with the asphalt. A metal pan (200 × 300 × 25 mm) was fabricated and placed on a kettle near the discharge pipe. Fifty milliliters of the masking agent was poured into the pan and the asphalt was heated for application on four consecu-tive days.

On each day a small amount of the masking agent, 10 mL, was added to the pan to replace material that had evaporated. On each of the 4 days, the outside temperature was below 5°C with negligible to moderate winds. As the project was a commercial building having a substantial amount of pedestrian traffic (individuals entered and left a store in the vicinity of the asphalt equipment), there was an opportunity to ascertain the reaction of the public and the effectiveness of the odor cover.

It was noted that on days 1, 2, and 4, the masking agent fragrance completely overwhelmed the odor of the asphalt. On each of these days a slight to moderate wind prevailed. On day 3 the winds were negligible and the odor of the asphalt was easily detected both on the ground outside the store and in the store itself. The ineffectiveness of the masking agent in periods of still air would indicate that wind may be an important factor in the dispersion of the asphalt fumes and the diminution of their fragrance. When questioned, the individuals entering and leaving the store indicated their general preference for the masking agent aroma over that of the asphalt. Observations on the roof, however, revealed that the agent had no effect on the asphalt odor at the location of application by this method.

A similar test was carried out with the use of an asphalt tanker located in the contractor's storage yard. Again, a metal pan was placed on top of the tanker near the pump outlet. The effects of the masking agent on the asphalt were monitored for 3 days, during which the office staff of the contractor were asked to comment on any detectable odor. On the first and third days, outside temperatures were below −4°C, accompanied by slight winds. On the second day the maximum temperature had reached −2°C and the winds were calm. The masking agent was observed to have completely veiled the fragrance of asphalt during days 1 and 3. On the second day, with no noticeable wind, the asphalt odor was easily detected in both the yard and the adjacent building. When questioned, the occupants of the office indicated that they found the fragrance of the masking agent to be generally effective but that it became increasingly unpleasant during the progress of the trial.

D. Field Trial No. 4

The fourth set of field trials were carried out in Nova Scotia at two separate locations, a small retail mall and a telephone utility office. At the mall site, approximately 250 mL of masking agent was added to the asphalt being heated in a portable kettle at roof level. The masking agent was reported as effectively masking the aroma of the asphalt for a period of approximately 3 h by the workers on the roof. At the utility office location a similar quantity was added to the asphalt in a heating kettle at ground level. When the building occupants were questioned on their reaction to the fragrance of the heating emissions, they responded that the aroma was somewhat pleasing. The contractor was unable to provide information as to the amount of time that the fragrance masked the asphalt odors.

V. CONCLUSIONS

Roofing with heated asphalt has often been viewed as unpleasant and disagreeable by some members of the public exposed to the process of heating the bitumen. Much of this negative perception is a consequence of the powerful and easily recognized fragrance of hot asphalt. The characteristic aroma that accompanies hot asphalt roofing applications has prompted numerous complaints, which in turn have resulted in costly interruptions to both the building and the roofing operations.

The limited laboratory and field testing clearly demonstrates that the "annoying" odor of heated bitumen can be effectively veiled, or suppressed, by the introduction of commercially available masking agents (fragrance masks). The testing revealed that small quantities of masking agent A, when mixed directly with the hot asphalt, effectively masked the asphalt odor, although changes in the physical characteristics of the bitumen were observed in the course of the laboratory testing. Field trials indicated that the most efficient method of using the masking agent was by heating it separately from the asphalt. This method of introducing the masking agent into the roofing operations would have no effect on the properties of the asphalt, as direct contact would be avoided.

The responses of individuals exposed to the aroma of the fragrance mask were generally favorable. However, in one instance, the positive reactions appeared to diminish in proportion to the duration of exposure.

The availability of masking agents represents an opportunity for roofing contractors to selectively enhance the comfort of individuals exposed to the offensive odors of hot process roofing. Discriminating use of these substances can reduce the level of annoyance associated with hot asphalt roofing and potentially reduce the number of costly complaints received by roofing contractors. Additional research is needed to verify that there are no potential harmful health consequences from exposure to the masking agent or from the fragrance and asphalt mixture. Further field tests are required to better establish the most effective and economical method of introducing the cover odor material into the roofing site operations.

VI. RECOMMENDATIONS

Based on the laboratory results and field trials, the following recommendations are made:

1. Establish a research program consisting of an analysis of all available data and information related to the health and safety effects of exposure to the masking agent.
2. Contact asphalt heating and roofing equipment suppliers and manufacturers to explore the feasibility of altering or adding to current equipment design to permit the efficient use of the masking agent at the asphalt heating and application source.
3. Conduct further testing on the effects of the fragrance on the performance properties of the roofing asphalt.

GLOSSARY OF TERMS

Exudate: A substance released or oozed through pores, incisions, etc. With particular reference to asphalt, it is an oily substance that separates out from the asphalt when the asphalt is in direct contact with an incompatible material.

Fallback: A reduction of bitumen softening point related to contamination, incompatibility, or overheating. Also referred to as softening point drift.

Flash point: The temperature at which vapor from asphalt ignites or flashes in air but does not remain burning.

Thiol: Any class of sulfur-containing organic compounds with the formula RSH, where R is an organic group. Also called mercaptan.

Slippage: Sliding movement, usually down a slope between plies of felt along the bitumen film separating them.

Softening point: An index of bitumen fluidity. The temperature at which a thermoplastic material changes from solid to liquid.

REFERENCES

1. L. Flynn, Public/Private Partnership Studies Ways to Reduce Asphalt Fumes. *Roads Bridges*, **33**(5): 34 (1995).
2. E. Thompson and R. Haas, Fumes from Asphalt and Leachates from Pavements: Are They Harmful? Proceedings of the 37th Annual Conference of Canadian Technical Asphalt Association, November 1992, Victoria, British Columbia, pp. 47–71 (1992).
3. E. Thompson and R. Haas, Health and Environmental Effects of Fumes from Asphalt Paving, Proceedings of the 1991 Annual Conference of the Transportation Association of Canada, Winnipeg, Manitoba, September 15–19, 1991, Vol. 1, pp. A37–A55 (1991).
4. G. Rinck, D. Napier, and D. Null, Exposure of Paving Workers to Asphalt Emissions (When Using Asphalt-Rubber Mixes), Asphalt Rubber Producers Group (1991).

5. K. O. Gunkel, Study of Paving Asphalt Fumes, National Asphalt Pavement Association, Special Report, issue no. 134 (1989).

6. Asphalt Institute, Asphalt Hot-Mix Emission Study, report 75-1 (1975).

7. V. P. Puzinauskas and L. W. Corbett, Report on Emissions from Asphalt Hot Mixes, Asphalt Institute, RR-75-1A (1975).

8. D. Saylak, L. E. Deuel, J. O. Izatt, C. Jacobs, and R. Zahray, Environmental and Safety Aspects of the Use of Sulfur in Highway Pavements. Vol. II: Field Evaluation Procedures (Final Report, June 1978–February 1982), report no. FHWA/RD-80/192 (1982).

9. L. L. Laster, Atmospheric Emissions from the Asphalt Industry (Final Report), report EPA-650/2-73-046 (1973).

10. T. J. Partanen, P. Boffetta, P. R. Heikkilae, R. R. Frentzel-Beyme, D. Heederik, M. Hours, B. G. Jaervholm, T. P. Kauppinen, H. Kromhout, S. Langard, O. B. Svane, M. Bernstein, P. A. Bertazzi, M. Kogevinas, M. A. Neuberger, B. Pannett, and J. Sunyer, Cancer Risk for European Asphalt Workers. *Scand. J. Work Environ. Health*, **21**(4): 252, (1995).

11. T. Norseth, J. Waage, and I. Dale, Acute Effects and Exposure to Organic Compounds in Road Maintenance Workers Exposed to Asphalt, *Am. J. Ind. Med.*, **20**(6): 737, (1991).

12. E. S. Hansen, Cancer Mortality in the Asphalt Industry: A Ten Year Follow Up of an Occupational Cohort. *Br. J. Ind. Med.*, **46**(8): 582, (1989).

13. A. Sivak, K. Menzies, K. Beltis, J. Worthington, and A. Ross, Assessment of the Cocarcinogenic/Promoting Activity of Asphalt Fumes, National Institute for Occupational Safety and Health (1989).

14. R. Ferguson, Health Hazard Evaluation Report GHETA 82-253-1301, NIOSH, HETA-82-253-1301 (1983).

15. Industrial Hygiene Survey Report at Asbestos Roofing Company, Report IWS-48.17 (1977).

16. A. G. Apol and M. T. Okawa, Health Hazard Evaluation Determination Report NIOSH-HE-76-55-443 (1977).

17. M. T. Okawa and A. G. Apol, Health Hazard Evaluation Determination Report NIOSH-77-56-467 (1978).

18. Criteria for a Recommended Standard—Occupational Exposure to Asphalt Fumes, NIOSH, DHEW/PUB/NIOSH-78-106 (1977).

19. R. M. Chase, G. M. Liss, D. C. Cole, and B. Heath, Toxic Health Effects Including Reversible Macrothrombocytosis in Workers Exposed to Asphalt Fumes, *Am. J. Ind. Med.*, **25**(2), 279, (1994).

20. L. Machado, P. W. Beatty, J. C. Fetzer, A. H. Glickman, and E. L. McGinnis, Evaluation of the Relationship Between PAH Content and Mutagenic Activity of Fumes from Roofing and Paving Asphalts and Coal Tar Pitch, *Fundam. Appl. Toxicol.*, **21**(4): 492, (1993).

21. T. Norseth, J. Waage, and I. Dale, Acute Effects and Exposure to Organic Compounds in Road Maintenance Workers Exposed to Asphalt, *Am. J. Ind. Med. (US)*, **20**(6): 737, (1991).

22. G. A. Carson, Health Hazard Evaluation Report HETA 83-198-1646 (1986).

7

Propagation and Healing of Microcracks in Asphalt Concrete and Their Contributions to Fatigue

Dallas N. Little, Robert L. Lytton, and Devon Williams
Texas A&M University, College Station, Texas

Y. Richard Kim
North Carolina State University, Raleigh, North Carolina

I. INTRODUCTION

Load-induced flexural stresses in asphalt concrete pavements induce microcracks, and microcracks may also be present in the form of voids or discontinuties. These microcracks propagate under load and coalesce to form larger macrocracks, which, in turn, result in pavement fatigue cracking and ultimately pavement cracking distress. The rate and extent of fatigue crack damage in asphalt pavements are controlled by the growth (propagation) and healing (rebonding during rest periods between loads) of microcracks. This chapter discusses the role of microcrack propagation and healing in the fatigue cracking process of asphalt concrete pavements.

This chapter reviews pertinent literature supporting microcrack growth and the very important phenomenon of microcrack healing. Historical evidence for healing and documentation of the importance of healing are traced to early work on polymers and extended to recent work on asphalt concrete mixtures. The development in understanding of the effects of healing on asphalt concrete mixtures is charted through three landmark research efforts at Texas Transportation

Institute. These research efforts were funded under separate grants or contracts by the National Science Foundation, the Air Force Office of Scientific Research, and the Federal Highway Administration. The important effects of microcrack healing have been corroborated by important work at North Carolina State University and Delft University in The Netherlands. This research is referenced and discussed. Using this historical background of documented research as a foundation, this chapter presents a fundamental model of microcrack damage and healing and explains how this model provides the basis for a clear and rational understanding of crack-induced damage in asphalt concrete mixtures.

II. ROLE OF MICROCRACKING AND HEALING IN ASPHALT CONCRETE FATIGUE

A. Mechanisms of Fatigue

Fatigue is the result of a crack initiation process followed by a crack propagation process. During crack initiation, microcracks grow from microscopic size until, as some research indicates, a critical size of about 7.5 mm. In crack propagation, a single crack or a few larger cracks grow until the pavement layer breaks in two. At that point, others of the larger microcracks begin to propagate and coalesce to complete the disintegration process.

Fatigue is a two-stage process: (a) microcrack growth and healing and (b) macrocrack growth and healing. There is no reason to expect that the two obey different laws, and the available evidence demonstrates that they do not, but instead they are governed by the same Paris law. The Paris law coefficients A and n are known to depend strictly on more fundamental properties. The chief among them are the compliance and tensile strength (mechanical properties) and the adhesive and cohesive surface energy density (chemical and thermodynamic properties).

Both microcracks and macrocracks can be propagated by tensile or shear stresses or combinations of both. Thus, in a pavement structure, microcracks form and grow in any location where sufficiently large tensile or shear stresses or a combination of both occurs. Thus, in a pavement structure, microcracks can form and grow in any location where tensile or shear stresses are generated by traffic or environmental stresses. The location of such zones relative to the placement of a tire is shown schematically in Figure 1. Microcrack zones can be introduced into the pavement by thermal stresses due to a drop in air temperature as shown schematically in Figure 2. These cracks may grow, reach critical size, and propagate due to either a significant decrease in temperature or smaller repeated daily decreases in temperature. The same pattern of microcracks may develop because of contractile stresses exacerbated by the embrittlement of the asphalt pavement through aging. Any tensile or shear stress applied to a field where

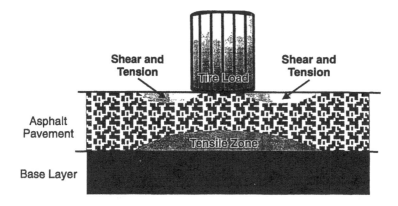

Figure 1   Schematic location of shear and tensile stress zones in the asphalt concrete layer under a tire.

microcracks exist may cause them to grow, to reach critical size, and then to propagate as macrocracks.

    Paris' law states that the "crack speed," dc/dN, depends upon the size of the J-integral or its elastic equivalent, the stress intensity factors due to tension (K_I) and shear (K_{II}). The distribution of K_I and K_{II} varies with depth and may be calculated using finite element methods. The K_I and K_{II} values induced by traffic are shown in Figure 3, for cracks initiating at the surface and at the bottom of the asphalt layer. Surface initiation is more common in thick asphalt layers (>200 mm) and bottom initiation is come common in layers of intermediate thickness (50–200 mm).

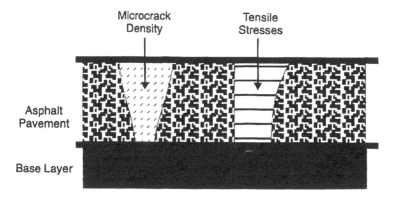

Figure 2   Schematic of microcrack density due to thermal stress.

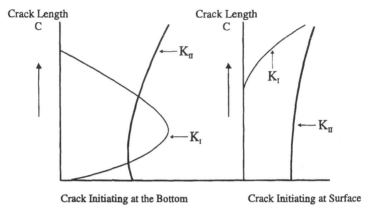

Crack Initiating at the Bottom   Crack Initiating at Surface

Figure 3   Stress intensity factors as a function of crack length in an asphalt concrete layer.

The number of traffic load cycles, N_f, to cause a crack to penetrate through the full depth of the pavement surface layer is the sum of the number of load cycles for crack initiation, N_i, and the number of load cycles required for the macrocrack to propagate to the surface, N_p.

$$N_f = N_i + N_p$$

Both N_i and N_p obey Paris' law as modified to include both fracture and healing. The actual number of load cycles required in each process is calculated by following the growth of a crack. Not only does the stress-intensity factor change with crack length, but also the values of the Paris law coefficients A and n for both fracture and healing vary depending upon whether the crack is momentarily growing along the surface of an aggregate (adhesive fracture) or out in the bitumen surrounding the aggregate (cohesive fracture) or is temporarily arrested by an object blocking its path (crack arrest).

B.   Microfracture Fatigue Relationship

The beam fatigue tests that were performed in the Strategic Highway Research Program A-003A contract [1] provided an excellent opportunity to observe the two phases of crack growth: initiation and propagation. The tests included both constant-strain and constant-stress tests, which made the process of separating the two phases of crack growth simpler.

In the constant-strain fatigue test, no crack propagation occurred and only the damage due to the formation and growth of distributed microcracks was observed. This was observed principally with two measurements: the stiffness of the beam fatigue sample decreased, and the rate of change of dissipated energy per load cycle continued to change with increasing number of load applications. Numerous measurements of the beam deflections and the dissipated energy per load cycle permitted the use of a finite element model of microfracture damage together with a systems identification method to determine the fracture properties of the mixture being tested.

In the constant-stress test, crack propagation did occur after a period of crack initiation. A clear separation of the two phases was observed in the plot of the rate of change of dissipated energy per load cycle versus the number of load cycles. At the beginning of the crack propagation phase, the rate of change of dissipated energy increased rapidly above the rather steady increase in the rate of change that characterized the previous crack initiation phase.

The fracture properties that were computed from the constant-*strain* fatigue tests were successfully used to compute the growth of microcracks and the rate of change of dissipated energy per load cycle in the constant-*stress* fatigue tests during the crack initiation phase. The same fracture properties were successfully used to compute the growth of the visible crack in the crack propagation phase.

C. Microfracture Model for Finite Element Analysis

A microfracture model of the crack initiation process was developed for use in finite element computations. Specifically, the model was used to back-calculate the Paris law parameters and some microcrack length and density parameters so as to match the measured changes in dissipated energy and stiffness of the beam fatigue samples. Subsequent applications of the same microfracture model were made to analyze the measured results of direct tensile fatigue tests.

In this model, an energy balance approach is used for the prediction of stiffness reduction and rate of change of dissipated energy due to microcrack growth as a function of the parameters A and n of Paris' law, the surface energy density Γ for the formation of crack faces, and the distribution of microcrack lengths. A small element of a homogeneous material having a microcrack of length $2c$ is shown in Figure 4. In the energy balance approach, this small element is represented by an equivalent uncracked element of reduced stiffness, whose strain energy is equal to the strain energy of the cracked element.

It is noted that both elements are subjected to the same stress state consisting of a normal stress, $\sigma_y = \sigma$, and a shear stress, $\tau_{xy} = \tau_{yx} = \tau$. For the small element with a center crack of length $2c$, the strain energy, U_c, per unit volume is given by:

$$U_c = \frac{\sigma^2 + 2(1 + \nu)\gamma^2}{2E} \left\{ 1 + 2\pi \left(\frac{c}{bl}\right) \left[\frac{4\gamma E}{\pi[\sigma^2 + 2(1 + \nu)\pi^2]} - c \right] \right\} \quad (1)$$

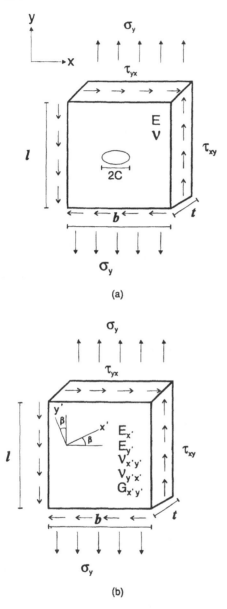

(a)

(b)

Figure 4 Elements used to derive microfracture properties by the energy balance approach.

where E = elastic modulus of the cracked element
   ν = Poisson's ratio of the cracked element
   b = width of the element
   l = element height

and the other terms are as defined previously.

Because of the expected orientation of the microcracks with increasing load repetitions, the equivalent uncracked finite element is modeled as an orthotropic material to account for the anisotropy that is induced in the cracked element with microcrack growth. In this formulation, it is assumed that the modulus parallel to the orientation of the microcracks remains unchanged with increasing load repetitions, so that the stiffness, $E_{x'}$, is equal to the original modulus, E, of the material. However, the modulus perpendicular to the microcracks, $E_{y}' = E'$, decreases with increasing load repetitions. It is also assumed that the shear modulus, $G_{x'y'} = G'$, at a given load cycle is related to the reduced Young's modulus E and shear modulus G through the relationship:

$$\frac{G'}{G} = \frac{E'}{E} \tag{2}$$

Using these assumptions, the following equation for strain energy U per unit volume for the equivalent uncracked small element is derived:

$$U = \frac{1}{2} \left[\frac{(F_1 + F_2)}{E'} + \frac{F_3}{E} \right] \tag{3}$$

where F_1, F_2, and F_3 are given by the following relationships:

$$F_1 = [\cos \beta(\sigma \cos \beta - 2\pi \sin \beta)]^2 \tag{4}$$

$$F_2 = 2(1 + \nu) \left(\frac{\sigma}{2} \sin 2\beta + \gamma \cos 2\beta \right)^2 \tag{5}$$

$$F_3 = [\sin \beta (\sigma \sin \beta + 2\gamma \cos \beta)]^2 \tag{6}$$

$$- \nu \sin^2 2\beta \left[\frac{1}{2} (\sigma + 2\gamma)(\sigma - 2\gamma) + \sigma\gamma \frac{(1 - \tan \beta)(1 + \tan \beta)}{\tan \beta} \right]$$

where β = angle of orientation of the microcracks.

Equating (1) and (3) above, the following relationship is derived for determining the reduced modulus, E', of the equivalent uncracked small element:

$$\frac{E'}{E} = \frac{F_1 + F_2}{F_4\{1 + 2\pi (c/bl)[(4\Gamma E/\pi F_4) - C]\} - F_3} \tag{7}$$

where $F_4 = \sigma^2 + 2(1 + \nu)\tau^2$.

If, instead of one crack, there are m microcracks, and m_i is the number of cracks of length c_i, then the ratio, m_i/m, is the probability of having a crack of length c_i, or

$$\frac{m_i}{m} = p(c_i) \tag{8}$$

In this case, E'/E becomes

$$\frac{E'}{E} = \frac{F_1 + F_2}{F_4 \{1 + 2\pi \, (m/bl) \int cp(c) \, [(4\Gamma E/\pi F_4) - C]\} - F_3} \tag{9}$$

This equation implies that the damage due to microcrack growth can be modeled through a stiffness reduction based on an energy equivalence. It also implies that all microcracks act independently, a condition which requires them to be some distance apart. The conditions for microfracture independence to be satisfied have been studied by Kachanov [2] and are considered to be valid for the asphalt concrete crack initiation phase.

The microcrack density (m/bl), in Equation 9, is assumed to be a function of the maximum stress σ_{max} in the load history of the material according to the relations:

$$\left(\frac{m}{bl}\right) = d \, (\sigma_{max} - \sigma_0)^q \qquad \text{if } \sigma_{max} > \sigma_0$$

$$= 0 \qquad \text{if } \sigma_{max} \leq \sigma_0 \tag{10}$$

where d and q are model coefficients and q is probably greater than 2.

The shape of this equation is illustrated in Figure 5. The term σ_0 is a threshold stress below which no microcrack growth occurs and no damage is done to the materials.

As indicated previously, microcracks vary in size, which results in a distribution of microcrack lengths. This distribution changes as the microcracks grow with repeated load repetitions as illustrated in Figure 6. In the microfracture model developed, the distribution of microcracks, $p(c)$, is assumed to follow a Weibull distribution given by the relation

$$p(c) = \lambda\gamma \, (\lambda c)^{\gamma - 1} e^{-(\lambda c)^\gamma} \tag{11}$$

where c = microcrack length
 γ = a shape parameter
 λ = a scale parameter

The Weibull parameters γ and λ are further assumed to change with load repetitions according to the following equations:

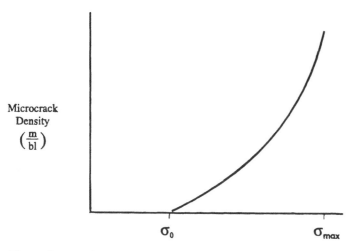

Figure 5 Relation between microcrack density and stress state.

$$\gamma = pe^{-(r/N)^s} \tag{12}$$

$$\lambda = ae^{-(f/N)^g} \tag{13}$$

where N is the load cycle and $p,\ r,\ s,\ a,\ f,$ and g are model coefficients. The foregoing equations are used to evaluate the microcrack distribution parameters γ and λ for any given load cycle, N. From these parameters, the integration in the denominator of Equation 9 is accomplished as follows:

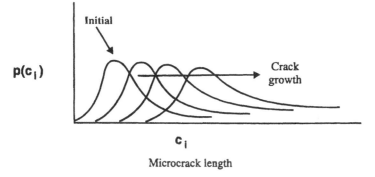

Figure 6 Probability density function of microcrack lengths.

$$\int cp(c) \left[\frac{4\Gamma E}{\pi F_4} - c \right] dc = \frac{4\Gamma E}{\pi F_4} \int cp(c)\, dc - \int c^2 p(c)\, dc \qquad (14)$$

Assuming that the microcrack lengths are distributed according to Equation 11, the $\int cp(c)\, dc$ becomes

$$\int cp(c)\, dc = \frac{1}{\lambda} \left[I\left(1 + \frac{1}{\gamma}, t_{max}\right) - I\left(1 + \frac{1}{\gamma}, t_0\right) \right] = \bar{c} \qquad (15)$$

where

$$I\left(1 + \frac{1}{\gamma}, t_{max}\right) = \text{incomplete gamma function with parameters } \left(1 + \frac{1}{\gamma}\right), t_{max}$$

$$I\left(1 + \frac{1}{\gamma}\right), (t_{max})_0 = \text{incomplete gamma function with parameters } \left(1 + \frac{1}{\gamma}\right) t_0$$

$$\bar{c} = \text{mean microcrack length}$$

$$t_{max} = (\lambda c_{max})^\gamma$$

$$t_0 = (\lambda c_0)^\gamma$$

$$c_{max} = \text{maximum microcrack length}$$

$$c_0 = \text{initial microcrack length}$$

Likewise, the integral, $\int c^2 p(c)\, dc$ is evaluated from the following expression:

$$\int c^2 p(c)\, dc = \frac{1}{\lambda^2} \left[I\left(1 + \frac{2}{\gamma}, t_{max}\right) - I\left(1 + \frac{2}{\gamma}, t_0\right) \right] = \overline{c^2} \qquad (16)$$

where the parameters are as defined previously. Equation 16 gives the expression for the mean of the squared microcrack lengths, $\overline{c^2}$. Thus, Equation 9 can be rewritten as

$$\frac{E'}{E} = \frac{F_1 + F_2}{F_4 \{1 + 2\pi(m/bl)\, [(4\Gamma E/\pi F_4) - \bar{c} - \overline{c^2}]\} - F_3} \qquad (17)$$

The stiffness reduction, E'/E, can also be related to the dissipated energy. The dissipated energy W per load cycle is the area enclosed by the load-displacement curve with each load cycle. From fracture mechanics, the change in dissipated energy per load cycle, dW/dN, is related to the J-integral through the relation

$$\frac{dw/dN}{t\, dc/dN} = J \qquad (18)$$

where t is the thickness of the material and dc/dN is the rate of crack growth. The rate of crack growth dc/dN is assumed to be given by the equation

$$\frac{dc}{dN} = A(K_I^2 + K_{II}^2)^{n/2} \qquad (19)$$

where K_I^2 and K_{II}^2 are given respectively by $\sigma^2 \pi c$ and $\tau^2 \pi c$ and are directly proportional to J^I and J^{II}.

From Equations 18 and 19, the following relationship may be derived between the stiffness reduction, E'/E, and the change in dissipated energy per load cycle, dW/dN.

$$\frac{E'}{E} = \frac{F_1 + F_2}{F_4\left\{1 + 2\pi\left(\dfrac{m}{bl}\right)\left[\dfrac{4\Gamma}{\pi} \dfrac{At\pi\, \overline{c^{(2+n/2)}}\, \sigma^2 + \gamma^{2n/2}}{dW/dN} - \overline{c^2}\right]\right\} - F_3} \qquad (20)$$

This relation is used in a finite element program to match the measured reduction in stiffness and change of dissipated energy by adjusting the values of

A and n (Paris' law coefficients)
d and q (microcrack density)
P (microcrack length distribution shape parameters)

No other microcrack length distribution parameters were found to be sensitive to changes in stiffness and dissipated energy.

The ranges of these five values, which were determined by back-calculation from both the constant-strain and constant-stress beam fatigue tests, are presented in Table 1.

English units were use in describing the microcrack size, modulus, and dissipated energy quantities, and the tabulated values of A and d are compatible with these. It is noted that the back-calculated values of A in the tables are based on the J-integral (following Equations 18 and 19).

Table 1 Back-Calculated Microfracture Parameters

Microfracture parameter	Constant-strain beam fatigue test	Constant-stress beam fatigue test
A	$0.6 \times 10^{-7} – 4.1 \times 10^{-5}$	$0.3 \times 10^{-6} – 39.5 \times 10^{-5}$
n	$1.19 – 12.12$	$1.60 – 10.11$
d	$0.21 \times 10^{-4} – 20.1 \times 10^{-4}$	$0.5 \times 10^{-5} – 16.6 \times 10^{-4}$
q	$2.34 – 2.79$	$2.21 – 2.79$
p	$0.13 – 3.40$	$0.10 – 6.40$
N (no. of samples)	17	22

In order to get corresponding values based on stress-intensity factors, the numbers given in the tables need to be divided by $E^{n/2}$, where E is the mixture stiffness at the appropriate test temperature.

Based on the results of the analysis of the controlled strain beam fatigue data, the following prediction equations for the fracture parameters were generated:

$$\log A = -2.605104 + 0.184408 AV$$
$$-4.704209 \log AC - 0.00000066E \qquad (21)$$
$$R^2 = 0.89, \quad RMS = 0.26$$

$$n = 12.244643 - 3.810569 \log A - 13.988526 \log \bar{\sigma} \qquad (22)$$
$$R^2 = 0.85, \quad RMSE = 1.41$$

$$\log d = -2.007118 - 0.201037n - 0.007647\bar{\sigma} \qquad (23)$$
$$R^2 = 0.84, \quad RMSE = 0.28$$

$$\log p = 0.979364 + 0.404126 \log A + 338.493411d + 0.005881\bar{\sigma} \quad (24)$$
$$R^2 = 0.83, \quad RMSE = 0.02$$

$$q = 2.649760 - 0.146474 \log p - 5429.318170A - 0.193129 \log AC \quad (25)$$
$$R^2 = 0.81, \quad RMSE = 0.05$$

where E = resilient modulus of the mixture (psi)
   AV = air voids (%)
   AC = asphalt content (%)
   $\bar{\sigma} = 1/3(\sigma_1 + \sigma_2 + \sigma_3)$, the mean principal stress (psi)

The observations of the number of loads cycles to complete the crack initiation phase in the constant-stress tests led to the development of the following equations for N_i, the number of load cycles for crack initiation:

$$\log_{10} N_8 = b_0 + b_1 + b_2\sigma_m + b_3[(\sigma_m)^2 + 2(1 + \mu)(\gamma_{oct})^2]E$$
$$+ (b_4 \log_{10}\sigma_m + b_5 \log_{10}E)(\%AC) \qquad (26)$$
$$+ b_6[(\sigma_m)^2 + 2(1 + \mu)(\gamma_{oct})^2]/E + b_7 \log_{10}\sigma_m (\%air)$$
$$+ [b_8(\sigma_m/E) + b_9 \log_{10}\sigma_m](\sigma_m/E)$$

where N_i = number of load cycles to crack initiation
   σ_m = mean principal stress, psi
   τ_{oct} = octahedral shear stress, psi
   E = asphalt concrete modulus, psi
   $\% AC$ = asphalt content by weight percent
   $\% air$ = air voids content, percent
   μ = Poisson's ratio

$b_0 = 4.415936$
$b_1 = -5.421 \times 10^6$
$b_2 = 1.11 \times 10^7$
$b_3 = -8.51796 \times 10^{-11}$
$b_4 = -0.838837$
$b_5 = 0.314813$
$b_6 = 3.089278$
$b_7 = -0.114846$
$b_8 = 35{,}787{,}201$
$b_9 = -12{,}144$
$b_{10} = 40.8396$

Figure 7 shows a comparison of the predicted N_i from Equation 26 with the back-calculated N_i as determined by the methodology presented previously. The fit is quite good with an R^2 of 0.847 and a standard error of the estimate (SEE) of 0.305, and the number of observations upon which the equation is based is 60.

It must be recognized that the rate of loading used in these fatigue tests was 50 Hz with no rest period between load applications. Thus, with this high

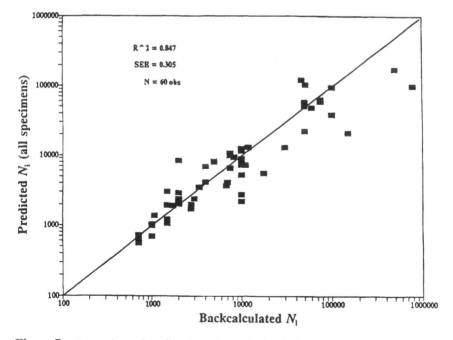

Figure 7 Comparison of predicted and backcalculated N_i.

rate of loading the beam material was acting essentially as an elastic material, rather than a viscoelastic material, and had no opportunity to heal. It is healing that provides a large portion of the "shift factor" that relates laboratory fatigue test results to the field fatigue observations.

D. The Shift Factor

Support for a shift factor between laboratory and field fatigue results can be found among many researchers. This underprediction of fatigue cracking based on laboratory data can be as high as 100-fold. Finn et al. [3] demonstrated that laboratory-derived phenomenological fatigue relationships for asphalt concrete used at the AASHTO road test required a shift factor of about 13 to match actual or observed fatigue in the road test sections. Raithby and Sterling [4] performed uniaxial tensile cyclic tests on beam samples sawed from a "rolled carpet" of asphalt concrete. They observed strain energy recovery during rest periods of up to three times longer than the loading periods. These rest periods resulted in longer fatigue lives by a factor of 5 or more than for the samples tested without rest periods.

Work at the Texas Transportation Institute during the past 10 years has demonstrated and quantified the significant effect of rest periods on fatigue life of asphalt concrete and on the recovery of strain energy. Balbissi and colleagues [5] found that rest periods increased fatigue lives by 10- to 17-fold and measured substantial strain energy recovery levels following rest periods. Kim [6] used the concept of pseudostrain to calculate the magnitude of pseudostrain energy that can be recovered following rest periods of various lengths. Kim found the pseudostrain energy recovery to be substantial and dependent on the length of the rest period, the temperature of the sample during the rest period, and the chemistry and rheological nature of the binder. The fact that pseudostrain energy was used by Kim to evaluate the "healing" effects of rest periods is significant, as the use of pseudostrain allows the time-dependent, viscoelastic effects to be separated from healing or rebonding of microcracks in damage areas.

Healing of microcracks during rest periods is the most significant factor affecting the difference between fatigue lives of continuously loaded laboratory fatigue asphalt concrete samples and actual pavements in which loading of the asphalt concrete is interrupted by rest periods of varying lengths. The length of the rest period can be significant even for highly trafficked pavements. This is because the period of significance is that between the passage of design or critical loads with the capacity to do the most damage.

To illustrate the significance of rest periods in highly trafficked parameters, consider an urban interstate with eight lanes (four in each direction) with an average daily traffic (ADT) of 200,000 (25,000 in each lane). However, during evening, low-traffic periods (10:00 p.m. until 4:00 a.m.) only about 10% of the traffic is applied [2500 vehicles over a period of 6 h (21,600 s)]. Now assuming a heavy truck traffic of 15% of ADT yields 375 damaging axles over a 21,600-s period or a rest period of about 58 s between axle loads. As another extreme, consider a moderately trafficked (four-lane) pavement (26,000 ADT) with 10% heavy truck. The average rest period is 173 s.

The difference between the time at which cracking is manifested in the pavement and the time at which it is predicted from cyclic laboratory fatigue tests without rest periods is due not only to the effects of healing but also to the effects of field residual stresses and aggregate dilation effects that occur in the field but cannot be duplicated in the laboratory. However, the most influential effect by far is that of microdamage healing [1]. In Strategic Highway Research Program (SHRP) study A-005, which resulted in the Superpave performance prediction models, Lytton et al. [1] used the following relationship to compute the healing shift factor between laboratory fatigue predictions and field expectations:

$$SF = 1 + \frac{n_{ri}}{N_0} a \left(\frac{t_i}{t_0}\right)^h$$

In this equation n_{ri} is the number of rest periods, N_0 is the number of loading cycles in which failure occurs based on cyclic laboratory tests without rest periods, t_i is the length of the period between load application in the field, t_0 is the length of the rest period between loading cycles in the laboratory test, and a and h are regression values and are directly related to a term established by Kim [6] called the healing index. The healing index was defined by Kim as the difference between the dissipated pseudostrain energy measured after a rest period and the dissipated pseudostrain energy measured before the rest period divided by the dissipated pseudostrain energy after the rest period. This can be expressed mathematically as follows:

$$\text{Healing index (HI)} = \frac{\phi_{\text{after}} - \phi_{\text{before}}}{\phi_{\text{before}}} = \frac{at_r^h}{1 + at_r^h} \qquad (27)$$

In this relationship ϕ represents pseudostrain energy, a and h regression constants, and t_r the period of rest. The relationship at the far right is that defined by Lytton in A-005. It represents a number that rises from 0 to 1 as the length of the rest period increases. For mixtures that heal well the number will rise faster than for poorly healing mixes. The exponent h is a material property and is independent of the period of rest. The exponent can be measured in the laboratory or can be inferred by back-calculation from fatigue cracking data observed in the field. Table 2 shows values of a and h determined in the calibration runs with SHRP A-005 pavement performance prediction models. In this case, the coefficient a depends on the actual traffic rate on each of the pavement test sections.

Table 2 Healing Coefficients Inferred from Field
Fatigue Data

	Healing coefficients	
Climatic zone	*a*	*h*
Wet, no freeze	0.09	0.843
Wet, freeze	0.059	0.465
Dry, no freeze	0.05	0.465
Dry, freeze	0.07	0.479

To illustrate the significance of the shift factor and the effects of healing, the *a* and *h* values from Table 2 can be used together with typical traffic data to calculate the magnitude of a healing shift factor. If this is done for a very highly trafficked urban interstate with 10% heavy trucks and 200,000 ADT (with eight lanes, four in each direction) and considering the *a* and *h* values for the wet–no freeze (Houston, Texas) region, the shift factor computed from Equation 27 is approximately 2.79. A similar shift factor calculated for a four-lane highway of moderate traffic (25,000 ADT, 10% heavy trucks) is approximately 6.73. Finally, the shift factor for a low-volume highway of 5000 ADT with 10% heavy trucks is approximately 13.4.

E. Historical Evidence of Healing in Polymers and Asphalt

Bazin and Saunier [7] introduced rest periods for asphalt concrete beam samples that had previously been failed under uniaxial tensile testing. They reported that a dense graded asphalt concrete mix could recover 90% of its original tensile strength after 3 days of recovery at 25°C (77°F). The researchers then performed cyclic fatigue tests. In these tests an asphalt concrete beam was loaded cyclically until fatigue failure occurred. The beam was then allowed to rest and recover and a cyclic load was again applied until fatigue failure reoccurred. The ratio of number of cycles to failure after the rest period to number of cycles to failure before the rest period was evaluated. This ratio was over 50% after 1 day of rest and with a 1.47 kPa (0.213 psi) pressure used to press the crack faces together. This research clearly showed the evidence of healing even though the rest periods and pressures were not necessarily realistic in terms of their ability to duplicate pavement conditions.

Despite the relatively small amount of historical research in the area of asphalt concrete healing, the mechanism of healing within polymeric materials has been intensely studied. Prager and Tirrell [8] described the healing phenomenon:

When two pieces of the same amorphous polymeric material are brought into contact at a temperature above the glass transition, the junction surface gradually develops increasing mechanical strength until, at long enough contact times, the full fracture strength of the virgin material is reached. At this point the junction surface has in all respects become indistinguishable from any other surface that might be located within the bulk material—we say the junction has healed.

Wool and O'Connor [9] identified stages of the healing process that influence mechanical and spectroscopic measurements: (a) surface rearrangement, (b) surface approach, (c) wetting, (d) diffusion, and (e) randomization. Kim and Wool [10] introduced the concept of minor chains and described the diffusion model. de Gennes [11] explained the microscopic sequences in a reptation model related to the Kim and Wool diffusion and minor chains model. The term "reptation" was defined as a chain traveling in a snakelike fashion, due to thermal fluctuation, through a tubelike region created by the presence of neighboring chains in a three-dimensional network. de Gennes explained that the wiggling motions occur rapidly, that their magnitudes are small, and that on a time scale greater than that of the wiggling motions, a chain, on average, moves coherently back and forth along the centerline of the tube with a certain diffusion constant, keeping its arc length constant.

Macromechanically, the most common technique used to describe the healing properties of polymers is to measure fracture mechanics parameters of a specimen that has healed. The fracture properties often used to evaluate healing potential are energy release rate, G_I, stress intensity factor, K_I, fracture stress, σ_f; and fracture strain, ϵ_f. These properties are dependent on the duration of the healing period, temperature, molecular weight, and pressure applied during the healing period.

Kim and Wool [10] used the critical energy release rate, G_{IC}, to define the portion of a chain that escapes from the tubelike regions defined earlier and influences the healing process through reptation-type interaction. Their model predicted that:

$$G_{IC} = t^{0.5} M^{-0.5}$$

where t is the duration of the healing period and M is the molecular weight. They also proposed the following experimental relationship:

$$\frac{\sigma_{fh}}{\sigma_0} = \frac{t^{0.25}}{M^{0.75}}$$

where σ_{fh} is the fracture healing strength and σ_0 is the original strength.

The temperature dependence of healing mechanisms has been reported by many researchers: Jud and Kausch [12], Wool [13], and Wool and O'Connor [9].

An increase in the test temperature shifts the recovery response to shorter times. Wool [13] constructed master healing curves by time-temperature superposition. Researchers in the area of polymer healing have also reported on the restoration of secondary bonds between chains of microstructural components and that Van der Waals forces or London dispersion forces play a very important role in healing [14]. Surface forces, electrostatic forces and hydrogen bonding have been reported to induce adhesive healing [14]. It has also been pointed out that adhesive forces and the bulk viscoelastic properties of the "hinterland" adjacent to the interface are the most important factors in the adhesion of elastomers (14).

According to other healing research (DeZeeuw and Pontente [15], Bucknall et al. [16]), the orientation and interpenetration of the flowing material influence the strength of the crack healing effect and this flow is dependent on healing temperature, contact period, and the extent of melt displacement.

In order to understand the healing mechanism of asphalt concrete, it is helpful to keep the healing models developed for polymers in mind. Petersen [17] claimed that the association force (secondary bond) is the main factor controlling the physical properties of asphalt. That is, the higher the polarity, the stronger the association force, and the more viscous the fraction, even if molecular weights are relatively low. Petersen [17] also presented a vivid description of the effect of degree of peptization on the flow properties as follows:

> Consider what happens when a highly polar asphaltene fraction having a strong tendency to self-associate is added to a petrolene fraction having a relatively poor solvent power for the asphaltenes. Intermolecular agglomeration will result, producing large, interacting, viscosity-building networks. Conversely, when an asphaltene fraction is added to a petrolene fraction having relatively high solvent power for the asphaltenes, molecular agglomerates are broken up or dispersed to form smaller associated species with less inter-association; thus, the viscosity-building effect of the asphaltenes is reduced.

Traxler [18] also suggested that the degree of dispersion of the asphalt components is inversely related to the complex (non-Newtonian) flow properties of the asphalt.

F.   Recent Developments in Research Affecting Fracture Healing

A significant breakthrough in the understanding of the effect of the composition of the asphalt on the healing of asphalt was made by Little et al. [19]. They found that healing was directly proportional to the amount of longer chain aliphatic molecules in the saturates and long-chained aliphatic side chains in the napthene aromatics, polar aromatics, and asphaltene generic fractions. They used the methylene-to-methyl carbon ratio (MMHC) as a quantifier of the nature of the long-chained aliphatic molecules and side chains. The MMHC is defined as the ratio of the number of methyl and methylene carbon atoms in independent aliphatic molecules or aliphatic chains attached to cycloalkanes or aromatic centers. Little and Prapnnachari [20] also found indications that the LETC settling test is a predictor of healing but did not have sufficient data to investigate this hypothesis thoroughly. However, this finding is in keeping with Petersen's view of the effects of peptization on flow properties.

Prapnnachari [21] continued some of the healing work and tried to relate fracture healing with relaxation spectra. He found evidence that the healing index and the relaxation spectra are indeed related. However, in their research, Little and Prapnnachari [20] discovered that the same methylene-to-methyl ratio determined by Little et al. [19] to be a significant indicator of the healing index is also indicative of the relaxation properties of the asphalt cement binders. Prapnnachari [21] used Fourier transform infrared (FTIR) spectroscopy to evaluate the asphalt cement while it was being stretched. Aliphatic appendages or side chains were found to influence significantly the relaxation properties of the binders tested. This finding, in conjunction with the finding of Little and Prapnnachari [20] with regard to the effect of MMHC on healing index, places added significance on further studies of the effects of aliphatic appendages on the various generic fractions.

Perhaps these findings of Little et al. [19] and Little and Prapnnachari [20] are verifications of Petersen's thoughts concerning the effect of association forces and dispersion on the flow and hence fracture healing properties of asphalt concrete. The extensive studies of Western Research Institute (WRI) in the Strategic Highway Research Program (SHRP) have provided a much more complete model of asphalt composition and microstructure. This advanced understanding now makes it possible to identify the compositional factors in the asphalt binder and in the mix that interact to influence healing.

A major breakthrough in the SHRP study by WRI was the development of a more accurate microstructural model of the asphalt cement. Whereas asphalt has often been popularly viewed in the past according to a micellar model, the new model pictures the microstructure of the asphalt as a continuous, three-dimensional association of polar molecules dispersed in a fluid of nonpolar or relatively low-polarity molecules. This model differs from the older micellar model in several ways. First, the micellar model postulates that the asphalt is a colloidal system with particles of a dimension very large on a molecular scale, representing agglomerations of many individual asphalt molecules. By contrast, the new model views the asphalt molecular structure as very small with no large-scale assemblages of molecules. Instead, the structure depends on instantaneous bonding among the wide variety of polar molecules dispersed in the asphalt.

The new model pictures the asphalt's molecular structure as continually forming and reforming as energy flows to and from the asphalt through the medium of external loading and temperature fluctuations. The new model stresses the importance of the polar asphalt molecules in mediating performance to a greater degree than does the micellar model. The principle class of molecules affecting performance might be the amphoteric materials—the ones that have at least one negative and one positive charge in the same molecule.

Thus this "latest" model of the asphalt microstructure is most certainly one that accommodates the phenomenon of healing. In this model, the viscoelastic properties of the asphalt and its response to load and temperature-induced stress result directly from the making and breaking of bonds between polar molecules. When the asphalt is subjected to stress, these secondary bonds are broken and re-formed continuously. The result is that the molecules move relative to each other. This is an excellent description of healing.

At first glance, the new three-dimensional model offered by SHRP may seem to contradict the strong positive relationship between MMHC, or the cumulative effect of aliphatic appendages, and healing. The three-dimensional model would seem to relate healing to continuous making and reforming of secondary (polar-activated) bonds, whereas the MMHC indicates a dominant influence of nonpolar aliphatics. However, upon more careful consideration, the effect of aliphatic appendages may be to act as tiny buffers or springs that inhibit agglomeration of the more polar fraction. This may help preserve a more dispersed structure and superior ability to flow across microfracture faces, or the aliphatic appendages may actually act in a reptation-type interaction with polar molecules in a synergistic fashion that has yet to be defined.

The latest SHRP studies do indeed offer encouragement that the healing phenomenon can be better understood by carefully considering the dispersed polar fluids (DPF) model in an experimental matrix of mechanical healing tests.

Kim et al. [22] used the elastic-viscoelastic correspondence principle (CP) to evaluate the hysteretic behavior of asphalt concrete. This principle simply states that one can reduce a viscoelastic (time-dependent) problem to an elastic (time-independent) problem merely by working in an appropriately transformed domain and substituting elastic moduli.

The nonlinear viscoelastic CP was developed by Schapery [23] in 1984. He suggested that the constitutive equations for certain nonlinear viscoelastic media are identical to those for the nonlinear elastic case, but stresses and strains are not necessarily physical quantities in the viscoelastic body. Instead, they are "pseudo" parameters in the form of a convolution integral.

This principle's application to the hysteretic stress-strain behavior of asphalt concrete was illustrated using actual data [22]. Repetitive uniaxial testing was performed under two completely different sets of testing conditions in order to demonstrate the applicability of the pseudostrain concept. The details of the testing conditions included the following:

Test 1	Test 2
Controlled-stress test	Controlled-strain test
Compressive loading	Tensile loading
Haversine waveform	Sawtooth waveform
0.2 s/cycle	1 s/cycle
Short rest periods (1,4,8,16 s)	Long rest periods (5.10,20,40 min)
Densely graded asphalt concrete (AC)	Sand-asphalt

When cyclic loading is applied to asphaltic mixtures, hysteresis loops are usually observed from the stress-strain curves with changing dissipated energy (i.e., area inside the stress-strain curve) as cycling continues. Based on the control mode (controlled stress versus controlled strain), two different trends are observed in the stress-strain behavior, as illustrated in Figure 8. It is noted here that the stress and strain levels in Figure 8 are low enough not to induce any significant damage in the specimens. Relaxation moduli are determined from both mixtures, and pseudostrains are calculated using the convolution integral. The data presented in Figure 8 are replotted in Figure 9 using the pseudostrains instead of the physical strain on the abscissa. The figures indicate that regardless of the control mode, the hysteretic behavior from cyclic loading disappears when pseudostrain is used, except for the first cycle, during which some minor adjustments occur in the specimen and test setup. It is also noted that the stress-pseudostrain behavior is linear.

The same approach is employed to investigate the ability of the CP to account for the relaxation effect during rest periods. Figure 10 presents the hysteresis loops observed in the controlled-strain and controlled-stress testing before and after rest periods. Again, the stress and strain levels are kept low so as not to induce any damage. A significant increase in the dissipated energy is observed after the rest period, irrespective of the control mode. Pseudostrains are calculated and plotted against the stress values in Figure 11. The use of pseudostrain successfully eliminates the effect of relaxation and results in a linear, elastic-like behavior between the stress and pseudostrain.

On the basis of the observations made from Figures 9 and 11, it can be concluded that, as long as the damage induced by the cyclic loading is negligible, the pseudostrain can be utilized to eliminate the time dependence of asphaltic mixtures from the hysteretic behavior, irrespective of mode of control, loading type (tension versus compression), loading rate, waveform, or length of rest period. It is this use of pseudostrain that makes it possible to evaluate the healing mechanism during rest periods separately from the relaxation.

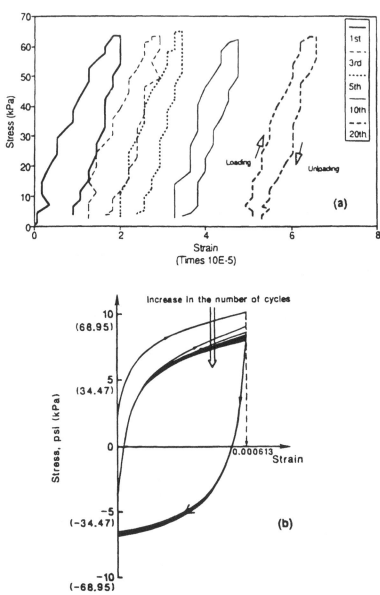

Figure 8   Hysteretic stress-strain behavior with negligible damage: (a) controlled-stress test; (b) controlled-strain test.

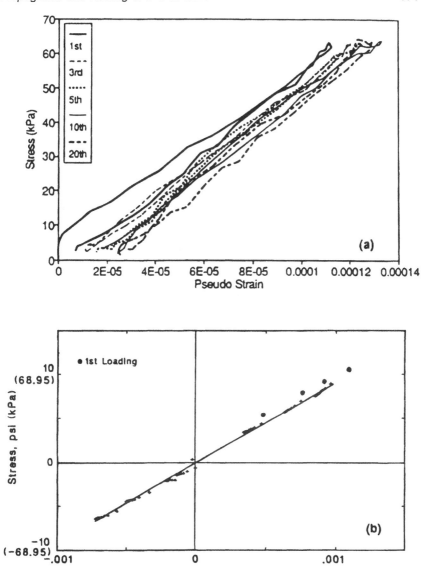

Figure 9 Application of CP to the data in Figure 1: (a) controlled-stress test; (b) controlled-strain test.

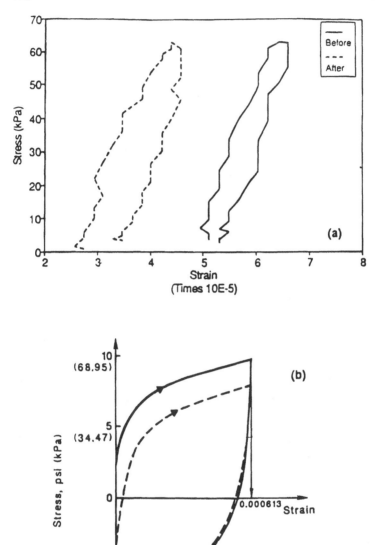

Figure 10 Effect of rest periods on stress-strain behavior (with negligible damage); (a) controlled-stress test; (b) controlled-strain test.

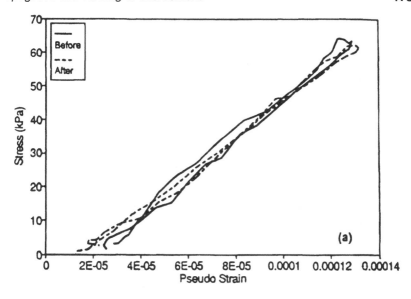

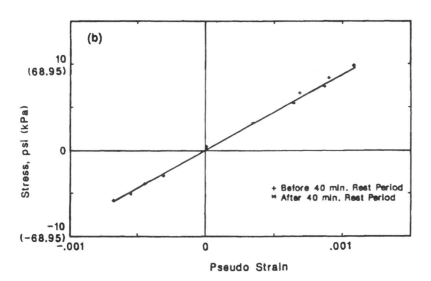

Figure 11 Application of CP to the data in Figure 10: (a) controlled-stress test; (b) controlled-strain test.

To evaluate fracture healing, the load amplitude is increased to a level at which significant damage growth can be expected during cyclic loading. Again, a significant increase in the dissipated energy is observed by comparing the stress-strain diagrams before and after the rest periods that could be resulting from either relaxation or healing or both. As mentioned previously, one cannot determine how much increase in the dissipated energy is related to healing merely by looking at stress-strain diagrams. Therefore, pseudostrains are calculated for both the controlled-stress and controlled-strain tests and plotted against the stresses in Figure 12. Different stress-pseudostrain behavior is observed in Figure 12 compared with that in Figure 11, where negligible damage occurred. Because the beneficial effect of the relaxation phenomenon has now been accounted for by using pseudostrain, changes in the stress-pseudostrain curves during the rest periods, as observed in Figure 12, must result from the fracture healing of microcracks.

One can determine the propensity for microcrack healing in different mixtures by measuring the difference in areas under the stress-pseudostrain curves before and after a rest period and normalizing it by the stress-pseudostrain area before the rest period. This method has been used successfully in differentiating the healing potentials of various binders.

Kim et al. [22] also used stress wave testing and impact resonance testing to identify healing after rest periods in laboratory asphalt concrete samples and in actual pavements. One of the most interesting aspects of this study was the elastic modulus evaluation of the asphalt concrete surface of a North Carolina highway, which was closed to traffic for a 24-h period. The stress wave test method measured stiffness of the asphalt concrete at middepth of the layer. Because of the impact nature of loading, the stiffness was in the glassy region (purely elastic), so the modulus change should not be affected by time-dependent relaxation. Therefore, the increase in modulus could be the result of microdamage healing within the asphalt concrete layer.

Figure 13 shows the hourly change in the elastic modulus of the pavement over a 24-h period. The figure illustrates the predictable decrease in modulus with temperature increase. But the most significant factor is the elastic modulus increase observed at a selected pavement temperature after the 24-h rest period. This modulus increase may be the result of microdamage healing.

III. FUNDAMENTAL CONCEPTS FOR DEVELOPMENT OF MICROSTRUCTURAL MODEL

A. General Background

Schapery [24] discovered in 1973 that what had been thought to be an empirical rule known as Paris' rule could, in fact, be derived from first principles applied

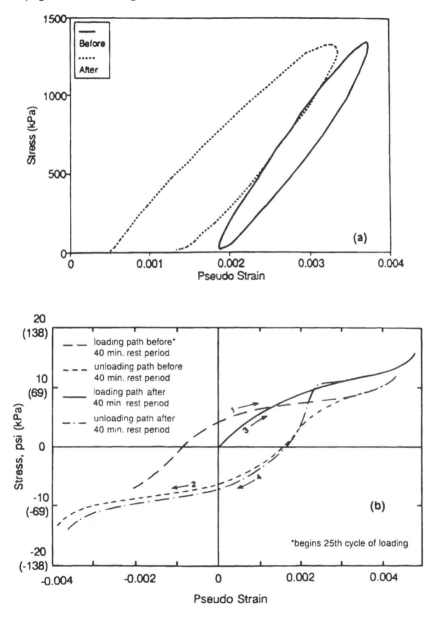

Figure 12 Stress-pseudostrain behavior before and after rest periods with significant damage: (a) controlled-stress test; (b) controlled-strain test.

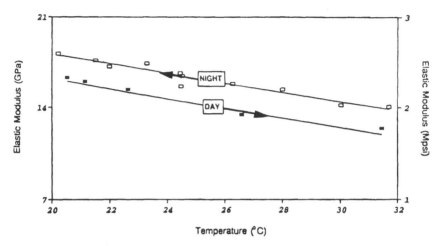

Figure 13   Elastic modulus versus pavement temperature from stress wave testing of a pavement in service.

to a viscoelastic material using a power law form of its creep compliance. Paris' law applied to fatigue crack growth is

$$\frac{dc}{dN} = A(K)^n \tag{28}$$

where c = crack length
   N = number of load repetitions
   A, n = empirical coefficients to be determined from physical repeated load test in the laboratory
   K = so-called stress intensity factor, which at first had to be calculated with an appropriate model

Schapery found that both A and n could be expressed in terms of fundamental material properties, namely:

1.   Power law creep compliance properties, D_0, D_1, and m
2.   Tensile strength, σ_t
3.   Surface energy density, Γ
4.   Waveform with time of the applied stress or strain, $w(t)$
5.   Length of time the stress or strain was applied, Δt

The most important property of these, and one that pervades the expressions for both A and n, is the slope of the curve of log creep compliance versus log time. In order to arrive at his results, Schapery had to consider the energy that is released in the fracture process zone ahead of the visible crack. In this zone, large plastic deformations and microcracking occur, both of which require energy to be expended. Schapery's work, both theoretical and experimental, advanced further and began branching and multiplying in several fruitful directions when he began to use the J-integral instead of the stress intensity factor, K. The two are related in lineraly elastic materials by the expression

$$J = \frac{K^2}{E} (1 - \nu^2) \quad \text{for plane stress conditions} \tag{29}$$

and

$$J = \frac{K^2}{E} \quad \text{for plane stress conditions} \tag{30}$$

However, the J-integral applies to any linear or nonlinear elastic material and can be measured directly in the laboratory because of its fundamental definition

$$J = \frac{\Delta W}{b \, \Delta C} \tag{31}$$

where ΔW = change of dissipated energy in advancing a crack by a length, Δc
$b \, \Delta c$ = change of crack area
b = width of sample

Schapery showed that the J-integral can be applied to viscoelastic materials through the use of the known correspondence principle and the extended correspondence principles, that he developed.

The correspondence principle states that the form of a relation between stress, strain, and energy for an *elastic* material is the same as the Laplace transform of the relation for the corresponding viscoelastic material. Schapery's next necessary advance was to demonstrate mathematically, using the thermodynamic principles of mechanics, that the correspondence principles could be generalized using the concept of what he termed pseudostrain. This permits, as discussed in Section 2.F, any time-dependent material to be transformed into an equivalent nonlinear elastic material, which is sometimes called a reference material. The transformation of the time-dependent strain in the actual material into a non-time-dependent pseudostrain in the reference material is done by using a memory integral, which is similar, mathematically, to the Laplace transform used in the correspondence principle. Pseudostrain is defined as ϵ_0 in the following equation:

$$\epsilon_0 = \frac{1}{E_R} \int_0^t E(t - \tau) \frac{\partial \epsilon}{\partial \tau} \, d\tau \tag{32}$$

where $E(t)$ = relaxation modulus of the material
   ϵ = time-dependent strain
   E_R = a reference modulus (dimensions FL^{-2})
   F = force
   L = length

The reference modulus makes the equation dimensionally correct and may be set at any constant value that produces a physically meaningful value of the pseudostrain. In tests with constant strain rate and constant maximum strain, the value of E_R is set to make the maximum pseudostrain equal to the maximum strain imposed on the test sample. This, in turn, permits the calculated pseudostrain energy on each load cycle to represent the actual total energy per unit volume that is expended on the material. Pseudostrain energy is the area under the stress-pseudostrain curve on each load cycle and is the energy per unit volume that is dissipated during that load cycle. Changes in this dissipated pseudostrain energy between load cycles define the *J*-integral that drives the growth of cracks in the material.

Theoretical and experimental work done in the SHRP A005 project by Lytton et al. [1] showed that the rate of growth of microcracks in beam fatigue samples that are loaded cyclically at relatively high frequency (50 Hz) with no rest completely explained the observed changes in dissipated energy and beam stiffness during the entire test. Companion beams were tested with constant center deflection (called constant-strain tests) and constant center load (called constant-stress tests). Inverse analysis of the constant-strain tests produced values of the Paris law coefficients A and n and three other parameters, p, d, and q, as explained earlier in this chapter. It was assumed, for the purpose of the analysis, that microcracks obey Paris' law when they grow. The values of A and n derived from the constant-strain tests were found to predict very well the growth of microcracks in the companion constant-stress tests. The cyclic loading rate in the beam fatigue tests was so high (50 Hz) that the material behaved essentially as an elastic material, not requiring the use of the pseudostrain energy in the analysis.

Two distinctively different parts of the crack growth process were observed consistently in the constant-stress tests. The first part consisted of a steady growth of the rate of change of dissipated energy per load cycle, dW/dN. In this part, microcracks grew in length until some reached a critical size, which was calculated to be 7.5 mm (0.3 inch). At that point, macrocrack growth began. In the second part of the crack growth process the rate of change of dissipated energy per load cycle accelerated as the single macrocrack demanded a much greater expenditure of energy to drive it to greater lengths. A graph of the rate of change of dissipated energy, dW/dN, versus the number of load cycles, N, is shown in Figure 14. These two parts of the crack growth process, illustrated in Figure 15, have been known to exist qualitatively for many years, during which they were termed the

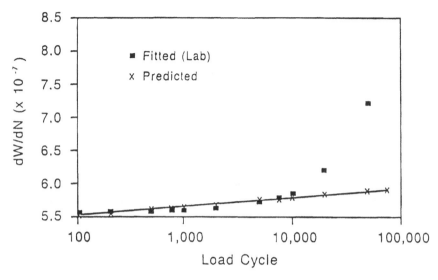

Figure 14 Sample illustration showing determination of N_i from dissipated energy data experiment conducted at 20°C (68°F).

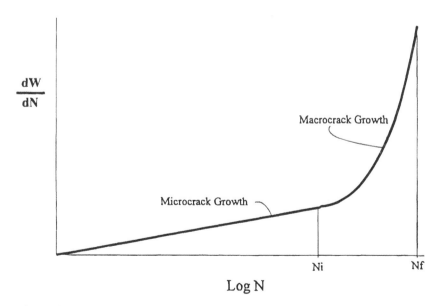

Figure 15 Illustration of the relationship of the parts of the crack growth process as related to the rate of change of dissipated energy per load cycle, *dW/dN*.

crack initiation and crack propagation phases. The distinction between the two parts is refined by focusing on the rate of change of dissipated energy, dW/dN, as shown in Figure 15. The Paris law coefficients, A and n, that predicted microcrack growth also predicted the growth of the macrocrack.

The coefficients p, d, and q mentioned previously also play a significant role in describing microcrack growth. Microcrack density varied considerably from one part of the fatigue beam to another. The strain energy was dissipated more rapidly in areas of high stress in the center of the beam than in areas that were under lower bending and shearing stresses. This was modeled by use of a finite element program that was capable of representing different anisotropic moduli in each of the several hundred elements used to represent a single beam. The density of microcracks in each element was found to be proportional to the level of principal tensile stress in each element as given by

$$\frac{\text{Microcracks}}{\text{Unit area}} = d(\sigma - \sigma_0)^q \tag{33}$$

d, q = microcrack density coefficients
σ_0 = treshhold stress below which microcracks did not form
σ = principal tensile stress in an element

The value of σ_0 appeared to vary slightly from one bitumen to another but generally to be around 34 kPa (5 psi). The value of p has to do with the distribution of microcrack lengths. A Weibull probability density function is assumed to describe the distribution of microcrack lengths. These lengths change with the number of load repetitions, and, if necessity, the Weibull distribution must also change. It was found by experimentation that the Weibull shape parameter, γ, changed but that the scale parameter, λ, did not with load repetitions. The shape parameter was modeled as a Gumbel cumulative distribution as follows:

$$\gamma = pe^{-(\lambda/N)^s} \tag{34}$$

This expression allows the Weibull shape parameter γ to grow larger as the number of load cycles, N, increases. The parameter p governs the rate of growth and is found to differ between bitumens. Knowledge of the values of p, d, and q permits the computation of the critical size of microcracks at crack initiation. The successful application of Schapery's and Paris' fracture laws to the observed micro- and macrocracking in all aspects of the behavior of the beam fatigue tests conducted in the SHRP A-003A project provided the motivation to develop the theory further and refine the understanding of the fundamental nature of the material properties it uses. The motivation leads to a return to what Schapery termed the fundamental relation of fracture mechanics.

B. Fundamental Relation of Fracture Mechanics

The fundamental relation of fracture mechanics as stated by Schapery is:

$$2\Gamma = E_R D(t_\alpha) J_\nu \tag{35}$$

where Γ = a surface energy density, i.e., per unit area of crack surface to separate a material from itself (units FL^{-1})

       E_R = reference modulus used, if necessary, in representing a nonlinear viscoelastic material as an equivalent nonlinear elastic (units FL^{-2})

       $D(t_\alpha)$ = tensile creep compliance of a material corresponding to the time, t_α, that is required for a crack to move through the distance α, which is the length of the fracture process zone ahead of the crack tip (units $L^2 F^{-1}$)

       J_ν = strain energy or pseudostrain energy J- integral; this is the rate of change of dissipated energy per unit of crack growth area from one tensile load cycle to the next (units FL^{-1})

Equation 44 is an energy balance: the energy given up on the right-hand side of the equation is taken up by the newly created crack surfaces on the left-hand side of the equation.

This "fundamental relation" describes the conditions of *cohesive* fracture. A similar fundamental relation of *adhesive* fracture is

$$\Gamma_1 + \Gamma_2 - \Gamma_{12} = E_{12} D(t_\alpha) J_\nu \tag{36}$$

The energy released on the right-hand side of Equation 44 is distributed among the terms on the left-hand side of the equation. The terms are, as applied to asphalt and aggregate interfaces,

Γ_1 = surface energy density of the asphalt
Γ_2 = surface energy density of the aggregate surface
Γ_{12} = energy of interaction between the two materials

This is actually a more general form of the fundamental fracture relation, because with cohesive fracture, $\Gamma_1 = \Gamma_2$ and $\Gamma_{12} = 0$.

There is a direct parallel between fracture and healing processes, healing being the reversal of fracture. The fundamental relations for both cohesive and adhesive healing processes are

$$2\Gamma_h = E_R D_h (t_\alpha) H_\nu \tag{37}$$

for cohesive healing, where

Γ_h, $D_h (t_\alpha)$ = surface energy density and creep compliance properties of the material during healing

H_v = healing H-integral representing the mechanical strain energy or pseudostrain energy needed to heal a fractured surface area:

and

$$\Gamma_{1h} + \Gamma_{2h} - \Gamma_{12h} = E_R D_h(t_\alpha) H_v \tag{38}$$

for adhesive healing, with similar meanings for the quantities Γ_{1h}, Γ_{2h}, Γ_{12h}, and $D_h(t_\alpha)$. The differentiation between fracture and healing properties is necessary because the compliances in fracture are tensile and those in healing are typically compressive. Also, the surface energies in fracture are for a dewetting process whereas for healing they are for a wetting process, which requires less energy than for dewetting.

The compliance of an asphalt-aggregate mixture is expressed in a power law form as

$$D(t) = D_0 + D_1 t_t^m \tag{39}$$

in tension and

$$D_h(t) = D_{0c} + D_{1c} t_c^m \tag{40}$$

in healing and compression. The time required the fracture or healing to traverse the length of the fracture process zone, α, is

$$t_\alpha = \frac{k_t \alpha}{\dot{f}} \tag{41}$$

in fracture and

$$t_\alpha = \frac{k_{th} \alpha}{\dot{h}} \tag{42}$$

in healing, where

k_t, k_{th} = constants that depend on the value of m, the slope of the log creep compliance versus log time curve; a common value of k_t is $1/3$ and k_{th} for healing is also approximately $1/3$

m_t, m_c = slopes of the tensile and compressive log creep compliance versus log time curves, respectively, for asphaltic concrete

$\dot{f}$ = fracture speed

$\dot{h}$ = healing speed

The crack speed, $\dot{c}$, is the difference between the fracture speed, $\dot{f}$, and the healing speed, $\dot{h}$:

$$\dot{c} = \dot{f} - \dot{h} \tag{43}$$

These relations lead to the basic equations or cohesive and adhesive fracture and healing as follows.

Fracture process

1. Cohesive fracture

$$2\Gamma_1 = E_R\left[D_0 + D_1\left(\frac{k_t\alpha}{f}\right)^{m_t}\right]J_\nu \tag{44}$$

2. Adhesive fracture

$$\Gamma_1 + \Gamma_2 - \Gamma_{12} = E_R\left[D_0 + D_1\left(\frac{k_t\alpha}{f}\right)^{m_t}\right]J_\nu \tag{45}$$

Healing process

1. Cohesive healing

$$2\Gamma_{1h} = E_R\left[D_{0h} + D_{1h}\left(\frac{k_{th}\alpha}{h}\right)^{m_c}\right]H_\nu \tag{46}$$

2. Adhesive healing

$$\Gamma_{1h} + \Gamma_{2h} - \Gamma_{12h} = E_R\left[D_{0h} + D_{1h}\left(\frac{k_{th}\alpha}{h}\right)^{m_c}\right]H_\nu \tag{47}$$

Combining the equations for fracture and healing speed to give the crack speed and then integrating over the time interval during which the load is applied, $(\Delta t)_l$, and over which healing proceeds, $(\Delta t)_h$, produce the following cohesive fracture equivalent of Paris' law:

$$\frac{dc}{dN} = \int_0^{(\Delta t)_l} \frac{k_t\alpha(D_1 E_R J_\nu)^{1/m_t}}{(2\Gamma_1 - D_0 E_R J_\nu)^{1/m_t}} \, dt \tag{48}$$

$$- \int_0^{(\Delta t)_h} \frac{k_{th}\alpha(D_{1h}E_R H_\nu)^{1/m_c}}{(2\Gamma_{1h} - D_{0h}E_R H_\nu)^{1/m_c}} \, dt$$

where $(\Delta t)_l$ = time during which the load is applied
   $(\Delta t)_h$ = time permitted for healing, during which the load is removed, or compressive stress is applied
   $dt = dN[(\Delta t)_l + (\Delta t)_h]$

Obviously, if there is no healing period or if $H_\nu = 0$, then there is no healing and the equation reverts back to the more familiar Paris law form of equation. Assuming that the J-integral dissipated energy is applied according to the normalized waveform with time, $w(t)$, as follows:

$$J_\nu = J_{\nu 0}w(t) \tag{49}$$

where $J_{\nu 0}$ = the maximum value of J_ν during the time interval $(\Delta t)_l$,

$$\frac{dc}{dN} = \left[\int_0^{(\Delta t)_t} \frac{k_t \alpha (D_1 \, E_R)^{1/m_t} w(t)^{1/m_t}}{[2\Gamma_1 - D_0 E_R J_{v0} w(t)]^{1/m_t}} \, dt \right] (J_{v0})^{1/m_t} \tag{50}$$

This is of the form

$$\frac{dc}{dN} = A[J_{v0}]^n \tag{51}$$

in which

$$n = \frac{1}{m_t} \tag{52}$$

with some possible additions due to the dependence of the fracture process zone size, α, on the size of the J-integral. The coefficient A is

$$A = \int_0^{(\Delta t)_t} \frac{k_t \alpha (D_1 E_R)^{1/m_t} w(t)^{1/m_t}}{[2\Gamma_1 - D_0 E_R J_{v0} w(t)]^{1/m_t}} \, dt \tag{53}$$

If D_0 is much smaller than D_1, as it commonly is, the expression in Equation 53 simplifies to

$$A = \left[\frac{k_t^{m_t} D_1 E_R}{2} \right]^{1/m_t} \int_0^{(\Delta t)_t} \frac{\alpha w(t)^{1/m_t} dt}{\Gamma_1^{1/m_t}} \tag{54}$$

This expression matches that proposed originally by Schapery in his theory published in 1973. It also illustrates the assumptions that were made in arriving at that expression, namely no healing and very small values of D_0.

In a similar way, the adhesive fracture equivalent of Paris' law is found to be:

$$\frac{dc}{dN} = \int_0^{(\Delta t)_t} \frac{k_t \alpha (D_1 \, E_R J_v)^{1/m_t}}{(\Gamma_1 + \Gamma_2 - \Gamma_{12} - D_0 E_R J_v)^{1/m_t}} \, dt$$
$$- \int_0^{(\Delta t)_h} \frac{k_{th} \alpha (D_{1h} \, E_R H_v)^{1/m_c}}{(\Gamma_{1h} + \Gamma_{2h} - \Gamma_{12h} - D_{0h} E_R H_v)^{1/m_c}} \, dt \tag{55}$$

Our laboratory measurements have shown the values of D_0 and D_{0h} to be very much smaller than D_1 or D_{1h}, which allows the preceding equation to be simplified to

$$\frac{dc}{dN} = [K_t^{m_t} D_1 E_R]^{1/m_t} \int_0^{(\Delta t)_t} \frac{\alpha [w(t)]^{1/m_t} \, dt}{(\Gamma_1 + \Gamma_2 - \Gamma_{12})^{1/m_t}} [J_{v0}]^{1/m_t}$$
$$- [K_{th}^{m_c} D_{1h} E_R]^{1/m_c} \int_0^{(\Delta t)_h} \frac{\alpha [w_n(t)]^{1/m_c}}{(\Gamma_{1h} + \Gamma_{2h} - \Gamma_{12h})^{1/m_c}} [H_{v0}]^{1/m_c} \tag{56}$$

$$\frac{dc}{dN} = A[J_{vo}]^n - A_h[H_{vo}]^{n_h} \tag{57}$$

This modification of Paris' law is for materials that heal and have time to heal between load applications.

The size of the fracture process zone, α, is estimated by the ratio of the J-integral (units FL^{-1}) to the area under the tensile stress-strain curve of the material (units FL^{-2}) out to the strain level where the tensile strength is reached. An equation for this estimation is

$$\alpha = \frac{J_v}{D_1 \sigma_t^2 I_1} \tag{58}$$

where $\quad \alpha_t$ = tensile strength of the material

I_1 = integral of the dimensionless stress-strain curve of the material out to the tensile strength, a number that varies between 1 and 2

A more general form of the relation that takes into account the hyperelastic nature of the large strains in the fracture process zone is

$$\alpha = \left(\frac{J_v}{D_1 \sigma_t^2 I_1}\right)^r \tag{59}$$

where the exponent, r, has a typical range from 0.5 to 1. Using this estimate for α in the expressions for cohesive fracture produces the following relations for the Paris law coefficients

$$A = \left(\frac{k_t^{m_t} D_1^{1-r} E_R}{2}\right)^{1/m_t} \int_0^{(\Delta t)_t} \frac{[w(t)]^{(1/m_t)+r} \, dt}{\Gamma_1^{1/m_t} \sigma_t^{2r} I_1^r} \tag{60}$$

and

$$n = r + \frac{1}{m_t} \tag{61}$$

In adhesive healing, the Paris law coefficients are

$$A_h = \left[k_{th}^{m_t}\left(\frac{D_{1h}}{D_1^r}\right) E_R\right]^{1/m_c} \int_0^{(\Delta t)_t} \frac{[w_n(t)]^{t+1/m_c} \, dt}{(\Gamma_1 + \Gamma_2 - \Gamma_{12})\sigma_t^{2r} I_1^r} \tag{62}$$

$$n_h = r + \frac{1}{m_c} \tag{63}$$

All of the values of material properties that make up these terms of the modified Paris law can be measured with material property tests that are much simpler than the fracture tests. They involve

1. Tensile and compressive creep compliance values, D_0, D_1, m_t, D_{0h}, D_{1h}, m_c
2. Tensile strength and the area under the tensile stress-strain or pseudostress-strain curve: σ_t I_1, r, E_R
3. Wetting and dewetting surface energy densities, Γ_1, Γ_2, Γ_{12}, Γ_{1h}, Γ_{2h}, and Γ_{12h}, for both cohesive and adhesive fracture and healing processes

The values of the J_{vo} and H_{vo} integrals may be estimated by the following method for tension and plane strain conditions:

$$J_{vo} = K_1^2(c) \mid D_t^*(w) \mid [1 - \nu_t^* (w)\overline{\nu_t^*}(w)] \tag{64}$$

and

$$H_{vo} = K_1^2(c) \mid D_c^*(w) \mid [1 - \nu_c^*(w)\overline{\nu_c^*}(w)] \tag{65}$$

where $\mid D_c'(w) \mid$ = magnitude of the tensile complex creep compliance
$\mid D_c'(w) \mid$ = magnitude of the compressive complex creep compliance
$\nu_t^*(w)$, $\overline{\nu_t}(w)$ = tensile complex creep Poisson ratio and its conjugate
$\nu_c^*(w)$, $\overline{\nu_c}(w)$ = compressive creep Poison ratio and its conjugate

The complex creep compliance components are frequency (or loading time) dependent and may be measured with frequency sweep tests as follows:

$$\mid D^*(w) \mid = [D^1(w)]^2 + [D^{11}(w)] \tag{66}$$

$$D_t^1(w) = D_0 + D_1\Gamma(1 + m_t)w^{-m_t} \cos\left(\frac{m_t\pi}{2}\right) \tag{67}$$

$$D_t^{11}(w) = D_1\Gamma(1 + m_t)w^{-m_t} \sin\left(\frac{m_t\pi}{2}\right) \tag{68}$$

$$D_t^1(w) = D_{0h} + D_{1h}\Gamma(1 + m_c)w^{-m_c} \cos\left(\frac{m_c\pi}{2}\right) \tag{69}$$

$$D_c^{11}(w) = D_{1h}\Gamma(1 + m_c)w^{-m_c} \sin\left(\frac{m_c\pi}{2}\right) \tag{70}$$

The loading time, Δt_l, and loading frequency are related to one another by

$$2\pi f = w = \frac{1}{2(\Delta t_l)} \tag{71}$$

The complex Poisson ratios $\nu^*(w)$ must be measured in tension and compression in order to complete the estimate of the J_{vo} and H_{vo} values. The values of the stress intensity factors, $K_I(c)$ for tension and $K_{II}(c)$ for shear change with the crack length, for both microcracks and macrocracks can be calculated using finite element methods capable of representing nonlinear, anisotropic elastic materials.

Measurements of the surface energies Γ, which are an important part of this fracture theory, may be made with equipment that is currently in use by surface chemists. The measurements are made in two parts: one part with the asphalt and the other with the aggregates. The asphalt surface energies are measured with a Wilhelmy plate apparatus and the aggregate surface energies may be measured very accurately with gas adsorption using a Cahn balance in an insulated, low-vacuum apparatus. This approach allows the surface energy components to be measured, taking into account both the polar (acid-base) and nonpolar components of surface energy. There is hysteresis in these Γ values, with more energy being required on dewetting (fracture) than on wetting (healing). The adhesive and cohesive properties of combinations of different asphalts and aggregates can be calculated from these components in the absence and presence of water, which gives a fundamental way of evaluating the stripping potential of the asphalt-aggregate mixture. The fracture and healing Γ values of a mixture can also be calculated for both adhesive and cohesive processes. The computations are made using the theory presented by Good and Van Oss [25].

C. Evidence of Microfracture Healing and Its Effects on Pavement Materials

The fracture mechanical relationships described in the preceding section define the crack growth and recession evinced by asphalt concrete materials in fatigue with periodic rests. The microcracks (and ultimately macrocracks) grow as the material is progressively and repeatedly stressed and then heal during the periods of rest at zero load and strain. The ultimate fatigue life of an asphalt concrete is derived from the relationship between these two competing processes and can certainly be lengthened by either minimizing the rate of crack growth or improving the healing properties of the mixture, or both.

An approximation to the progress of this material degeneration through crack growth can be made by back-calculating the average estimated crack lengths ($\bar{c}$) using a micromechanical model consisting of the fracture relationships presented in Section III.B. The model consists of a finite element representation of the test specimen, which is updated throughout a fatigue test, and its response is fit to actual laboratory stress-strain data. The resulting model parameters can then be used to develop a number of descriptive fracture parameters detailed in Section III.B, notably the shape characteristics of the assumed Weibull distribution of microcracks present in the specimen.

The average estimated crack length is the mean value of the Weibull distribution (Figure 6) of crack lengths generated in the back-calculation process. As a specimen is progressively fatigued and the Weibull distribution becomes more heavy tailed toward greater crack lengths, the mean necessarily shifts to a greater value. After the material heals, the calculations yield a less heavy-tailed distribution with a reduced mean value. Results for these average estimated crack lengths appear in Figure 16. Each individual line represents the change in the average estimated crack length throughout a group of 23 strain wave loading cycles. The fatigue period represented by these lines begins at the left side of the plot and moves to the right; then a rest period is introduced. The next load cycle after a rest period is the first cycle of the succeeding load period. This next load period is generally plotted just above the previous one, as the microcracks do not entirely heal during the rest period.

The healing of microcracks renders a material stronger than it was in its previous state of damage. The increased "strength" is made apparent by the upward shift and more steeply sloped hysteresis loop plots depicting stress-strain data from fatigue tests of asphalt concrete specimens. Shifts in hysteresis loops have been noted in different tests and stress states: cylindrical direct tension, direct tension overlay testers with and without notched beams, and flexural beam fatigue. Examples of this shift in fatigue data appear in Figures 17 and 18. The resistance or strength recovery of asphalt concrete mixtures via healing can be quantified by use of the healing index discussed in detail in Section II.D.

The healing index values vary with binder composition. Exactly which chemical fractions most greatly affect the healing tendencies of a given binder are currently under investigation, yet the hierarchy among binders established for a given test regime (stress or strain control, for instance) remains constant for that regime. This trend has been evident through strain-controlled overlay, reflection crack simulation, direct tensile, and beam fatigue testing at Texas Transportation Institute.

Additional work has revealed that the healing rate (as expressed by the healing index calculation) varies not only with the binder but also with the test temperature, length of rest permitted, and also possibly the damage state of the material. Figure 19 indicates the significantly increased rate of healing with increased temperature. This effect may be partially explained by the reduced binder viscosity at 40°C, which would allow the binder to flow more readily across fracture faces. The potential contribution of higher molecular activity at greater temperatures is currently under investigation. Figure 20 indicates the relation of healing index to longer periods between loads. The healing index value appears to increase in a logarithmic fashion with time for each binder, suggesting that even small variations in the rate of rest period application can exert an influence on asphalt concrete fatigue properties.

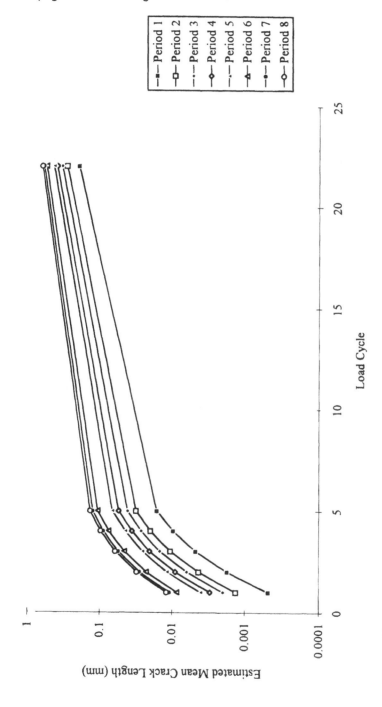

Figure 16 Back-calculated mean crack lengths of a cylindrical asphalt concrete specimen subjected to direct tensile fatigue loading. Note the dramatic shift in the mean value within a given loading period and the subsequent reduction of the mean value during the ensuing rest period. This reduction is made clear by the difference between the mean crack length values of the last load cycle of a given load period and the first cycle of the succeeding load period.

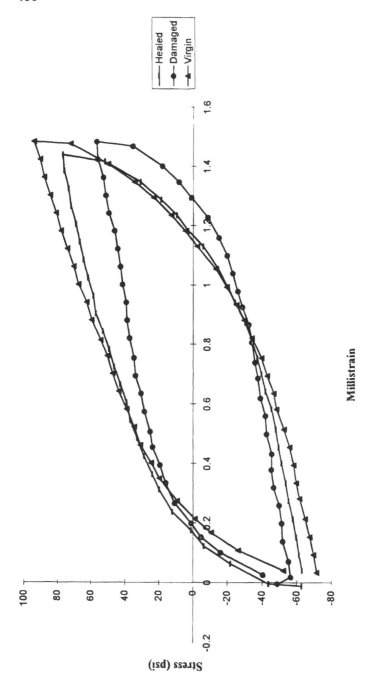

Figure 17 Overlay test data of a notched beam sample fatigued in direct tension. The "virgin" data represent the test's first load cycle; the "damaged" data represent the last load cycle applied before the introduction of the rest period; and the "healed" data were acquired from the first load repetition following the rest period. The slope and energy dissipation increase are evident when the "healed" hysteresis loop is compared with the "damaged" hysteresis loop.

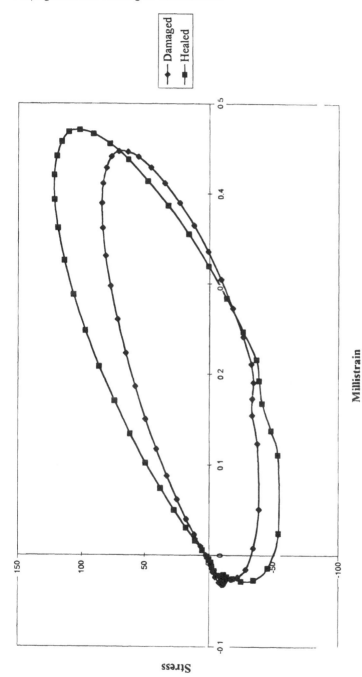

Figure 18 Comparison of a cylindrical specimen's response to a direct tension haversine strain wave before ("damaged") and after ("healed") a 2-min rest period. Again, the augmented slope and enclosed area of the "healed" hysteresis loop data delineate the increased strength and energy dissipation characteristics of the specimen after healing during the rest period.

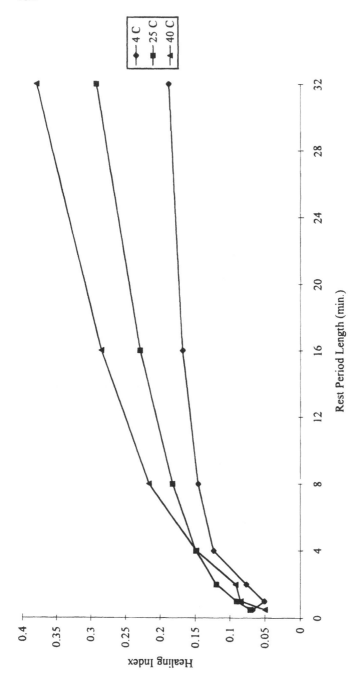

Figure 19 Comparison of healing response of identical AAM-1–based specimens to similar strain-controlled fatigue test regimes. While the strain rates necessarily increased with temperature to provide legible data, the tests performed at higher temperature clearly exhibit greater degrees of healing.

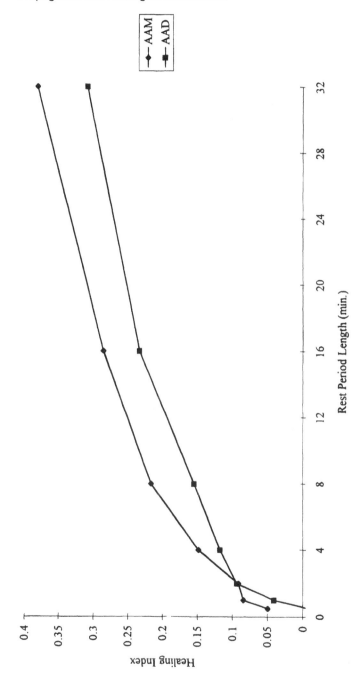

Figure 20 This comparison of healing index data from equal direct tensile fatigue tests of AAM-1 and AAD-1 mixtures illustrates the generally superior healing rate of the AAM-1 mixtures in strain-controlled testing. This relationship is noted in other strain-controlled test regimes as well.

The researchers believe that mixture type will also have a substantial influence on fatigue properties. High-binder-content mixes such as stone mastic mixes tend to be considerably more resistant to fatigue cracking than conventional dense graded mixes. Part of the reason is probably that the thicker films of binder promote cohesive fracture or formation of microcracks within the mastic as opposed to at the asphalt-aggregate interface. Such microcracks or fractures have a greater propensity to heal than would that which form at the asphalt-aggregate interface. The effect of mixture type will be evaluated in this continuing research. It is also probable that particles dispersed in the mastic may act as energy-absorbing crack arresters.

REFERENCES

1. R. L. Lytton, J. Uzan, E. G. Fernando, R. Roque, D. Hiltunen, and S. M. Stoffels. Development and Validation of Performance Prediction Models and Specifications for Asphalt Binders and Paving Mixes, Report SHRP-A-357, 1993.
2. M. Kachanov, Elastic Solids with Many Cracks and Related Problems, in *Advances in Applied Mechanics*, Vol. 30 (J. W. Hutchinson and T. Y. Wu, eds.), Academic Press, San Diego, pp. 260–438 (1994).
3. F. Finn, C. Sarf, K. Kulkarni, W. Smith, and A. Abdullah. The Use of Prediction Subsystems for the Design of Pavement Structures, in *Proceedings, Fourth International Conference on Structural Design of Asphalt Pavements*, pp. 3–38 (1977).
4. K. D. Raithby and A. B. Sterling. The Effect of Rest Periods on the Fatigue Performance of a Hot-Rolled Asphalt under Reverse Axial Loading, *Proc. Assoc. Asphalt Paving Technol.*, **39**: 134 (1970).
5. D. N. Little, A. H. Balbissi, C. Gregory, and E. Richey. Design and Characterization of Paving Mixtures Based on Plasticized Sulfur Binders—Engineering Characterization, Report DTFH-61-80-C-00048, 1984.
6. Y. S. Kim. Evaluation of Healing and Constitutive Modeling of Asphalt Concrete by Means of the Theory of Nonlinear Viscoelasticity and Damage Mechanics, Ph.D. Dissertation, Texas A&M University, December 1988.
7. P. Bazin and J. B. Saunier. Deformability, Fatigue and Healing Properties of Asphalt Mixes, in *Proceedings, Second International Conference on the Structural Design of Asphalt Pavements*, Ann Arbor, MI, pp. 553–569 (1967).
8. S. Prager and M. Tirrell, *J. Chem. Phys.*, **75**: 10 (1981).
9. R. P. Wool and K. M. O'Connor, *J. Polym. Sci. Polym. Lett. Ed.*, **20**: 1 (1982).
10. Y. Kim and R. P. Wool, A Theory of Healing at Polymer-Polymer Interface, *Macromolecules*, **16** (1983).
11. P. G. de Gennes, *J. Chem. Phys.*, **55**: 2 (1971).
12. K. Jud and H. H. Kausch, *Polym. Bull.*, **1**: (1979).
13. R. P. Wool, Material Damage in Polymers, in *Workshop on a Continuum Mechanics Approach to Damage and Life Prediction*, National Science Foundation, pp. 28–35 (1980).
14. B. J. Briscoe, *Polymer Surfaces*, Wiley Interscience, New York (1978).

15. K. DeZeeuw and H. Potente, *Soc. Plast. Eng.*, **23**: (1977).
16. C. B. Buchnall, I. C. Drinkwater and G. R. Smith, *Polym. Eng. Sci.*, **20**: 6 (1980).
17. J. C. Petersen, *Transportation Research Record*, 999 (1984).
18. R. N. Traxler, *J. American Chemical Society*, **5**: 4 (1960).
19. D. N. Little, Y. S. Kim, and F. C. Benson, Investigation of the Mechanism of Healing in Asphalt, Report to the National Science Foundation, Texas A&M University, 1987.
20. D. N. Little and S. Prapnnachari, Investigation of the Microstructural Mechanisms of Relaxation and Fracture Healing in Asphalt, Annual Report to the Air Force Office of Scientific Research, 1991.
21. S. Prapnnachari, Investigation of the Microstructural Mechanism of Relaxation in Asphalt, Ph.D. dissertation, Texas A&M University, 1992.
22. Y. R. Kim, S. L. Whitmoyer, and D. N. Little, *Transportation Res. Rec.*, 1454 (1994).
23. R. A. Schapery, *Int. J. Fracture*, **25**: (1984).
24. R. A. Schapery, A Theory of Crack Growth in Viscoelastic Media, Report 2, MM 2764-73-1, Texas A&M University, 1973.
25. R. J. Good and C. J. Van Oss, in *Modern Approach to Wettability: Theory and Applications* (M. E. Schrader and G. Loeb, eds.), Plenum Press, New York, pp. 1–27 (1991).

8

Polymer Modifiers for Improved Performance of Asphaltic Mixtures

Heshmat A. Aglan
Tuskegee University, Tuskegee, Alabama

I. INTRODUCTION

The durability of asphaltic pavements is greatly influenced by the environmental changes during the year, especially between summer and winter, and between day and night, when the daily average temperature change can be considerably large. In summer, the high temperature can soften the asphalt binder and consequently reduce the stiffness of the paving mixture. On the other hand, in winter the low temperature can stiffen the asphalt binder and reduce the flexibility of the paving mixture. As a result, thermal cracking may develop that adversely affects the performance and lifetime of the pavement. Thus, high-temperature stiffness and low-temperature flexibility are important properties that increase the lifetime of pavements.

Various elastomer and plastomer modifiers have been sought to address this problem. Polymer modifiers vary in function and effectiveness. Elastomers, which are at least to some extent derived from a diene chemical structure, toughen asphalt and improve temperature-viscoelastic properties. Plastomers, which come from nondiene chemicals, improve the high-temperature viscoelastic properties of softer asphalts that have good intrinsic low-temperature properties [1]. The properties of asphalt mixtures can be improved by selecting modifiers in the proper molecular weight range and mixing the modifiers with asphalt mixtures in an appropriate manner. In addition, these modifiers must have solubility parameters close to those of the asphalt mixtures. One of the critical factors that should be considered for better polymer-modified asphalt is the air void percentage in the total mix. The performance of the polymer-modified mixture will be improved as this percentage is reduced [2,3]. In general, the desired air void percentage depends on the load capacity of the transportation facility being designed. Lower air void percentages can be obtained by increasing both the modifier and the asphalt binder content [2] or by compaction.

Exploration of the combo-viscoplastic fundamental properties of asphalt pavements demands a focused and innovative study of their micro- and macromechanical phenomena relevant to their long-term performance. These phenomena encompass deformation processes induced by the applied mechanical forces accelerated by environmental challenges. In this regard, relevant micro- and macromechanical phenomena in pavement would involve localized irreversible deformation processes, e.g., microcracking of the binder and binder/aggregate interface. Interfacial microcracking is related to yet another fundamental phenomenon, adhesion of the binder to the aggregate surfaces. It is also essential to recognize that micromechanical processes in the mixture are controlled to a great extent by the viscoelastic character of the asphalt, which is influenced by the state of stress or strain, the processing conditions, and aging. Hence, it is fundamentally important to study the dependence of the viscoelastic character of the systems considered, particularly on the design parameters such as stress level and aging (exposed to increased temperature).

An innovative approach to characterizing micromechanical phenomena resulting from fatigue cracking and relating them to the design and durability of asphalt concrete mixtures has been developed [4,5]. In this approach crack propagation and its associated damage are considered an irreversible process and hence the general framework of the thermodynamics of irreversible processes is employed for modeling the fatigue crack propagation phenomenon. In the current study, invoking the modified crack layer model [6], the influence of some of the commonly used polymer modifiers on the fatigue crack propagation behavior of asphalt concrete mixtures is studied. Scanning electron microscopy was performed to explain the microstructural origin of fracture toughness of these modified mixtures. The general mechanical behavior of these modified mixtures is also evaluated.

II. THEORETICAL CONSIDERATIONS

A. Modified Crack Layer Model

The development of fatigue crack–resistant pavements necessitates a thorough understanding of the combo-viscoplastic behavior of the binders (asphalt) and the additives (modifiers), which have been shown to be the major constituents influencing the pavements' crack resistance. Recently, a methodology, the modified crack layer (MCL) approach, has been developed [6] to characterize the resistance of materials to fatigue crack propagation (FCP). The ability of this approach to discriminate the subtle effects introduced by different chemical structures and processing conditions in asphalts has been demonstrated [4, 6–8]. The MCL model essentially addresses the difficulties encountered in the general applicability of the crack layer model [9,10]. These difficulties are the identification and quantification of damage species associated with fatigue crack propagation in materials.

For stress control fatigue, the modified crack layer model is expressed as

$$\frac{da}{dN} = \frac{\beta' \dot{W}_i}{\gamma' a - J^*} \tag{1}$$

Where da/dN is the cyclic crack growth speed, a is the crack length, γ' is a candidate material parameter characteristic of the mixtures's resistance to FCP, and β' is the energy dissipative character of the mixture. The change in work $\dot{W}_i$ and the energy release rate J^* will be discussed in detail below.

The proposed J-integral [11] can be used to evaluate the crack tip elastic-plastic field. Invoking the deformation theory of plasticity, Rice [11] formulated mathematically the energy release rate for materials displaying an elastic-plastic response. For a cracked body subjected to a two-dimensional deformation field, the J-integral is expressed as

$$J = \int_\Gamma \left(w \, dy - T \frac{\partial u}{\partial x} \, ds \right) \tag{2}$$

where Γ is a closed contour surrounding a crack tip in a stressed solid, w is the strain energy density $w = w(x,y) = \sigma_{ij}\delta\epsilon_{ij}$, T is the traction vector perpendicular to Γ, u is the incremental displacement in the x direction, and ds is an element of Γ. Rice has also shown that the J-integral around a crack tip is the change in potential energy for a crack increment Δa. The path-independent J-integral proposed by Rice was viewed [12] as a measure of the crack tip elastic-plastic field and can be evaluated from the load-displacement curve associated with crack extension during monotonic loading. The deformation theory of plasticity upon which the J-integral is founded does not permit unloading; therefore the J-integral evaluated by Begley and Landes [12], strictly speaking, cannot be used to characterize the crack tip elastic-plastic field in pavement during FCP.

Little et al. [13] have modified the J-integral analysis to include unloading effects. On this basis, the energy release rate for an elastic-plastic crack tip field can be evaluated from the change in potential energy (the area above the unloading curve) with respect to the crack length divided by the thickness of the specimen. Little et al. stated that "this modified J-integral (J^*) should be suitable in a comparative analysis of energy release rates among various asphalt binders." This J^* will be evaluated experimentally from the hysteresis loops at various crack lengths and used with the modified crack layer theory in the current FCP analysis. Thus,

$$J* = \frac{1}{B}\left(\frac{dP}{da}\right) \tag{3}$$

where P is the area above the unloading curve and B is the specimen thickness. The quantity $\dot{W}_i$, which is the change in work, is measured directly as the area of the hysteresis loop at any crack length, H_i, minus the area of the loop just before crack initiation, H_0, divided by the specimen thickness, B [5]. Thus,

$$\dot{W}_i = \frac{H_i - H_0}{B} \tag{4}$$

In viscoelastic materials, $\dot{W}_i$ includes work expended on damage processes associated with crack growth and history-dependent viscous dissipation processes. Both processes are irreversible.

The parameters γ' and β', extracted from fatigue crack propagation experiments using the MCL model, will be employed to study the effect of additive type on the fracture resistance of AC-5 asphalt concrete mixtures.

B. Critical Energy Release Rate J_{Ic}

Invoking the linear elastic fracture mechanics (LEFM) concept, many workers [14–17] studied the fracture characteristics of various asphalt concrete mixtures. The J-integral, Equation 2, has been used by other workers [18–20]. The formula for computing the critical energy release rate J_{Ic} based on Equation 2 was given by Rice for one specimen and one crack length tested under monotonic loading as

$$J_{Ic} = \eta \frac{U}{B}(d - a_0) \tag{5}$$

where η is a constant, U is the area under the load-deflection curve up to the failure point (maximum load sustained by the specimen), d is the depth of the specimen, and a_0 is the initial notch depth. Equation 5 was studied by Sumpter and Turner [21]. These workers found that for length-to-depth ratios equal to 4 and notch-to-depth ratios between 0.5 and 0.7 the constant η in Equation 5 is equal to 2. Therefore Equation 5 can be written as

$$J_{Ic} = 2\frac{U}{B}(d - a_0) \tag{6}$$

The present research work is concerned mainly with the fatigue crack propagation behavior of various asphaltic pavements in view of the modified crack layer model. It is important to compare results obtained from Equation 6 with those obtained from the MCL model, Equation 1. This will demonstrate whether the J_{Ic} concept, which is based on static testing, can provide true information on the fracture resistance of asphaltic pavements. The consistency of the J_{Ic} criterion with the γ'

criterion (obtained from the MCL model) needs to be investigated. This is beyond the scope of the current work reported here and will be considered later.

III. EXPERIMENTAL

The following sections describe the materials and methodologies used to determine the optimum asphalt content and to prepare the test specimens to yield adequate data to study the fracture and fatigue crack propagation behavior of asphaltic mixtures. This includes three AC-5 mixtures modified with various additives.

A. Materials

1. Aggregate Type and Gradation

Crushed limestone aggregate and quartzitic sand were selected for the preparation of test specimens. The hardness and abrasion characteristics (Los Angeles abrasion test) of the aggregate are such that they meet the minimum standards specified by the Ohio Department of Transportation (ODOT), when used in the preparation of both Portland cement concrete and asphalt cement concrete.

The bulk aggregate was initially sieved to accumulate enough material within each size range specified by the gradation requirements and was stored in separate containers.

Gradation requirements corresponding to ODOT specification item 403 [22] were adopted in the preparation of all specimens. Item 403 is typically used as leveling or intermediate (asphalt concrete) course. Fatigue cracking of asphalt concrete pavement generally starts at the bottom of the concrete pavement and propagates upward toward the surface. Thus, it would be of more significance to study the crack propagation characteristics due to fatigue loading of the leveling course rather than of the surface course, which utilizes a finer gradation, such as ODOT item 404. Specimens following item 403 specification were prepared according to the details in Table 1.

Table 1　Gradation Requirements

Sieve size	Total passing (%)
1/2 inch	100
3/8 inch	95
No. 4	59
No. 16	28
No. 50	9
No. 200	0

2. Asphalt Cement

An AC-5 asphalt cement was used in the study. Typically the AC-5 is mixed with a modifier (polymer). The binder physical properties in Table 2 were determined by the supplier and reported to clients, including the Ohio Department of Transportation.

3. Modifiers

Three asphalt cement modifiers, Kraton, Elvax, and Novophalt, were studied. These modifiers belong to the general group of thermoplastic polymers and are claimed to improve the engineering characteristics of the asphalt cement and ultimately of the asphalt concrete. Their general characteristics and attributes are presented next.

Kraton D4463 (thermoplastic rubber of styrene-butadiene-styrene block copolymer) is supplied in pellets, which melt at the prescribed mixing temperature between 320 and 380°F. Mixing it with asphalt cement tends to improve the engineering properties of asphalt concrete.

Elvax (ethylene vinyl acetate) is supplied in free-flowing pellets that melt at a prescribed mixing temperature between 275 and 300°F. Mixing it with asphalt cement is claimed to increase the durability, toughness, tenacity, and resistance to cracking of the asphalt concrete.

Novophalt modified asphalt cement is a biphase binder consisting of a small percentage (4–6%) of polyolefin additives. The polyolefin phase is primarily low-density polyethylene. Virgin or recycled polyethylene can be incorporated in preparing the Novophalt. The exact chemical composition of Novophalt is proprietary. Its prescribed melting point is between 290 and 310°F.

B. Specimen Preparation

Before any fatigue testing specimens could be prepared, the optimum asphalt cement content of the asphalt concrete mixtures was determined. The procedure used for this purpose followed generally recognized and approved methods as explained below.

Table 2 Physical Properties of the Unmodified AC-5 Binder

Physical properties of unmodified binder	AC-5
Penetration at 77°F, ASTM D-5	204
Viscosity at 275°F, cSt, ASTM D-2170	201
Viscosity at 140°F, poise, ASTM D-2171	461
Flash (Clev. open cup), °F, ASTM D-92	595

1. Optimum Asphalt Content Determination

An optimum asphalt content determination was conducted using the Marshall method of mix design (ASTM D1559-89). Cylindrical specimens 4 inches in diameter by 2.5 inches in length were compacted with the standard impact hammer by applying 50 blows on each side. The unit weight was then determined, and the specimens were tested for their Marshall stability and flow. The consideration of the average (of three specimens) Marshall stability, flow, unit weight, percentage voids in mineral aggregate (VMA), and percentage air voids allows the determination of the optimum asphalt cement content. The optimum asphalt cement content for the AC-5 was determined to be 8%. In addition, the target asphalt concrete unit weight was determined to be 149 pounds per cubic foot (pcf) by the same procedure.

2. Beam Preparation

Asphalt concrete beams measuring 15 inches long by 2 inches thick by 3.5 inches deep were prepared using the following procedure. The aggregate, the asphalt cement, and the mold were preheated to 320°F before blending and compaction. Beams were compacted by sustaining a 2000 psi pressure through a steel plate for 1 min, in agreement with ASTM D3202-83.

Beams were allowed to cool off in the compaction mold and were usually tested 7 days after preparation and conditioning. Conditioning consisted of subjecting each beam to a constant temperature of 140°F for 1 day.

As previously indicated, all beams were prepared at an optimum asphalt content, determined according to the Marshall method of mix design of 8% of the total weight, and a target asphalt content unit weight of 149 pcf (determined by the same procedure). Beams were prepared using AC-5 separately, to develop comparative static and dynamic (fatigue) testing data for the eventual fracture mechanics–based analysis of asphalt concrete mixtures.

Polymer-modified beams were prepared by mixing asphalt cement AC-5 with Kraton, Elvax, or Novophalt. The bulk polymer was first heated to the compaction temperature of 320°F and then blended with the liquefied asphalt cement at the required percentages of binder. Mixing of the asphalt cement and polymer was performed with a high-speed electric mixer until a uniform blend was achieved. The mixture of AC-5 asphalt and Novophalt modifier was prepared by Pri Asphalt Technology, Tampa, Florida. The modified asphalt cement was left overnight at 150°F and raised to 320°F before mixing with the aggregate, as recommended by the supplier. Beams containing 6% of each polymer as a percentage of the total binder (AC-5) were prepared for static and dynamic (fatigue) testing. This is in order to study the effect of polymer type on the fracture and fatigue behavior of AC-5–modified asphalt concrete mixtures.

C. Testing Procedures

1. Static Testing

A Wykenham-Farrance frame was used to conduct four-point flexural static tests to evaluate the general mechanical properties such as ultimate strength and flexural modulus of typical specimens from each mixture prior to the fatigue tests. The frame was outfitted with an electronic load cell and linear variable differential transformer (LVDT) to provide continuous load and deformation readings, respectively. Data were displayed on an X-Y plotter and were also recorded by a microcomputer containing an analog-to-digital converter. This allowed the display of results at a desired scale, as well as the automatic calculation of areas under the recorded curves. All necessary software for data acquisition and reduction was developed and tailored to meet the project needs. Static testing was conducted at a rate of loading of 4 mm/min. A four-point bend specimen similar to those tested under fatigue loading was used with a support span of 10.2 inches and a midspan (between the inner load points) of 3.4 inches (Figure 1a). Because both load and deflection were measured during the static test, it is

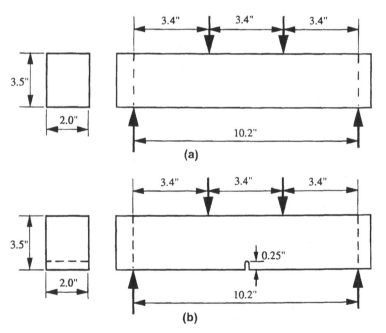

Figure 1   Specimen geometry for static and fatigue testing. (a) Unnotched specimen, (b) notched specimen (fatigue).

possible to calculate the elastic modulus of asphalt concrete by known elasticity-based formulas, as well as other parameters of interest in further analyses.

2. Fatigue Testing

A fatigue testing frame was designed and assembled that is capable of applying repeated loads of equal magnitude and duration. Loads were applied through the use of air cylinders, which are activated by an electronic signal generated by a timer that activates the independent solenoid valves. The fatigue testing machine also contains a load cell and an LVDT for load-deformation measurement. A microcomputer and an X-Y plotter were used to collect data from loading loops. Four-point bend fatigue testing on notched beams of each mixture was conducted. The same specimen geometry as used in the static testing was used for the fatigue testing except that an initial straight notch with a total depth of 0.25 inch was inserted at the middle of the specimens with a 0.156-inch saw with a round tip of radius 0.094 inch. The geometry of the fatigue test specimens is shown in Figure 1b. The tests were conducted at a constant frequency of about 0.5 Hz under load control with a constant maximum load of 65 lb. Each cycle consisted of a 0.2-load duration. The specimens were cycled from zero to the maximum load until catastrophic failure (complete separation) occurred. All tests were conducted at 70°F for the additive type and percent additive studies. Multiple identical specimens (three from each mixture) were tested under the same set of experimental conditions. A hysteresis loop was recorded at 1/4-inch intervals of crack growth. Software was developed to digitize graphical data and to calculate pertinent areas within the load-deflection curves obtained during fatigue testing. These areas are considered important in the current analysis of asphalt concrete under fatigue loading. A traveling video camera equipped with a zoom lens was used to monitor the crack propagation and capture any damage events associated with the fracture process. In order to determine the crack speed, the crack length was monitored with respect to the number of cycles. This monitoring is a well-established procedure and many techniques are available to measure the crack length in FCP experiments. In the current experiment, an accurate transparent flexible scale was attached to the specimens to measure the crack length.

3. Scanning Electron Microscopy

Scanning electron microscopy (SEM) analysis of the effect of additive modifier percentage was performed. The SEM samples were taken from fractured beams of both Kraton- and Elvax-modified AC-5 mixtures. They were compared with SEM samples taken from fractured beams from unmodified AC-5 mixtures. Focus was placed on binder-rich areas as well as aggregate surfaces.

IV. RESULTS AND DISCUSSION

A. Effect of Additive Type

The effect of additive type on the fracture resistance of modified AC-5 asphalt concrete mixture has been studied. The effect of the three modifiers (Kraton, Elvax, and Novophalt) on the static flexural behavior (ultimate strength and modulus) was evaluated. In addition, fatigue crack propagation tests were conducted to evaluate the effect of the additives on the fracture toughness.

The load versus the load point displacement response for a typical unnotched specimen of each of the three modified AC-5 mixtures is shown in Figure 2. The experimental conditions as well as the additive loading was kept the same (6% of the asphalt cement by weight) for each of the three asphalt mixtures. The load-deflection curve for the Novophalt-modified mixture exhibits an initial linear response followed by another more compliant linear response. A deviation from linearity is observed prior to the maximum load. After the maximum load is reached the curve decays rather sharply in comparison with the curves for the Elvax- and Kraton-concrete mixtures. Average values of the flexural modulus and ultimate flexural strength for the Kraton-, Elvax-, and Novophalt-modified AC-5 mixtures are given in Table 3. It is seen from Table 3 that the 6% Kraton

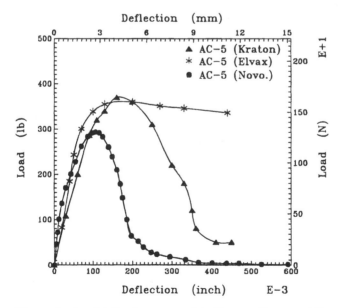

Figure 2   Load-deflection curves of the Kraton-, Elvax-, and Novophalt-modified AC-5 asphalt concrete mixtures (6% additive of the asphalt cement by weight).

Table 3 Mechanical Properties of Modified AC-5 with Novophalt, Elvax, and Kraton Additives

Mixture	E (psi)	Ultimate bending strength (psi)
AC-5 + Novophalt 6%	16,560	118
AC-5 + Elvax 6%	10,142	139
AC-5 + Kraton 6%	7,200	151

mixture has the highest ultimate strength and the lowest flexural modulus. The superiority of the Kraton-modified mixture is evident from Table 3.

The effect of additive type on the fatigue resistance of the modified asphalt concrete mixtures is examined in view of the modified crack layer theory. The crack length *a* versus the number of cycles *N* for typical specimens of the Novophalt-, Elvax-, and Kraton-modified AC-5 mixtures is shown in Figure 3. It is observed from Figure 3 that the Novophalt is inferior with respect to both crack initiation and propagation when compared with both the Elvax- and Kraton-modified mixtures. The crack speed was calculated at intervals of crack length from the slope of each curve in Figure 3.

The energy release rate *J**, evaluated on the basis of the potential energy principle, is plotted against the crack length in Figure 4. The value of *J** for the Kraton-modified specimens is always higher than its counterpart for the Novophalt- and Elvax-modified specimens at the same crack length. Again, a higher value of *J** indicates the fracture resistance superiority of the Kraton-modified AC-5 to the Novophalt- and Elvax-modified AC-5 mixtures. The value of *J** at each crack length will be employed in the present analysis to extract γ' and β' for each mixture from Equation 1. It is evident from Figure 4 that the Novophalt-modified mixture is inferior to both the Kraton- and Elvax-modified mixtures.

The change in work W_i versus the crack length for typical beams tested from the Novophalt-, Elvax-, and Kraton-modified mixtures is shown in Figure 5. The value of W_i for the Kraton-modified mixture is in general higher than that for the Elvax- and Novophalt-modified mixtures. Thus, more work has been expended on damage formation and history-dependent dissipative processes within the active zone of the Kraton-modified mixture. These relationships between W_i and the crack length will also be employed in the current analysis (Equation 1) to evaluate γ' and β'. The results for a typical specimen from each mixture are plotted, on the basis of the modified crack layer theory, in Figure 6. It is evident from Figure 6 that the experimental points for all of the mixtures make a good straight line, where γ' is the intercept and β' is the slope. The average values of γ' and β' based on three identical specimens of each mixture are shown in Table 4. The value of γ' for the Kraton-modified AC-5 mixture is higher than that for the Novophalt- and Elvax-

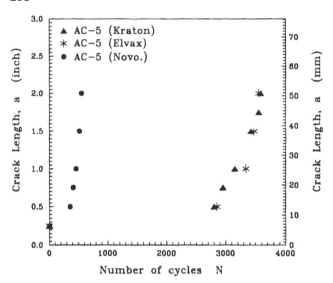

Figure 3 Crack length versus the number of cycles of the Kraton-, Elvax-, and Novophalt-modified AC-5 asphalt concrete mixtures.

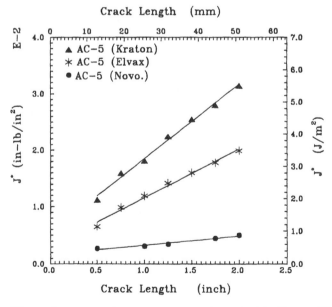

Figure 4 Energy release rate versus crack length of the Kraton-, Elvax-, and Novophalt-modified AC-5 asphalt concrete mixtures.

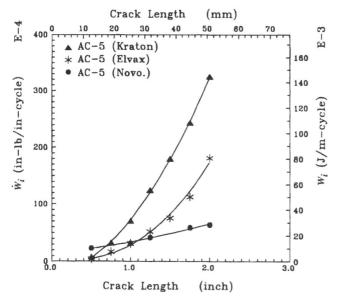

Figure 5   The change in work, W_i, versus the crack length of the Kraton-, Elvax-, and Novophalt-modified AC-5 asphalt concrete mixtures.

modified mixtures. The increase in γ', in the case of the Kraton mixture, makes it more resistant to crack propagation. i.e., tougher than the other mixtures. This is clear in Figure 7, where the theoretically predicted crack speed is plotted versus the energy release rate, using the values of γ' and β' for each mixture. Also in Figure 7 the experimental results are plotted together with the theoretical results. Again, the theory describes the experimental results very well.

B.   Fracture Surface Morphology

Scanning electron microscopic analysis was performed to study the effect of additive type on the fracture surface morphology of modified asphalt mixtures. Samples were cut from the portions of the specimens ahead of the initial notch that were fractured under fatigue loading. The polymer loading (% polymer) was kept the same, about 6% by weight of asphalt in each mixture. The results of the SEM analysis were also compared with those generated for the AC-5 unmodified mixture. It should be mentioned that only Kraton- and Elvax-modified mixtures are discussed here, because the Novophalt-modified mixture did not show any change in the morphological fracture features compared with the unmodified AC-5 mixture. The micrographs in Figure 8a and b show the fracture surface morphology of the unmodified AC-5 mixture at $100\times$ and $2000\times$. It is seen in

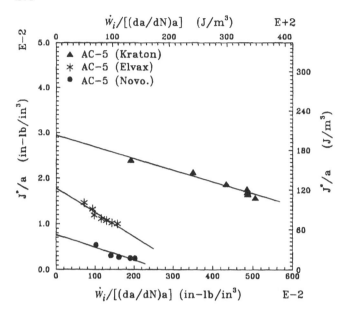

Figure 6   Fatigue crack propagation behavior of the Kraton-, Elvax-, and Novophalt-modified AC-5 asphalt concrete mixtures plotted in the form of the modified crack layer model to obtain γ' and β'.

Figure 8a that the binder is smooth with no evidence of strong cohesiveness or resistance to separation. At 2000×, the micrograph of Figure 8b reveals lack of adhesion of the binder to the aggregate in the unmodified AC-5 mixture. The micrographs in Figure 9a and b are for an Elvax-modified mixture. At 100× (Figure 9a), evidence of some cohesiveness in the binder is displayed. This is in the form of ridges in the asphalt binder-rich areas. It is believed that these features are due to the increased resistance of the matrix to separation. The micrograph in Figure 9b attests to the better adhesion of the Elvax-modified asphalt binder to the aggregates. The difference is clear when Figure 9a and b

Table 4 γ' and β' for Novophalt-, Elvax-, and Kraton-Modified AC-5 Mixtures

Mixture	γ' (in-lb/in^3)	β'
AC-5 + Novophalt 6%	0.83×10^{-2}	3.10×10^{-3}
AC-5 + Elvax 6%	1.79×10^{-2}	2.97×10^{-3}
AC-5 + Kraton 6%	2.13×10^{-2}	2.61×10^{-3}

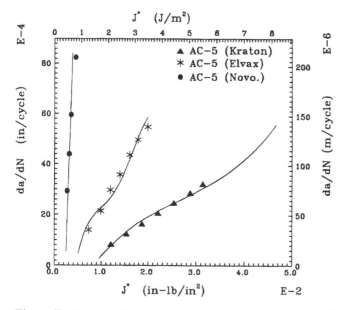

Figure 7 Theoretically predicted fatigue crack propagation speed based on the modified crack layer model (together with the experimental data) for the Kraton-, Elvax-, and Novophalt-modified AC-5 asphalt concrete mixtures.

and Figure 8a and b are compared. The micrograph in Figure 10a displays a binder-rich area of the Kraton-modified AC-5 mixture at 100×. A considerable increase in the size and the frequency of the ridges is evident in Figure 10a. This attests to the improvement in the resistance of the Kraton-modified AC-5 mixture to fatigue crack propagation. This is in comparison to Figures 8a and 9a. The micrograph in Figure 10b reveals a considerable improvement in the adhesion properties of Kraton-modified asphalt binder to the aggregate. From the comparison of Figure 8a and b, Figure 9a and b, and Figure 10a and b, it appears to be the case that both Elvax and Kraton have improved the cohesion and adhesion properties of the AC-5 asphalt binder. Kraton-modified AC-5 binder appears to be superior to the Elvax-modified and the unmodified AC-5 binder.

V. CONCLUDING REMARKS

A methodology has been developed to characterize the resistance of polymer-modified asphalt concrete mixtures to fatigue crack propagation. A theoretical development with an experimental counterpart constitutes the main component of this methodology. Experimental techniques have been developed to evaluate parame-

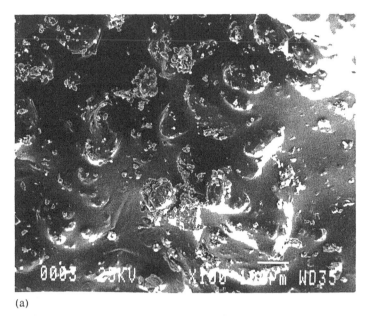

(a)

(b)

Figure 8 (a) Fracture surface morphology of an asphalt-rich area in the unmodified AC-5 mixture at 100×. (b) Morphology of fine aggregate particles on the fracture surface of the unmodified AC-5 mixture at 2000×.

(a)

(b)

Figure 9   (a) Fracture surface morphology of an asphalt-rich area in the Elvax-modified AC-5 mixture at 100×. (b) Morphology of fine aggregate particles on the fracture surface of the Elvax-modified AC-5 mixture at 2000×.

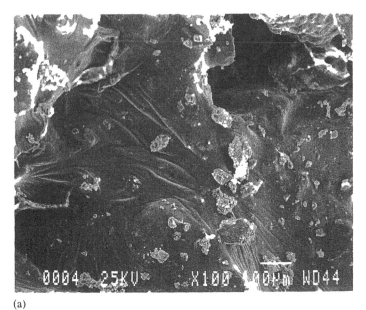

(a)

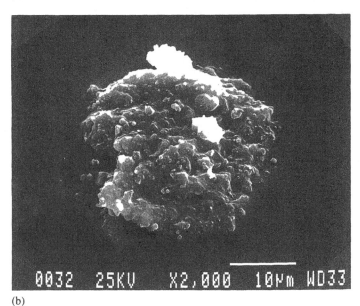

(b)

Figure 10 (a) Fracture surface morphology of an asphalt-rich area in the Kraton-modified AC-5 mixture at 100×. (b) Morphology of fine aggregate particles on the fracture surface of the Kraton-modified AC-5 mixture at 2000×.

ters controlling the fatigue crack propagation process, namely the crack speed, the crack driving force, and the change in work expended on damage formation and history-dependent viscous dissipative processes. The modified crack layer model, derived from the thermodynamics of irreversible processes, has been employed to examine the resistance of various polymer-modified asphalt concrete mixtures to fatigue crack propagation. The model has been utilized to extract the specific energy of damage, γ', a candidate material parameter characteristic of an asphalt mixture's resistance to crack propagation (fracture toughness), and a coefficient β', which reflects the dissipative character of the mixture.

Analysis of controlled-stress fatigue experiments conducted at room temperature (70°F) on Kraton-, Elvax-, and Novophalt-modified AC-5 asphalt mixture revealed the fracture resistance superiority of the Kraton (SBS)-modified asphalt mixture to those modified with Elvax and Novophalt. This is based on the value of the specific energy of damage. It was also found from static flexural tests that the Kraton-based mixture possesses relatively higher flexibility and higher strength at 70°F.

Scanning electron microscopic analysis was performed on the fracture surface of the polymer-modified and unmaodified AC-5 mixtures. Profound ridges in binder-rich areas of the Kraton-modified mixture were observed. These microstructural features were not profound in the case of the Elvax-modified mixture and not at all obvious in the case of the unmodified AC-5 mixtures. Improvement in the adhesion between the binder and the aggregate and better cohesion within the binder result in microstretching of the binder producing these ridges on the fracture surface. It is this mechanim by which the Kraton-modified AC-5 asphalt concrete mixtures acquire their resistance to fatigue crack propagation.

ACKNOWLEDGMENT

This work was sponsored by the Army Corps of Engineers, Waterways Experiment Station, Vicksburg, Mississippi.

REFERENCES

1. Key Facts About Polymer Modified Asphalt, *Better Roads*, July: 39 (1989).
2. F. P. Narusch, Alaska Experience with Rubberized Asphalt Concrete Pavements, 1979–1882, Division of Design and Construction, Alaska Department of Transportation and Public Facilities, Central Region, 1982.
3. D. N. Little, Performance Assessment of Binder-Rich Polyethylene Modified Asphalt Concrete Mixtures (Novophalt), presented at the 70th Annual Meeting of the Transportation Research Board, Washington, DC (1991).
4. H. Aglan, I. Shehata, L. Figueroa, and A. Othman, Structure-Fracture Toughness Relationships of Asphalt Concrete Mixtures, in *Transportation Research Record 1353*, TRB, National Research Council, Washington, DC, pp. 24–30 (1992).

5. H. Aglan and L. Figueroa, Structure-Fracture Toughness Relationships of Asphalt Concrete Mixtures Under Fatigue Loading, Year-end technical report, U.S. Army Corps of Engineers, Waterways Experiment Station, 1992.
6. H. Aglan, Evaluation of the Crack Layer Theory Employing a Linear Damage Evolution Approach, *Int. J. Damage Mech.*, **2**: 53 (1993).
7. H. Aglan and L. Figueroa, A Damage Evolution Approach to Fatigue Cracking in Pavements, *J. Eng. Mech.*, **119**(6): 1243–1259 (1993).
8. H. Aglan, M. Motuku, A. Othman, and L. Figueroa, Effect of Dynamic Compaction on the Fatigue Behavior of Asphalt Concrete Mixtures, presented at the Transportation Research Board 72nd Annual Meeting, Washington, DC, 1993.
9. C. Chudnovsky, Crack Layer Theory, in *10th U.S. Conference on Applied Mechanics,* (J. P. Lamb, ed.), Houston, ASME, p. 97 (1986).
10. H. Aglan, A. Chudnovsky, A. Moet, T. Fleischmann, and D. Stalnaker, Crack Layer Analysis of Fatigue Crack Propagation in Rubber Compounds, *Int. J. Fract.*, **44**: 167 (1990).
11. J. R. Rice, A Path Independent Integral and the Approximate Analysis of Strain Concentrations by Notches and Cracks, *J. Appl. Mech.*, **35**: 379 (1968).
12. J. A. Begley and J. D. Landes, The J-integral as a Fracture Criterion, ASTM STP 514, p. 1 (1972).
13. D. N. Little, J. W. Button, et al. Investigation of Asphalt Additives, FHWA/RD-D87/001, Texas Transportation Institute, College Station, 1986.
14. K. Majidzadeh, E. M. Kauffmann, and D. V. Ramsamooj, Application of Fracture Mechanics in the Analysis of Pavement Fatigue, *Proc Assoc. Asphalt Paving Technol.*, **40**: 227–246 (1971).
15. K. Majidzadeh, D. V. Ramsamooj, and A. T. Chen, Fatigue and Fracture of Bituminous Paving Mixtures, *Proc. Assoc. Asphalt Paving Technol.*, **38**: 495–518 (1969).
16. M. Herrin and A. G. Bhagat, Brittle Fracture of Asphalt Mixes, *Proc. Assoc. Asphalt Paving Technol.*, **37**: 32–55 (1968).
17. D. L. Monismith, R. G. Hicks, and Y. M. Salam, Basic Properties of Pavement Components, Report FHWA-RD-72-19, FHWA, U.S. Department of Transportation, 1971.
18. A. A. Abdulshafi and K. Majidzadeh, J-Integral and Cyclic Plasticity Approach to Fatigue of Asphaltic Mixtures, in *Transportation Research Record 1034,* TRB, National Research Council, Washington, DC, pp. 112–123 (1985).
19. H. Aglan, Evaluation of the Crack Layer Theory Employing a Linear Damage Evolution Approach, *Int. J. Damage Mech.*, **2**: 53–72 (1993).
20. D. N. Little and K. Mahboub, Engineering Properties of First Generation Plasticized Sulfur Binders and Low Temperature Fracture Evaluation of Plasticized Sulfur Paving Mixtures, in *Transportation Research Record 1034,* TRB, National Research Council, Washington, DC, pp. 103–111 (1985).
21. J. D. G. Sumpter and C. E. Turner, Method for Laboratory Determination of J_{Ic} Cracks and Fracture, ASTM STP 601, American Society for Testing and Materials, pp. 3–18, 1976.
22. Ohio Department of Transportation Specification, *Construction and Materials Specifications,* State of Ohio, Columbus, p. 112 (1989).

9

Effect of SBS Polymer Modification on the Low-Temperature Cracking of Asphalt Pavements

Robert Q. Kluttz
Shell Chemical Company, Houston, Texas

Raj Dongré
EBA Engineering, McLean, Virginia

I. INTRODUCTION

Since the Strategic Highway Research Program (SHRP) binder specifications were provisionally adopted by AASHTO [1], the authors have been evaluating them with particular attention to polymer-modified asphalt binders (PMAs).

The SHRP binder specification made a radical change in the philosophy of grading asphalt binders, moving from a collection of property-based requirements (viscosity, penetration, etc.) to a unified temperature-based scheme. Rutting is addressed by setting a minimum on $G^*/\sin \delta$, the inverse loss compliance, at the highest pavement temperature. Thermal cracking is controlled by setting a maximum on stiffness and a minimum on creep rate (which is assumed to be related to relaxation) at the lowest pavement temperature. Fatigue cracking is addressed by setting a maximum on $G^* \sin \delta$, the loss modulus, at an intermediate temperature.

Modifiers can improve both low- and high-temperature performance of binders. As shown in Table 1 [2], Styrene-butadiene-styrene (SBS) improves both rut resistance in a wheel tracking test with hot mix and low-temperature cracking resistance of the binder. These results agree with years of field experience with modified binders.

Table 1 Effect of SBS on Performance Properties

Binder	Cycles to 5 mm rut depth[a]	Critical cracking temperature[b] (°C)
AC5	25,300	−30
AC20	54,600	−24
AC5 + 6% SBS	118,000	−52

[a] At 60°C in Mesquite, NV aggregate.
[b] Development of visible cracks in a plaque poured onto concrete cooled at 5°C per hour.

In the past 3 years there has been extensive work on correlating PG binder grading results with mix performance and field performance [3]. Generally, the results have been encouraging. However, the majority of the work has focused on unmodified binders rather than PMAs. In particular, questions have been raised about the low-temperature specification as it relates to polymer-modified binders. The focus of this work is on low-temperature properties of polymer-modified binders as they relate to the PG grading system.

A.   Current Low-Temperature Specification

Thermal cracking occurs during rapid temperature drops. This is due to differences in thermal expansion coefficients, asphalt having a much higher expansion coefficient than most other materials. As the asphalt cools, it shrinks more rapidly than the surrounding, constraining environment, building up stress.

There are three basic ways to control thermal cracking in asphalt. One is to assume that the induced stress is proportional to the modulus of the asphalt. Hence, limiting the stiffness of the asphalt will control the buildup of stress. This is incorporated in the bending beam requirement of $S(60) \leq 300$ MPa [4]. Another way is to maximize the rate of stress relaxation. This factor is included with the bending beam requirement of $m(60) \geq 0.30$. The third mechanism for controlling cracking is increasing the strength of the asphalt. If the breaking stress of the asphalt is increased, it can tolerate faster and lower temperature drops. This property is inherently beyond the scope of test methods such as the bending beam rheometer (BBR) and the dynamic shear rheometer (DSR), which do not measure failure properties, which is the reason for the recently developed Superpave direct tension test [5].

Data on the low-temperature cracking performance of pavements indicate that if the asphalt binder stiffness exceeds 300 MPa at 7200 s loading time, cracking is observed [4]. To control this, laboratory testing to determine stiffness at 7200 s loading time and the lowest pavement temperature (generally below −6°C) is necessary. The time and temperature requirement is not practical for routine testing in a specification setting. Therefore, a +10°C shift is used in the laboratory along with 60 s loading time to determine the stiffness (cracking potential) of the binder. Both bending beam rheometer and direct tension tests are run at 10°C above the specification temperature with a 60 time scale. With typical time-temperature shifts this translates to a 2-h time scale at the desired temperature. Relative rates are shown in Table 2. The rates for the bending beam rheometer and actual pavement contraction are quite close, but the rate for direct tension testing is more than an order of magnitude higher. One would expect that such a large difference in rate might require a change in the temperature shift to reflect field rates accurately.

The current specification requires that the bending beam test be run first. The binder passes at the test temperature if both $S(60) < 300$ MPa and $m(60) > 0.30$. If the binder passes the m value requirement but fails in stiffness [300 $< S(60) < 600$], then direct tension may be done. The binder must have an elongation of greater than 1% at an extension rate of 3.75% per minute (1 mm/min).

What is needed, therefore, is that binder stiffness, relaxation, and fracture properties should be considered equally in predicting thermal cracking. A scheme to accomplish that will be discussed in the following. First, a case will be made for a need to incorporate failure (strength) properties to accommodate SBS polymer–modified asphalts in the current binder specification. Then an approach to accomplish this will be proposed.

II. A CASE FOR SBS POLYMER–MODIFIED ASPHALTS

A. Experimental

1. Sample Preparation

Blends of SBS and asphalt were prepared at 180°C (356°F) using a Silverson L4R high-shear mixer. The end point was determined visually and then the blending was continued for 15 min more.

Table 2 Comparative Strain Rates

Test method	Elongation rate
Direct tension at 1 mm/min	3.8%/min
BBR at $S(60) = 300$ MPa	0.1%/min
Asphalt concrete cooled at 5°C/h[a]	0.1%/2 h

[a] The 10°C temperature shift compensates for the 120-fold time decrease from field to laboratory test.

2. Binder Tests

When blending was complete, samples were poured for testing according to American Society for Testing and Materials (ASTM) procedures for traditional paving binder testing. Sample plaques for rheological testing were poured onto release paper and stored at ambient temperature.

PG grading tests were conducted according to AASHTO MP1 utilizing a Bohlin dynamic shear rheometer, a Cox rolling thin-film oven, a Rainhardt pressure aging vessel (PAV), and a Cannon bending beam rheometer.

Direct tension tests were run on a modified Satec direct tension tester with a computer-controlled T2000 frame. Initial tests were run at constant crosshead speed. Where indicated below, the hardware was further modified so that the strain was controlled using the laser response in real time with a feedback loop.

During initial screening, very erratic results were obtained in the direct tension test, especially on modified asphalts. This was traced to voids in the aphalt created during the pressure release after PAV. The following procedure was developed to eliminate these voids.

The Standard Practice for Accelerated Aging of Asphalt Binder Using a Pressure Aging Vessel, AASHTO PP1, leaves the residue with a high concentration of dissolved gas. As the pressure in the aging vessel is reduced, this gas forms a myriad of tiny bubbles dispersed in the binder. If the viscosity of the binder is too high, these bubbles will not migrate to the surface. They can remain entrained in the sample even during heating, stirring, and pouring to form specimens for DSR, BBR, or direct tension (DT) testing. The presence of these bubbles can cause minor errors in DSR or BBR measurements and premature failure during direct tension testing. This problem is particularly serious for direct tension testing, in which the bubbles function as crack initiation sites. The following vacuum degassing procedure was developed to remove these entrained bubbles.

The vacuum oven used in this work was a Napco model 5851 vacuum oven, although any oven of suitable volume, pressure integrity, and temperature range may be used. An oven with a viewing window is desirable so that the process and completion of degassing can be observed.

The vacuum oven was preheated to 100°C. After the pressure on the PAV was released and the pans were removed (AASHTO PP1 section 10.15), the pans were transferred to the vacuum oven. The vacuum valve was opened to lower the pressure below 200 mm Hg. This equates to a standard vacuum gauge reading greater than 25. The pans were held under vacuum for 2 h. The vacuum was released, the sample pans were removed, and section 10.16 of AASHTO PP1 was followed. The degassing time may be shortened by using higher temperatures, but this might cause unwanted additional aging.

B. Results

The initial step of this work was to test unmodified and modified binders to see if the current specification accurately reflects measured critical cracking temperatures and known field performance. To this end, a Wood River AC5 binder was modified with 3%, 4%, and 5% SBS polymer. The four binders were graded for low temperature using the bending beam rheometer and then were tested on the direct tension tester to determine the 1% failure strain temperature. The same binders were used to prepare hot mix specimens [6], which were used for thermal stress restrained specimen testing [7]. The results are shown in Figure 1.

For the unmodified asphalt, bending beam and direct tension gave the same specification value. This is a critical observation. The basis for existence of the direct tension test is to detect nonstandard fracture characteristics. If a binder exhibits standard fracture properties, then BBR and DT should give the same results. Thus the results seem to indicate that the basic principles are sound.

The bending beam tests, limited by the *m* value, show no improvement as polymer is added. This is quite reasonable, as an added elastic modifier could only be expected to increase the elasticity of the asphalt and hence reduce relaxation or *m* value.

The direct tension tests, however, show significant improvement, which agrees with the earlier critical cracking tests. Although there is a significant offset between the mix test and the two binder tests, the TSRST results clearly follow the trend of the direct tension test more closely than that of the bending beam rheometer test. Whereas the specification, as it currently exists, says that the

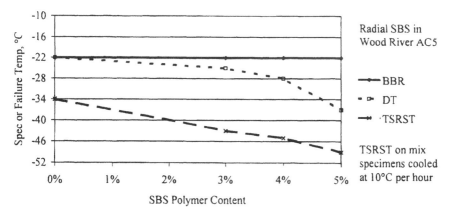

Figure 1 SBS-modified Asphalt: bending beam versus direct tension and thermal stress restrained specimen tests.

binder is not improving, the thermal stress restrained specimen test (TSRST) results differ and the direct tension results standing alone agree.

At least part of the reason for this observation is shown in Figure 2. As the polymer content increases, viscoelastic properties may not change significantly but the breaking strength of the binder more than doubles. This change will not be seen in the viscoelastic properties determined by BBR, and, because the m value is limiting, the direct tension results, although positive, are not considered.

At this point a procedure was developed to calibrate the strain rate, comparing the crosshead response to the laser response. The first results are shown in Figure 3. The actual strain rate is far below the commanded strain rate. At 15 s, the point at which 1% elongation should be reached, the strain rate is nearly an order of magnitude lower than it should be. An extensive hardware upgrade was implemented to achieve full closed-loop control of the strain rate. The results, shown again in Figure 3, were satisfactory.

After the upgrade, more comparative testing was done. Some representative results are shown in Figure 4. Running the direct tension test at a true strain rate of 1 mm/min now gives failure points considerably higher than obtained with the bending beam for both modified and unmodified asphalts. Reducing the strain rate from 1 to 0.16 mm/min, the estimated strain rate before the upgrade, brings the data back into line.

The failure characteristics of asphalts were explored. As shown in Figure 5, the ductile-brittle transition is very sharp, with elongation to break jumping from 0.5% to greater than 5% in as little as 1°C. The sharpness of the transition will be useful for specification purposes. A relatively large precision and bias can be tolerated because they will have only a slight effect on the transition temperature.

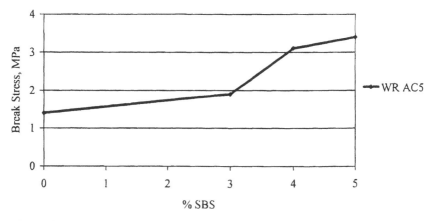

Figure 2   Effect of SBS on breaking stress.

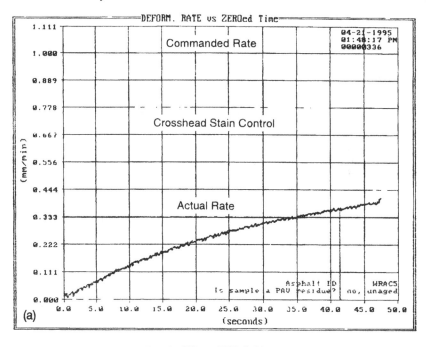

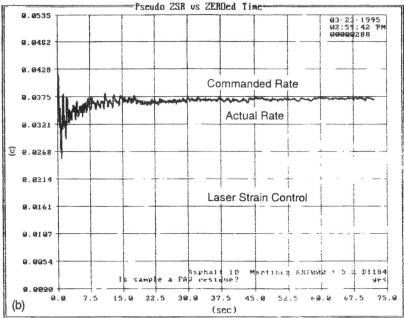

Figure 3 Strain control from crosshead versus laser.

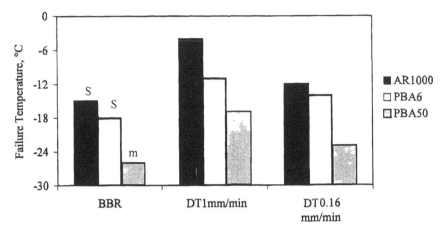

Figure 4   Test results with new hardware on unaged asphalts.

A simple study was conducted on the effects of strain rate on the temperature at which 1% failure strain is achieved. Typical results are shown in Figures 6 and 7 and two significant points are illustrated. In both cases, addition of SBS lowers the temperature at which 1% strain at break is achieved. This is in contrast to the results of the BBR test. Also, the slopes of the lines are not the same for the California Valley asphalt. The slope translates to the temperature shift as a function of strain rate. This variability in the time-temperature shift indicates the

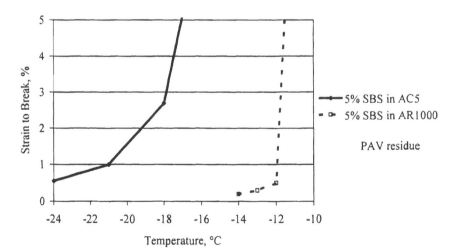

Figure 5   Effect of temperature on strain to break.

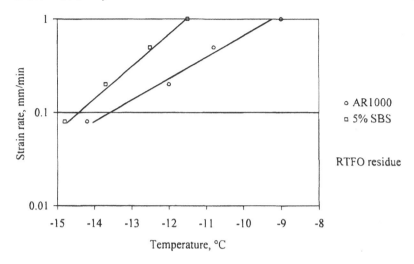

Figure 6 Effect of strain rate on direct tension failure temperature.

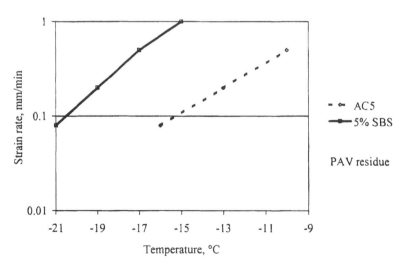

Figure 7 Effect of strain rate on direct tension failure temperature.

importance of running the test at as low a temperature and as low a strain rate as reasonably possible.

The effect of strain rate on the breaking stress was determined for several binders. As shown in Figure 8, within the limited rates tested, there is little effect. The importance of this observation will be discussed later.

Several binders are compared in direct tension versus bending beam in Figure 9. There is generally an offset of about 6° dropping the strain rate from 1 to 0.08 mm/min. Thus the test could be run at 1 mm/min and change the specification offset from 10°C to 16°C. However, the offset varies from about 3°C to about 7°C. As mentioned above, this variability makes it desirable to run the test at the lowest reasonable strain rate and temperature. Note that the offset between BBR and DT varies considerably for different binders. For unmodified AR1000 direct tension, even at the low rate, is worse than BBR. For the modified AC5, direct tension is much better than BBR. This highlights the importance of using the data from BBR and DT correctly to predict thermal cracking.

To validate these predicted low temperatures, commercial binders with several years of excellent field performance were tested. Results are shown in Figure 10. The highly modified Alaska and North Dakota binders offer an interesting contrast. Both perform well in extreme climates, which is reflected in the direct tension results. The bending beam results are quite different, very good for the North Dakota binder and poor for the Alaska binder. These SBS-modified binders all fail BBR on *m* value, which, with the current specification, would preclude the direct tension test.

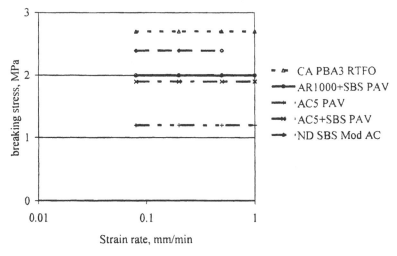

Figure 8 Effect of strain rate on breaking stress.

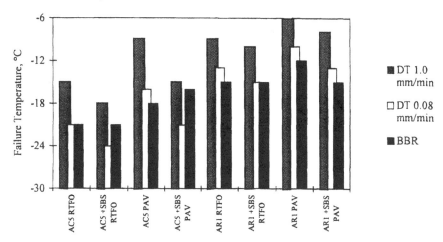

Figure 9 Correlation between bending beam and direct tension at low and high strain rates.

By conducting a series of measurements, a plot of failure stress as a function of temperature can be developed as shown in Figure 11. At the ductile-brittle transition, the breaking stress can be replaced by the yield stress to give a continuous function.

At this point the results were examined to try to develop a reasonable plan for modifying the specification. A simple recommendation to lower the strain

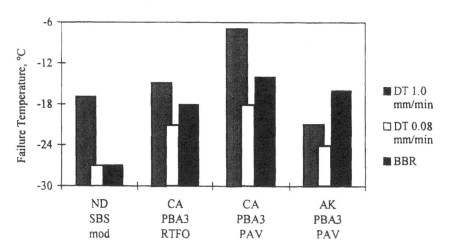

Figure 10 Commercial binders with field performance data.

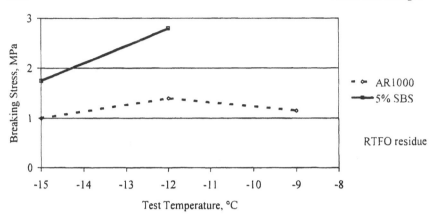

Figure 11 Effect of temperature on breaking stress.

rate and elongation to break requirement would be straightforward but inelegant. The specification would still be purely empirical and thus would be very difficult to justify without large amounts of data. Also, there is a wealth of information in the direct tension test. Some of the parameters and information available include strain rate, strain failure criterion, failure stress energy to break, and temperature offset. It seems reasonable to explore the use of some of this additional information in the specification. If possible, it is desirable to base the specification on a fundamental material property.

III. PROPOSED SCHEME TO MODIFY THE BINDER SPECIFICATION

The simplest model, which has often been used to describe thermal cracking [8], is shown in Figure 12. As temperature drops, thermally induced stress builds. When the thermally induced stress exceeds the breaking stress of the material, cracking occurs. This temperature is designated T_c, the critical cracking temperature.

The importance of the m value is shown in Figure 13. If the m value is low, high stresses build even at temperatures above the critical cracking temperature. These stresses also relax slowly. Binders with a low m value will be susceptible to low-temperature fatigue cracking well above their critical cracking temperature. High m value, on the other hand, means the pavement is unlikely to crack until the critical cracking temperature is closely approached or passed.

The effects of polymer modification on fracture properties, shown in Figure 14, are easily seen in this model. If the breaking stress is increased, the critical

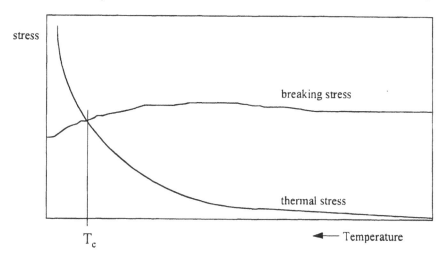

Figure 12   Simple thermal cracking model.

cracking temperature will be lower irrespective of the linear viscoelastic properties, *S* and *m*, of the binder.

A similar model is used as the basis for the Superpave Indirect Tensile
Test (IDT) for thermal cracking [9].

The proposal is to set the Superpave low-temperature specification at the
critical cracking temperature, T_c, of the binder. The T_c can be calculated using

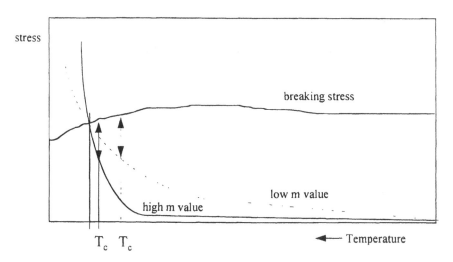

Figure 13   Effect of *m* value.

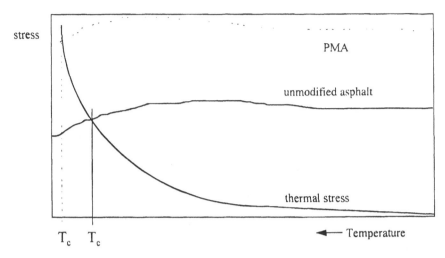

Figure 14   Effect of polymer modification on thermal cracking.

the simple model in Figure 12. Breaking stress is measured with the direct tension test. The thermally induced stress can be calculated via a convolution integral from the relaxation modulus and shift factor function. These, in turn, can be determined with data from the bending beam test.

This procedure will allow exact calculation of critical cracking temperatures based on fundamental material properties. However, this procedure is much too extensive and time consuming, requiring multiple tests at multiple temperatures, to be used as a specification test. Instead, the proposal is to determine T_c for a number of neat and modified asphalts. The critical cracking temperature will then be fit to a simple empirical model, for example:

$$T_c = f(T_{BBR}, T_{shift}, m, \sigma_c)$$

where T_{BBR} is temperature from BBR where $S(60) = 300$ MPa, T_{shift} is a constant offset to compensate for higher rate, m is a relaxation term to account for induced stress curve shape and thermal fatigue cracking, and σ_c is the breaking stress from the direct tension test. These parameters can be readily determined at a single temperature to generate a pass-fail criterion.

For unmodified asphalts, the current specification utilizing the bending beam rheometer is recognized to give good predictions of thermal cracking. This will allow us to simplify the process by breaking it into two parts. Since BBR is sufficient for standard asphalts, the model will be fitted using only BBR parameters:

$$T_c = f(T_{BBR}, T_{shift}, m) = \text{BBR specification temperature}$$

Binders with nonstandard fracture characteristics will then be included to modify the simplified model with terms from the direct tension test.

$$T_c = f(T_{BBR}, T_{shift}, m) + f(\sigma_c)$$

The accuracy of the model will naturally depend on the assumptions used in developing master curves and shift factor functions. Optimization of this procedure is still in progress and is beyond the scope of this report. BBR and dynamic shear rheometer (DSR) data will be used as raw data for the master curves. The most straightforward and rational approach is to use the modified Williams, Landel, Ferry (WLF) approach to determine shift factors. The resulting master curve raw data will be fitted to the Christensen-Anderson model [10] as a mathematical form. This, in turn, will be fitted to a Prony series to simplify calculations in the convolution integral. A preliminary example of this procedure, using actual data, is shown in Figure 15. The critical cracking temperature and hence the specification temperature would be −24°C.

IV. CONCLUSIONS

Although the current Superpave low-temperature specification gives reasonable predictions for most asphalts, it does not adequately reflect the performance of asphalts, especially modified asphalts, which do not have standard fracture properties. In particular, the current prescribed rate for the direct tension test is too high, being more than an order of magnitude higher than the rate for the BBR test and the time-shifted rate of actual pavement deformation.

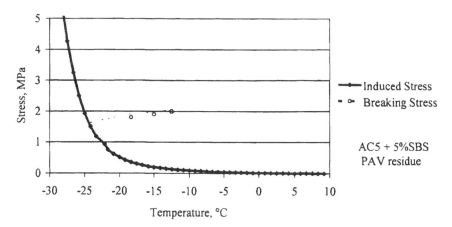

Figure 15 Calculated induced stress versus measured breaking stress.

Rather than simply adjusting the elongation rate and hoping for the best, a simple model has been developed that is based on fundamental material properties. This model, while too complex for specification testing, will be fitted to a simple empirical formula for the core Asphalt Materials Reference Laboratory (AMRL) asphalts and then expanded to include modified asphalts with unusually large (or small!) breaking stress or breaking energy. This empirical model will be validated by comparison with actual field performance of binders in severe low-temperature climates. So far, the data set includes binders from California, Pennsylvania, North Dakota, Texas, and Alaska.

ACKNOWLEDGMENTS

We thank Mark Bouldin of Applied Paving Technology for his support in carrying out the thermal stress restrained speciment tests. Furthermore, we thank David Cushing and Dr. King Him Lo of Shell Chemical Company and Professor David Anderson of Pennsylvania State University for their contributions to this work.

NOMENCLATURE

δ	phase angle, degrees
$G*$	complex modulus, MPa
$G* \sin \delta$	shear loss modulus, MPa
$G*/\sin \delta$	inverse shear loss compliance, MPa
m	$m(60)$
$m(t)$	slope of $\log(S(t))$ versus $\log(t)$ determined at a specific time
SBS	styrene-butadiene-styrene block copolymer
S	$S(60)$
$S(t)$	stiffness determined at a specified time, MPa
σ_c	critical breaking stress, MPa
t	time, s
T	temperature, °C or °F
T_{BBR}	temperature at which one of the bending beam rheometer failure parameters is met
T_c	critical cracking temperature
T_{shift}	temperature shift factor to account for differing test rates

REFERENCES

1. *AASHTO Provisional Standards,* American Association of State Highway and Transportation Officials, Washington, DC (1995).
2. *Shell Chemical Publication SC;1494:93,* Shell Chemical Company, Houston (1993).

3. See, for example (a) F. Finn, C. Monismith, G. Hicks, and R. Leahy, Validation of SHRP Binder Specification Through Mix Testing, *Proc. Assoc. Asphalt Paving Technol.,* **62**:565 (1993); (b) R. Cominsky and P. Teng, Post SHRP Validation of SHRP Binder and Mixture Specifications, *Proc. Assoc. Asphalt Paving Technol.,* **62**:639 (1993); (c) P. S. Kandhal, R. Dongré, and M. Malone, Prediction of Low Temperature Cracking of Pennsylvania Project Using Superpave Binder Specifications, *Assoc. Asphalt Paving Technol. Prepr.,* Session IV (1996).

4. D. A. Anderson, *Binder Characterization and Evaluation,* Vol. 3, *Physical Characterization,* Strategic Highway Research Program Report SHRP-A-369, National Research Council, Washington, DC (1994).

5. R. Dongré, J. D'Angelo, and S. McMahon, Development of the New Superpave Direct Tension Test, presented at the 1997 TRB Meeting, Washington, DC.

6. M. G. Bouldin and R. Q. Kluttz, Evaluation of the Low Temperature Properties of Straight and Polymer Modified Asphalt, presented at the Petersen Asphalt Research Conference, Laramie, WY 1994.

7. D. H. Jung and T. S. Vinson, *Low Temperature Cracking; Test Selection,* Strategic Highway Research Program Report SHRP-A-400, National Research Council, Washington, DC (1994).

8. T. S. Vinson and D. H. Jung, Low Temperature Cracking Resistance of Asphalt Concrete Mixture, *Proc. Assoc. Asphalt Paving Technol.,* **62**:54 (1993).

9. (a) R. L. Lytton, R. Roque, J. Uzan, D. R. Hiltunen, E. Fernando, and S. M. Stoffels, Performance Models and Validation of Test Results, Final Report to Strategic Highway Research Program; Asphalt Project A-005, 1994; (b) R. Roque and W. G. Buttlar, The Development of a Measurement and Analysis System to Accurately Determine Asphalt Concrete Properties Using the Indirect Tensile Mode, *Proc. Assoc. Asphalt Paving Technol.,* **61**:304 (1992).

10. D. Christensen and D. Anderson, Physical Properties of Asphalt Cement and Their Relationship to Microstructural Model, *Proc. Assoc. Asphalt Paving Technol.,* **62**:475 (1993).

10

Rheological Properties of Polymer-Modified Asphalt Binders

Peter E. Sebaaly

University of Nevada, Reno, Nevada

I. INTRODUCTION

As the nation's highway system ages, rutting, low-temperature cracking, and fatigue failures have become a common occurrence. Conventional asphalt cements like AC-20 and AR-4000, although they showed good performance in the past, are showing increased signs of unsatisfactory performance when used in pavements subjected to high traffic volumes, high tire pressures, and severe climatic conditions. In order to combat such failures, the state highway agencies (SHAs) have been using polymer-modified asphalt binders under high traffic loads and severe environmental conditions (i.e., hot and cold climates).

The rutting problem can be traced to the changes that have taken place in the tire type and inflation pressure. The traditional bias-ply has been replaced by radial tire that can be used at much higher tire inflation pressures for better fuel economy and improved axle load-carrying capacity. Because of this increase in contact pressure, pavements are being subjected to stresses higher than those for which they were designed. These circumstances, combined with high summer temperatures, have led to many pavement failures due to rutting.

Low-temperature cracking is another type of common failure caused primarily by the environment. As its temperature drops, the pavement shrinks, causing tensile stresses in the pavement; in the meantime loading stresses are superimposed that result in high total tensile stresses.

One alternative that considerably slows down the mentioned failures is the use of polymer-modified asphalt binders in the mixtures. It is anticipated that by polymer modifying the asphalt, improved properties will be attained compared with those of conventional asphalts. The improved properties include better binder stiffness at high temperatures and more flexibility at low temperatures.

A. Polymers

Polymers are added to asphalt cements to improve one or all of the following mixture properties: resistance to rutting, resistance to thermal cracking, improved moisture and temperature susceptibility, and increased fatigue resistance.

Polymers can be further subdivided into two main types as follows [1]:

1. Rubber
   (a) Natural latex
   (b) Synthetic latex
   (c) Block copolymers
   (d) Reclaimed rubber
2. Plastic

Rubber has been used for a number of years in various forms to improve certain properties of mixes. Rubber in the form of latex has been used to improve the asphalt-aggregate bond. Rubber has also been used to improve stiffness and flexibility in order to improve rutting resistance and low-temperature susceptibility of mixtures.

Plastics have also been used to improve the temperature susceptibility and stiffness of asphalt mixes. Some examples of plastics are polyethylene, polypropylene, ethyl vinyl acetate, and polyvinyl chloride.

B. Literature Review

The use of polymers in binders can be traced back to the early 1950s in Europe [2]. During the early years research on polymer-modified binders and mixes was sparse. However, in the past decade or so the interest in alternatives such as polymers has increased as social and technological advances require more durable pavements.

The earliest research available on polymer-modified asphalts was published by Gregg and Alcocke [3] in 1954. They evaluated asphalt binders modified with a latex made from styrene-butadiene polymer. They concluded that the addition of polymers to asphalt binders improved or reduced the temperature susceptibility as measured by viscosity and penetration. Similarly, Thompson [4] in 1964 concluded that the addition of neoprene to asphalt binders improved their toughness and aging characteristics. He reported that the characteristics of the polymerized asphalt mix depend entirely on the nature and source of the binder and the type of polymer.

King et al. [5] conducted research on the high-temperature performance of polymer-modified asphalt cements. In their research they took Pen graded asphalt cements (ACs) ranging from 40/50 to 180/200 grades with polymer contents ranging from 0 to 2% for each grade. Various conventional consistency tests and dynamic mechanical analysis (DMA) were conducted on these ACs. These asphalts were then mixed with the same aggregate and tested for rutting on the Hamburg wheel tracking device and the French wheel track rutting tester. The rutting results were compared with the rutting predicted from the conventional AC tests and the DMA, which led them to following conclusions:

1. Classic consistency tests for asphalts do not adequately predict results from laboratory rutting simulators when polymerized asphalts are used and these consistency results also do not correlate with the DMA results. Therefore there is a tendency to overpredict the stiffness and hence the expected benefits of the modified binder.
2. There is a good correlation between results from DMA and rutting simulators. However, the choice of temperature and frequency is critical if DMA results are to be used to predict high-temperature pavement performance. The experimental DMA parameters should not vary too much from the extreme conditions that are to be encountered on the actual pavement.

Stock and Arand [6] conducted research on low-temperature cracking of polymerized binders at Braunschweig University in partnership with British Petroleum. They studied the effects of adding synthetic rubber polymer additives to asphalts. They concluded that the addition of polymers always improves the low-temperature performance of the binder, but the effectiveness of the additive can be dependent on the characteristics of the base asphalt.

II. EVALUATION AND ANALYSIS OF BINDER PROPERTIES

A total of 11 binders were evaluated in this research. All of them were classified as AC-20Ps, which are polymer-modified AC-20 binders. All 11 binders were tested against the Nevada Department of Transportation (NDOT) specifications for AC-20P binders as shown in Table 1. The current NDOT specification is based on viscosity, ductility, toughness, and tenacity properties of the binder.

The objective of this research is to evaluate and compare the rheological properties of the polymer-modified binders. Even though all 11 binders were classified as AC-20Ps on the basis of their viscosity and ductility properties, their rheological properties may be significantly different.

A. Rheological Properties

Rheology is the science of the deformation and flow of material, whether in liquid, melt, or solid form, in terms of the material's elastic and viscoelastic

Table 1   Nevada DOT Specifications for AC-20P Binders

	Specification		
Property	Minimum	Maximum	Allowable tolerance
Viscosity at 140°F, poise	2300		12%
Viscosity at 275°F, cSt	600	3000	2%
Flash point, °F, COC	450		15°F
Ductility at 39.9°F, cm	50		
Toughness, inch-pounds	110		
Tenacity, inch-pounds	75		
Tests on RTFO residue			
Viscosity at 140°F, poise	3000		12%
Ductility at 39.9°F, cm	25		
Loss on heating, %		0.5	

properties. The science of rheology can be complex, but rheology testing itself need not be complicated. Both rational measurements and parameters are needed to obtain the rheological behavior of binder, which would serve as the basis for an effective performance-based binder specification. Basic rheological properties of asphalt binders include the following:

G': The storage modulus (elasticity) of the asphalt binder.

G'': The loss modulus (viscous loss) of the asphalt binder.

G^*: The complex modulus, which is the amount of energy to deform the asphalt binder.

δ: The phase angle, which is the difference between the phase of the sinusoidally varying input quantity and the phase of the output quantity, which also varies sinusoidally at the same frequency. In the case of binder testing, the input quantity represents the applied strain and the output quantity represents the resulting stress.

Figure 1 is a graphical representation of the relationship among the foregoing variables. In obtaining these rheological properties, sinusoidal shear strains γ are applied to the binder samples. At cold testing temperatures (below 34°C), the strains are kept constant at 1% and increased to 12% at higher test temperatures (above 52°C). Keeping the strain constant throughout a given test allows the sample to remain in the linear viscoelastic range. Although no material is perfectly linear under all conditions, linear viscoelastic characterization has been found in the past to best represent the rheological behavior of asphalt binders.

Because most rheological properties are time dependent, it is essential to test the binder at a constant frequency, ω. This study evaluated the binders at a frequency of 10 rad/s. If the stress-strain behavior of the binder is completely elastic, the resulting stress will be in phase with the applied strain, as illustrated

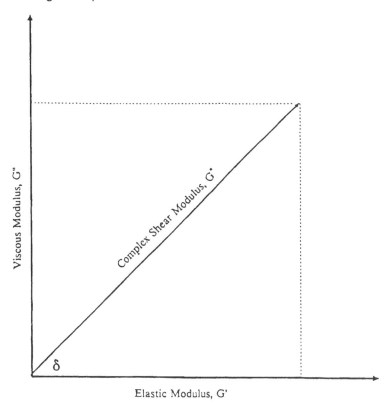

Elastic Modulus, G'

Figure 1   Relationship among the various rheological properties of asphalt binders. Storage (elastic) modulus $G' = (G^*)\cos(\delta)$; loss (viscous) modulus $G'' = (G^*)\sin(\delta)$; loss tangent $\tan(\delta) = G''/G'$.

in Figure 2. Otherwise, when the response is completely viscous, the stress response will be 90° out of phase with the applied strain. In general, the phase angle (δ) can very from 0° to 90°, which indicates the amount by which the resulting stress is out of phase with the applied strain.

This research effort evaluated the rheological properties of the polymer-modified binders at the high and intermediate temperatures in terms of their G^* and δ and at the low temperature in terms of their bending stiffness $S(T)$. The dynamic shear rheometer (DSR) was used to evaluate G^* and δ and the bending beam rheometer (BBR) was used to evaluate $S(t)$ for all binders. Both of these tests are described below.

1. Dynamic Shear Rheometer

The DSR system used in this research was developed by Rheometrics and is referred to as the Rheometrics Asphalt Analyzer (RAA). The instrument can

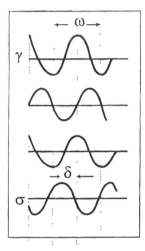

Strain Input (at Frequency ω)

Stress Response of Ideal Viscous
Fluid (90° Phase Shift)

Stress Response of Ideal Elastic
Solid (0° Phase Shift)

Stress Response of Viscoelatic
Material (δ Phase Shift)

Figure 2   Relationship among stress, strain, and phase lag.

apply a precise oscillatory, steady, or step shearing strain to the test sample and precisely measure the sample's stress response.

One of the most important aspects of obtaining repeatable rheological data while conducting binder testing with the RAA is the sample and equipment preparation. The parallel plate configuration with the temperature sensor inside the upper plate was used in the test as shown in Figure 3. The size of the plates (i.e., 8 or 25 mm in diameter) varies depending on the test temperature. More details of sample and RAA preparation are summarized in Ref. 7.

2. The Bending Beam Rheometer (BBR)

The BBR is a "creep" test device that applies a constant load at the center of a simply supported asphalt binder beam for a selected period of time (Figure 4). During the loading time the deflection at the center of the beam is continuously measured. The asphalt beam is 127 mm long, 12.7 mm wide, and 6.3 mm thick, supported at both ends on metal supports that are 100 mm apart. The Cannon BBR was used in this research. The data generated from the test include the time history of load and deflection. The analysis of the data provides the stiffness [$S(t)$] values for selected loading times. This research effort evaluated the $S(t)$ at 60 s loading time.

3. Binder Aging

Because asphalt binders have a tendency to become brittle with age, the evaluated binders were aged prior to BBR testing and the DSR testing at intermediate temperatures.

Elevation View   Plan View

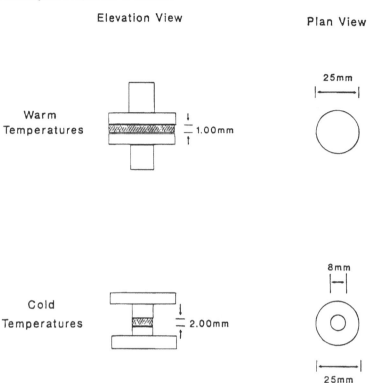

Figure 3   Parallel plate test configurations.

To accomplish similitude between laboratory and field binders, two types of aging process were performed. For short-term aging, the rolling thin-film oven test (RTFOT) is used to represent aging or hardening of the binder that occurs during the mixing and laydown process. To simulate long-term exposure in the field, the pressure aging vessel (PAV) is used [7].

After short-term aging through the RTFOT, the asphalt binder is aged using the PAV for 20 h under a constant pressure of 2.07 MPa and temperature between 90 and 110°C. The temperature of the test varies depending on the climate in which the asphalt will be used. The vessel can hold 10 thin trays with 50 ± 0.5 g of binder per tray.

The RTFOT-PAV–aged binders are then tested in the DSR to evaluate their G^* and δ at intermediate temperatures and in the BBR to evaluate their $S(t)$ at low temperatures.

Asphalt Beam Plan View

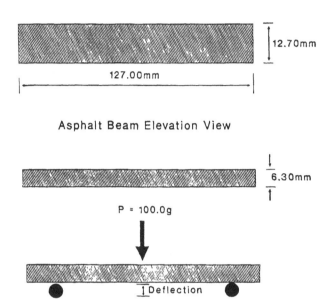

Figure 4   Bending beam test configurations.

B.   Data Analysis

The objective of this research effort was to evaluate the variability in rheological properties of asphalt binders within one grading group, namely the AC-20P binders. First, the rheological properties of the binders were measured at high temperatures of 52, 58, and 64°C to evaluate their resistance to permanent deformation. Second, the rheological properties of the binders were measured at intermediate temperatures of 16, 19, and 22°C to evaluate their resistance to fatigue cracking. Third, the rheological properties of the binders were measured at low temperatures of -10 and -20°C to evaluate their resistance to low-temperature cracking.

In this study, 11 binders were graded as AC-20Ps on the basis of their conventional viscosity and ductility properties. Figures 5 and 6 show the distributions of the G^* and δ values for the unaged binders, respectively. Figures 7 and 8 show the distributions of G^* and δ for the aged binders, respectively. Figure 9 shows the distribution of $S(t)$ for the aged binders. The data in these figures show that there are some differences in the properties of the AC-20P binders in their unaged and aged stages.

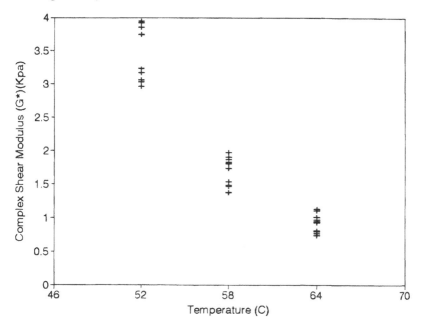

Figure 5 Distribution of the complex shear modulus for unaged polymer-modified binders.

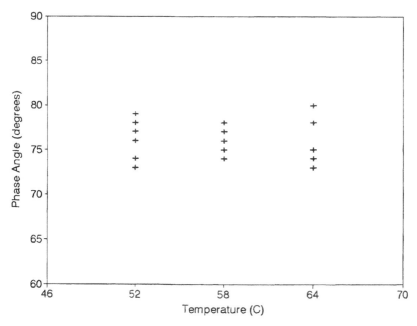

Figure 6 Distribution of the phase angle for unaged polymer-modified binders.

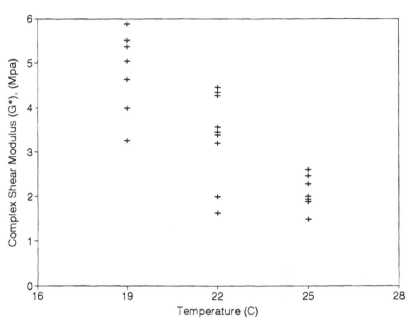

Figure 7   Distribution of the complex shear modulus for aged polymer-modified binders.

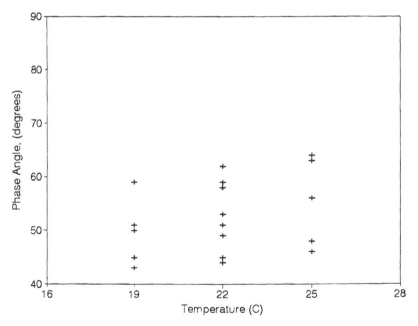

Figure 8   Distribution of the phase angle for aged polymer-modified binders.

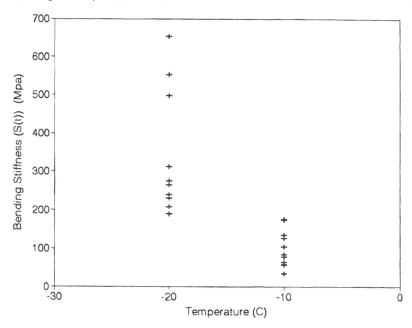

Figure 9   Distribution of the bending stiffness for aged polymer-modified binders.

In order to determine the magnitude of the variation, the mean, standard deviation, and coefficient of variation (CV) were evaluated for each of the measured properties. Table 2 summarizes the statistical data for all of the tested binders. It is commonly assumed that CV values below 10% indicate very low variability and CV values above 10% indicate high variability. The data in Table 2 indicate that the phase angles δ of the unaged binders are relatively uniform; when the binders are aged, their δ values become significantly different. On the other hand. G^* varies significantly among the binders in both the unaged and aged stages. The $S(t)$ shows extremely high variability among all binders.

III. CONCLUSIONS

Based on the data generated from this experiment, the following conclusions can be drawn:

1. Even though asphalt binders are graded within one grading group on the basis of their viscosity and ductility properties, they can have significantly different rheological properties.
2. The variation in the rheological properties becomes more significant as the binders age and as the testing temperatures drop. This indicates

Table 2 Statistical Data for the Measured Properties of All AC-20P Binders

| | Unaged binders | | | | | | Aged binders | | | | | | | |
| | 64°C | | 58°C | | 52°C | | 25°C | | 22°C | | 19°c | | -10°C | -20°C |
	G^* (kPa)	δ (deg)	G^* (kPa)	δ (deg)	G^* (kPa)	δ (deg)	G^* (MPa)	δ (deg)	G^* (MPa)	δ (deg)	G^* (MPa)	δ (deg)	$S(t)$ (MPa)	$S(t)$ (MPa)
Average	0.93	76	1.7	76	3.5	76	2.1	58	3.4	55	5.4	51	99	376
Standard deviation	0.13	3	0.20	1.4	0.44	1.8	0.38	8	0.94	7	1.1	7	48	190
Coefficient of variation (%)	14	4	12	2	13	2	18	14	28	13	21	14	48	50

that the polymers used to modify these binders exhibit different aging characteristics and have significantly different low-temperature properties.

3. It is expected that these 11 binders will have significantly different field performances in term of resisting fatigue and low-temperature cracking. This further emphasizes that grading asphalt binders on the basis of their viscosity and ductility properties alone may not guarantee uniform field performance.

ACKNOWLEDGMENT

The author would like to thank the Nevada Department of Transportation, Materials Division, for providing the funds and materials for this research project.

REFERENCES

1. H. Von Quintus and J. A. Epps *Highway Materials Engineering,* National Highway Institute Course, Publication FHWA-HI-90-008 (1990).
2. F. L. Roberts, P. S. Kandhal, E. R. Brown, D. Lee, and T. W. Kennedy, *Hot Mix Asphalt Materials, Mixture Design and Construction,* 1st ed., NAPA Education Foundation, MD (1991).
3. L. E. Gregg and W. H. Alcocke, Investigation of Rubber Additives in Asphalt Paving Mixtures, *Proc. Assoc. Asphalt Paving Technol.,* **23**: 28 (1954).
4. D. C. Thompson, Rubber Modifier, in *Bituminous material* (A. Holberg, ed.), Vol. 1, Interscience, New York, p. 375 (1964).
5. G. N. King, P. Harders, and P. Chaverot, Influence of Asphalt Grade and Polymer Concentration on the High Temperature Performance of Polymer Modified Asphalt, *J. Assoc. Asphalt Paving Technol.,* **61** (1992).
6. A. F. Stock and W. H. Arand, Low Temperature Cracking in Polymer Modified Binder, *J. Assoc. Asphalt Paving Technol.,* **62** (1993).
7. Federal Highway Administration, Office of Technology Applications, *SHRP Asphalt Binder Specification and Equipment Training,* Phoenix, AZ (1992).

11

Current and Potential Applications of Asphalt-Containing Materials

Arthur M. Usmani

Usmani Development Company, Indianapolis, Indiana

I. INTRODUCTION

Building and construction products are very important, since shelter is a basic human need. These products are made from wood, asphalt, gypsum, glass, Portland cement, concrete, brick, and synthetic polymers.

Bitumen from petroleum or crude oil is known as asphalt. Bitumen is a black or dark brown mastic-like thermoplastic material. The most important sources of bitumen are petroleum or coal deposits. The natural product is called gilsonite or pitch. This mineral was formed by an old weathered petroleum flow at the surface of the earth from the larger petroleum molecules left behind. Gilsonite, with substantial reserve, is available from Utah. The deposit at Lake Trinidad was a principal source in the past.

Asphalt is left behind after more valued components, e.g., gasoline and kerosene, distill from the crude oil. The quality and amount of asphalt left behind distillation depend on the source of the crude oil used in the refinery. The largest petroleum reserves are in the Middle East. Most crudes, e.g., Saudi Arabian and Kuwaiti, produce good asphalt [1].

As a construction material, asphalt finds major applications in paving and roofing. About 75% of asphalts find application in paving, 15% in roofing, and the remaining 10% in coatings, adhesives and batteries [2].

249

Currently, we use about 70 billion pounds of asphalt in the United States. Much more asphalt would find use if other applications were fully in advanced stages through research efforts. Asphalt is a polymer-like material and the most widely used thermoplastic material. It is also inexpensive, around $0.06/lb. Polyolefins, specifically linear low-density polyethylene (LLDPE), low-density polyethylene (LDPE), and high-density polyethylene (HDPE), are the next most widely used thermoplastics [3]. They are much more expensive, around $0.40/ lb. Engineering thermoplastics are a dollar or more per pound.

In this chapter, I will describe currently practiced asphalt applications, polymer and rubber modifiers, and an outline of developing newer applications.

II. PAVING

The asphalt type used in paving depends on the intended application. The types include straight asphalts known as asphalt cements (ACs), asphalt emulsions, cutback asphalts made by dissolving ACs in suitable solvents, and road oils.

Asphalt cements find application in paving well above 50% of the total volume. These asphalts come in a wide rage of viscosities from 250 P (25 Pas) to 4000 P (400 Pas) at 60°C; for example, AC-20 has a viscosity of 200 Pas. Over 90% of the roads in the United States are asphaltic concrete, also called as blacktop. Asphaltic concrete consists of sand, aggregate (rock), and usually polymer-modified asphalt. The heated mixture coats the aggregates. It is applied to the prepared roadbed and pressed while hot to compact before the mixture cools and sets.

Asphalt cutbacks are useful as road sealers and coatings. The cutbacks set up depending on the evaporation rate of the solvent. To repair minor road damage, aggregate and sand are mixed with cutback asphalt and used as cold patches. The chuck holes are filled up with cold patch and cured by solvent evaporation. Asphalt cutbacks applied to foundation as a coating film retard water vapor transmission. Because of restriction on volatile organic compounds (VOCs), use of cutback asphalt will decease in the future.

Asphalt emulsions are stabilized by either anionic or cationic surfactants. Hot asphalt when mixed with water and surfactant in a colloid mill produces asphalt emulsion. The particle size can be as small as 3 μm. The emulsion particles do not flocculate because of the repulsion barrier provided by electrostatic forces [4]. Cationic asphalt emulsions are flocculated by metal or silicate. Most rocks are silicate based and hence cationic emulsions find use in subbase stabilization. Anionic emulsions flocculate due to water evaporation and hence are useful in making cold patch compounds. Road oils are very fluid asphalts. They find application in dirt road and sand stabilization.

III. ROOFING

Asphalts are useful for waterproofing roofs and inferior foundations. Roofing is by far the most important waterproofing application of asphalt. Roofing slopes are low slope (under 2-inch rise per foot, also called a 2-in-12 slope) and steep slope. Membrane roofing systems find application on slope of up to 6 in 12. Shingles are never useful under 2 in 12 and are generally used over 4 in 12.

The current roofing market size, including installation costs, is above $12 billion for commercial and over $5 billion for residential. It should grow by 5% annually.

A. Shingles

Asphalt shingles are most popular in residential roofing because of their low cost. The market share for shingle is about 65%. Reinforcements for shingles are fiberglass mat and organic felt, with the former dominating the market by 85%.

The fiberglass mats are made from thin, around 16 μm, fiber randomly dispersed and bonded by a thermoset. These mats are thin, in the range of 80 to 120 g/m^2. The mat is coated on both sides on a production coating line by a highly filled asphalt composition. To the top surface, granules of the desired color are adhered, and sand or any other inexpensive release agent is applied on the bottom side to eliminate tackiness. Finally, an asphalt-based adhesive as a strip serves to seal the shingles on the roof.

Organic felt shingles started about 100 years back. The felt made from cellulosic materials looks like a porous paper. Rolls of the felt, on a production line, soak up very soft asphalt called saturant by several passes. The saturant-soaked felt is coated on both sides by a highly filled asphalt composition. The asphalt coating composition is about 50 to 60% calcium carbonate–filled, air-blown asphalt. The highly filled coating does not burn well and thus the shingle has good fire resistance. Granules, if attached to the top surface, enhance weathering and decoration. The bottom surfaced with sand reduces tackiness for easy handling. An asphalt-based adhesive, as a strip, is finally coated on the bottom side of the shingle. The adhesive helps to provide a secured roofing system and avoid wind damage.

B. Roof Membranes

Membranes are useful on large commercial buildings. Currently, three types of asphalt-containing materials are in use. These are modified bitumen (modbit) membranes, built-up roofing (BUR), and mopping asphalt (MA).

1. Modbit Membranes

Modbits were developed in Italy around 1960. In 1957, stereoregular isotactic polypropylene (IPP) production began using Ziegler-Natta catalyst. The early catalysts produced waste by-product, atactic polypropylene (APP). Finding an application for APP led to the discovery of modbit. Use of modbit began during mid-1970 in the United States. Today, there are more than 10 manufacturers of modbit, notably GAF and Bridgestone/Firestone, with large production capacities.

The basic components of modbit membrane are polymer- or rubber-modified asphalt compound and the reinforcement mat. Commercially, the asphalt modification uses either atactic polypropylene or styrene-butadiene-styrene (SBS). Roofing membranes made from polypropylene are known in the roofing industry as APP membrane. The SBS membranes use styrene-butadiene-styrene-modified compounds. The two most commonly used reinforcers are nonwoven polyethylene terephthalate (PET) and nonwoven fiberglass. New mats, e.g., PET/Nylon 6, that do not use thermoset binders are coming into the market. Typically these mats have weighs in the range 160 to 250 g/m². The mat contributes significantly to the cost of the membrane. The new trend is to use the lowest weight mat that one can run on the production line without significant necking.

Because of advances in highly efficient coordination catalysts, there is no atactic polypropylene by-product in the IPP manufacture. Instead of APP, polypropylene copolymers containing minor amounts of ethylene have been developed for the roofing and adhesive industries [5]. Major manufacturers of such copolymers are Rexene and Eastman Chemicals. Using about 25% APP or polypropylene copolymer produces a novel structure with outstanding properties.

Modification of asphalt with about 10% SBS elastomer also produces novel structures with outstanding properties.

On the production line, the mat goes in one or several dip tanks containing the compound. The excess is removed, followed by mat consolidation and chilling with one or several water trays. The membranes are cut to the desired size and wound.

There are three useful methods for attaching modbit to the roof. A mopping asphalt bonds the membrane. In torch grade, a torch melts the compound and the molten compound serves as the adhesive. Use of an asphalt-containing cold adhesive is a third method. This method is gaining acceptance in the United States [6].

2. BUR Roofing

Built-up roofing, with many types of reinforcer, has been in use for almost 100 years. Presently, fiberglass mat is most popular. Organic felt finds slight use. BUR roofing systems consist of several layers of reinforcer between several asphalt layers (asphalt/reinforcer 1/asphalt/reinforcer 2/asphalt/reinforcer and so on). Asphalt serves as an adhesive and waterproofer. The reinforcer provides strength to the BUR composite and prevents asphalt flow. A ply felt coated with filled asphalt, commonly called ply sheet, is used on each side when the roof must be nailed to the substrate instead of being adhered. A cap sheet, or a sheet with granules, when added, improves decorative functionality. The BUR market is still substantial in the United States. Modbit membranes will gain market share at the expense of BUR in the future.

3. Mopping Asphalt

Mopping asphalts are asphalts oxidized by blowing air onto hot asphalt. Oxidized asphalt has improved properties compared with asphalt. Grades of mopping asphalt are classified by softening point and penetration hardness.

Type	Softening point (°C)	Penetration at 25°C (dmm)	Remarks
I	68	18–50	Only for 0.25 in 12 slope
II	71–79	18–40	For less than 1 in 12 slope
III	85–93	15–35	For up to 3 in 12 slope
IV	96–107	12–25	Used in flashings and hot climates to keep roofing system from sliding.

IV. POTENTIAL APPLICATIONS

Besides traditional building materials, plastics are becoming important in this market. During the past 10 years, the growth in total plastic volume consumed for building and construction applications has been around 7%. Initially, plastics were used for decorative purposes, but now they are finding ever increasing use in structural and decorative applications. Plastics have superior strength, light weight, corrosion resistance, low cost, good insulating properties, and suitability for fabrication in any shape or size. The most serious problems of plastics in building materials are their flammability and smoke toxicity.

The properties that govern use of plastics as building materials are glass transition temperature, ease of conversion, heat deflection temperature, outdoor stability, and mechanical properties.

Much work on asphaltic compositions was done before or just when polymer science became a worthwhile discipline. Because of the low cost and polymer-like nature of asphalt, it is distinctly possible to develop newer products from asphalt by using the very many polymers that are now available. Major shortcomings of asphalt are low softening point, high temperature dependence, permanent deformation under load, insufficient hardness and toughness, and limited wear-out properties.

We overcome the shortcomings of asphalt by polymer modification or cross-linking. Introduction of entangled chains (interpenetrating polymer network) should also improve properties. Incorporation of polymers, [e.g., epoxy, polyurethane, polysulfide, unsaturated polyester, phenolic, polyester, rubber, thermoplastic olefinics (TPOs), thermoplastic elastomer (TPE), in situ polymerized styrene, in situ polymerized vinyls, and in situ polymerized allylic] in asphalt should improve properties for various applications.

The shortcomings of asphalt can be mitigated by modification with various polymer systems. Asphalt types and polymer types in the same polymer family should match for compatibility and development of unique structures.

It is distinctly possible to optimize the promising composition for various applications, e.g., hot moldings, cold moldings, mineral rubber, improved paving binders, highway coatings, roof coatings, waterproofing compositions, and corrosion-resistant coatings. A brief account of the approach now follows.

A. Methodology

We suggest using very soft asphalt (AC-2.5) to a hard propane-extracted (AC-30) to air-blown asphalt for polymer and chemical modifications. Compatibility of asphalt with a polymer can be determined after simple to intense mixing at elevated temperatures, around 200°C. Formation of structures can be determined by ultraviolet spectroscopy or scanning electron microscopy. Some polyblends will be compatible in all proportions, whereas others will show a narrow compatibility range. Both low (10, 20% polymer) and high (50% polymer) levels need investigation to develop newer products.

Thorough mixing requires an intensive mixer, e.g., Banbury. For roofing membrane, adhesive, and coating, an intensive mixer will be suitable equipment. In the case of molded products, asphalt, polymer (also rubber polymer), fiber, mineral filler, and other component of the formulation can be added to a Banbury for fluxing, blending, and chopping. The molded compound lends itself to fabrication by a variety of techniques. The latex modification of asphalt is straightforward; the latex is added to hot and stirred asphalt.

B. Polymer-Modified Asphalt

1. Epoxy-Asphalt

Epoxies react with curing agents or hardeners to cross-link. They offer a unique combination of properties and characteristics, e.g., low curing shrinkage, no by-products of cure, chemical and environmental resistance, very good adhesion to many substrates, and excellent mechanical properties.

Epoxies contain a highly reactive ring called an oxirane or epoxide group [7]. There are many types of epoxies, e.g., bisphenol A/epichlorohydrin resin (aromatic), epoxy novalac resin, aliphatic epoxy, cycloaliphatic epoxy, highly functional resin, and high-molecular-weight linear epoxy resin. Aliphatic polyamines and their derivatives cure epoxies at room temperature. Examples are diethylenetriamine (DETA) and polyamides. Acid anhydrides, e.g., hexahydrophthalic anhydride, are the second most commonly used curing agents. Aromatic epoxies are compatible and make asphalt rigid.

2. Polyurethane-Asphalt

Polyurethanes are formed by condensation of an isocyanate and a polyol [8]. The commonly used isocyanates are toluene diisocyanate (TDI), methylene diphenyl isocyanate (MDI), and polymeric diisocyanates (PMDIs). Polyols used with isocyanate are polymeric glycols based on polyesters, polyethers, or hydroxylated acrylic. Polyesters, e.g., polyethylene adipate or phthalate, have free hydroxyl groups that can react with the isocyanate. Polyethers are low-molecular-weight glycols usually based on polyalkylene oxides. Acrylics are usually low-molecular-weight resins containing a hydroxylated monomer, e.g., hydroxypropyl methacrylate. Tin compounds, e.g., dibutyltin dilaurate, added to polyols catalyze the formation of urethane linkages as well as cross-linking allophanate and biuret linkages for added rigidity.

3. Polysulfide-Asphalt

The highly reactive terminal mercaptans of liquid polysulfide polymers are converted into high-molecular-weight products at ambient temperature by oxidation or by reaction with other active polymers [9]. Polysulfides have excellent solvent resistance, predictable cure, good low-temperature performance, good weathering, and excellent adhesion to many substrates.

Polysulfide sealants have found wide acceptance in construction, aircraft, automotive, marine, and insulating glass applications. A typical liquid polysulfide, such as Thiokol LP-2 or LP-3, can be cured by 3% activated manganese dioxide. Asphalt and liquid polysulfide can be compounded with filler. The curative consists of MnO_2 plus cure modifiers.

4. Unsaturated Polyester-Asphalt

The usual unsaturated polyester resins result from the condensation of an aliphatic diol with an unsaturated dibasic acid or acid anhydride, specifically maleic anhydride [10]. The unsaturated polyesters are generally solutions in a monomer that serves to cross-link the resin chains during the curing reaction. The most commonly reactive monomer is styrene, which is usually present in amounts ranging from 25 to 45%. Since all the styrene is consumed in the curing reaction, the unsaturated polyester resin is a 100% solid coating.

Free radical initiators, e.g., organic peroxides or hydroperoxides, are useful catalysts in curing unsaturated polyesters. A 60% solution of methyl ethyl ketone peroxide in dimethyl phthalate produces room temperature cure. Cobalt naphthenate is an accelerator in curing.

Modification of asphalt with unsaturated polyester should be aimed at developing coatings. Room temperature and elevated temperature curing coatings can result from unsaturated polyester modification of asphalt.

5. Phenolic-Asphalt

Phenolics result from condensation of phenol and formaldehyde and have many outstanding properties, e.g., chemical resistance and durability. Phenolics are used as molding compounds and as bonding agents in plywood, laminated sheets, foundry sand bonding, grinding wheel and friction brake elements, and coatings. We have used phenolics successfully to bond bagasse and clay to produce roofing and building materials [11].

A phenolic powder resin can be slurried into molten asphalt and then allowed to convert to the desired form. Fillers, e.g., clays, calcium carbonate, and sand, can be useful in molding and coating applications.

6. Polyester-Asphalt

Polyesters are condensation polymers obtained by reacting a polyol and a dibasic acid [12]. Polyarylates are aromatic polyesters of phthalic acids and bisphenols. Polybutylene terephthalate (PBT) is obtained by condensation of dimethyl terephthalate and 1,4-butanediol. PBT is highly crystalline with a melting point of 440°F. Polyethylene terephthalate (PET) is widely used as fiber and film. Polyesters can be slurried into molten asphalt.

7. Rubber-Asphalt

Rubber modification of asphalt has been known for more than 100 years. The early interest was in caulking compositions and sealants. Current interest in this system is for paving and roofing.

Almost any type of rubber can be used, including ground rubber, vulcanized tire stocks, reclaimed (crumb) rubber, natural rubber, and various synthetic elastomers, e.g., SBS. The type of asphalt, type of rubber, method of rubber blending, and processing parameters determine the properties of rubberized asphalt. Rubber modification of asphalt usually produces softening point increase, hardness increase, reduction of cold flow properties, and improvement in low-temperature brittleness.

Rubber can be added to asphalt in many ways. The easiest form of addition is as rubber latex. Latex is added to molten asphalt above the boiling point of water. Controlling foaming is essential. With a heavy-duty mixer, rubber can be directly added, although it is better to break down the rubber. Thermoplastic elastomers (TPEs) are appearing on the market now. Although they are more expensive now, the potential for obtaining improved products is good.

Synthetic and crumb rubbers are used as binders for paving. The presence of rubber improves resistance to compacting under heavy traffic, eliminates rutting, and results in less "shoving" in areas that have stop-and-start traffic. Nitrile rubberized asphalt can be developed for building paving areas with jet fuel resistance in airports. In a similar vein, a host of industrial products can be developed using asphalt as a filler.

8. Thermoplastic-Asphalt

Polypropylated asphalt finds application in commercial roofing and is described elsewhere in this book. There is a need to investigate other polyolefinics. Newer grades of polypropylenes based on single-site catalysts are appearing on the market. These resins may be investigated to modify asphalt.

9. TPO-Asphalt

TPOs are unique in that they process as thermoplastics yet produce properties similar to those of thermosets. Recently, Dow, Himont, and others have introduced TPOs that could potentially modify asphalt. At present, they are more expensive than polypropylene. Also, these TPOs have high molecular weight. For asphalt modification, the molecular weight of TPOs must be lowered substantially. Roofing companies should work with manufacturers so that several grades of TPOs suitable for asphalt modification become available. Ultimately, TPO will dominate asphalt modification.

10. In Situ Polymerized Monomers on Asphalt

There is a distinct possibility that monomers, e.g., styrene, vinyl acetate, acrylates, and diallyl phthalate, can be grafted onto asphalt by using initiators, such as benzoyl peroxide and others that produce resonance-stabilized radicals. This will expand current applications and open up newer application areas for modified asphalt.

V. SUMMARY

We have described current applications of asphalt in paving and roofing. There is a distinct possibility that the use of asphalt in current applications can be increased by further research. In addition, we expect newer application developments in the future. A brief R&D approach has been presented, and investigators are encouraged to develop newer technologies.

REFERENCES

1. A. M. Usmani, Special Report, Research Institute, University of Petroleum and Minerals, Dhahran, 1982.
2. A. M. Usmani and W. B. Gorman, American Chemical Society Meeting, Rubber Division, Philadelphia, March 1995.

3. A. M. Usmani and I. O. Salyer, *J. Sci. Ind. Res.*, **38**: 555 (1979).
4. A. M. Usmani and I. O. Salyer, *J. Appl. Polym. Sci.*, **23**: 381 (1979).
5. N. Akmal and A. M. Usmani, in *Handbook of Polyolefins* (R. B. Seymour, ed.), Springer-Verlag, New York (1993).
6. A. M. Usmani, *Cold Adhesives*, Usmani Development Company, Indianapolis (1996).
7. G. Odian, *Principles of Polymerization*, Wiley, New York, 1981.
8. J. H. Saunders and K. C. Frisch, *Polyurethane: Chemistry and Technology*, Vols. 1 and 2, Wiley, New York (1962, 1964).
9. N. Akmal and A. M. Usmani, in *Handbook of Adhesive Technology* (A. Pizzi and K. L. Mittal, eds.), Dekker, New York (1994).
10. H. F. Payne, *Organic Coating Technology*, Vol. 1, Wiley, New York (1954).
11. A. M. Usmani and I. O. Salyer, in *Use of Renewable Resources for Polymer Applications* (C. E. Carraher and L. H. Sperling, eds.), Plenum, New York (1982).
12. R. B. Seymour and C. E. Carraher, *Polymer Chemistry*, Dekker, New York (1992).

12

Amorphous Polyalphaolefins (Apaos) as Performance Improvers in Asphalt-Containing Materials

Andrés Sustic
Rexene Corporation, Odessa, Texas

I. INTRODUCTION

A. Asphalts

Asphalt is defined by the American Society for Testing and Materials (ASTM) as "a dark brown to black cementitious material in which the predominating constituents are bitumens that occur in nature or are obtained in petroleum processing" [1]. The generic name bitumen, as used in the United States, refers to mixtures of hydrocarbons derived from natural or pyrogeneous origins, or combinations of both, and which are completely soluble in carbon disulfide, i.e., no insoluble minerals.

Asphalts are classified as either natural asphalts, which include native, rock, and lake asphalts, or artificial asphalts. Artificial asphalts include petroleum asphalts, which, in contrast to native asphalts, are mostly organic with only trace amounts of inorganic materials. Petroleum asphalts derive their characteristics from the nature of their crude origins, with some variation in composition and properties made possible by choice of manufacturing process.

Determination of the components of asphalts has always presented a challenge because of the complexity of the mixture, in which many different types of chemical species are present. For this reason, the composition of asphalts is usually investigated by analytical methods that separate the components into generic groups, such as by liquid chromatography, gel permeation chromatography, or ion exchange.

The structure of asphalt is viewed as a dispersion of highly associated, high-molecular-weight, polar aromatic hydrocarbons and heteroaromatics, called *asphaltenes*, dispersed in a continuous oily phase medium called *malthenes*. The asphaltenes are not soluble in the oils but are dispersed or peptized in the medium by lower molecular weight resins found in the malthenes. The asphaltenes are high-molecular-weight aggregates (molecular mass 1000 to 4000 daltons) as a result of associations that occur by stacking of highly conjugated aromatic rings and by polar attractions of nitrogen-, sulfur-, or oxygen-containing moieties. The associations of asphaltenes do not involve covalent bonding.

Asphalts can be classified as gel or sol types. A gel asphalt contains relatively large quantities of asphaltenes (20 to 35%), which are not well peptized and which associate to form an extensive network. This network is responsible for the asphalt's non-Newtonian behavior. A sol-type asphalt contains a lower amount of well-dispersed asphaltenes (5 to 10%). Sol-type asphalts exhibit Newtonian flow [2].

A road paved with a gel-type asphalt is more resistant to high-temperature deformation (rutting) than a road paved with a sol-type asphalt, but at low temperatures it can be more brittle and less resistant to cracking.

B. Asphalt Modification with Polymeric Modifiers

Approximately 95% of the highways in the United States are surfaced with an asphalt/mineral aggregate product. Not only are these roads subjected to many different environmental distresses, they are also subjected to a wide variety of mechanical stresses due to increasing vehicular axle loads and tire pressures. Building and maintaining these roads at a reasonable cost are further complicated because asphalt is produced and distributed from worldwide crude oil sources, resulting in a greater variability of the asphalt properties.

A desirable asphalt mixture is one that is strong and durable, yet flexible. It should resist permanent deformation at high temperatures and thermal cracking at low temperatures. In order to reduce the potential for permanent deformation and thermal cracking, mixture tensile strength and flexibility must be simultaneously increased.

Continuing improvement in oil refining technologies and the energy crisis of the early to mid-1970s caused a change in the quality and grade of the oil being refined and in the extent and efficiency of the refining operations. The resultant residual bitumen or "bottoms," more devoid of the lower molecular weight oils, which have a plasticizing effect on bitumen, became harder and less flexible.

One way to achieve a strong and flexible asphalt mixture is to use asphalt additives such as polymers. Polymer-modified asphalt cements increase the level of field performance of asphalt concrete. Polymer modifiers improve thermal cracking resistance [3], provide resistance to permanent deformation [4], and improve resistance to moisture damage [5]. Furthermore, previous research on polymer additives to asphalt cement revealed that polymers decrease the temperature susceptibility of asphalt cements [6] and increase the tensile strength of asphalt.

Extensive research has been conducted on asphalt additives. For example, Terrel and Walter [7] have summarized various studies and reports on the performance of paving grade asphalt cement modified by rubbers and plastics, compared with the performance of unmodified asphalt cements. This research showed that rubber additives, which include natural and synthetic latex, styrene-butadiene rubber (SBR), etc., increase the mixture's cohesion, lower the temperature susceptibility, and improve aggregate retention along with increasing elasticity of the modified asphalt cement. Similarly, asphalt cements modified with either polyethylene, polypropylene, poly(ethylene-co-vinyl acetate), or ethylene-propylene rubbers show increased stability and stiffness modulus, increased resistance to permanent deformation (rutting), and improved lower temperature flexibility. In general, polymer additives improve the base asphalt characteristics at both low and elevated temperatures.

For certain applications such as modified bitumen roofing or pavement modification, the use of polyolefinic modifiers has become very common because the reinforcing effect of the modifiers results in improved physical and mechanical properties of the final product. Typically, in pavement modification, the quantity of polyolefinic modifier is limited to less than 7 to 8 wt %, whereas in modified bitumen roofing, 15 to 25 wt % of modifier is added to the bitumen.

An important property of polyolefins is their chemical inertness. Unsaturated elastomers have one great disadvantage. After hot mixing, unsaturation may lead to severe oxidation, which can negatively affect the mixture's physical properties.

Button, Little, and colleagues [2, 3, 8] evaluated several different asphalt additives to increase pavement flexibility. Their results showed that each additive lowered the temperature susceptibility of the mixtures but the influence of the additives depended on the physical properties and the chemical composition of the asphalt cement. More generally, when a given quantity and type of polymer was added to an asphalt product, the resulting physical properties of the product varied with asphalt source. These authors found that APAOs, because of their lower molecular weights, had less effect than styrene-butadiene-styrene (SBS), polyethylene (PE), and styrene-butadiene rubber (SBR) on viscosity at 135°C (275°F). This result is a positive finding because it indicates that, by comparison with other polymer-modified asphalts, lower temperatures may be used to bring the APAO-modified asphalts to the appropriate mixing and compaction viscosities.

Since most polymers, including APAOs, are not really soluble or miscible in asphalt, compatibility is a difficult term used to describe the ability of a polymer to be incorporated into an asphalt and remain in suspension during static, hot storage. A polymer that is not compatible with asphalt will physically separate into lumps. Polymers containing aromatic moieties, such as SBR or SBS, will be more "compatible" with asphalts of higher aromatic content.

Polyolefins have low Hildebrand solubility parameter (an expression widely used for correlating polymer-solvent and polymer-polymer interactions) values because of their highly saturated aliphatic nature [9]. Therefore, it is unlikely that they can absorb significant amounts of oils from the asphalt and they tend to be at best slightly soluble in asphalt, even at high temperatures. Given that the chemical composition of the asphalt is significantly polar, the asphalt/polyolefin blends will tend to segregate into two separate phases, driven by the difference in density between the dispersed and continuous phases. Evidence from several tests indicates that APAO, when dispersed in asphalt, exists as microscopic but discrete particles that will segregate during static, hot storage.

There are a few remedies to this problem. One of them is to keep the size of the dispersed particles small enough so that Brownian motion compensates for the density difference. This can be accomplished by proper agitation. Another approach is to increase the viscosity of the continuous phase and also of the dispersed phase. This can be accomplished by adding small quantities of a high-molecular-weight polyolefin like polyethylene or polypropylene and/or a thickening inorganic material like kaolin, talc, or calcium carbonate [10].

C. Coordination Polymerization

As described in the patent literature, first-generation Ziegler-Natta catalysts are formed by allowing a metal alkyl such as triethylaluminum, the cocatalyst, to react under an inert atmosphere with a transition metal salt, such as titanium trichloride, the procatalyst [11]. Organic and/or inorganic compounds are added to modify some aspects of the polymerization. For example, hydrogen is added to control molecular weight. These first-generation catalysts have a low productivity, i.e., kilograms of polymer produced per gram of catalyst, and also have a moderate degree of stereospecificity, typically producing 10–15 wt % of atactic polypropylene (aPP) as a by-product of the polymerization reaction.

For the so-called solution polymerization processes using an organic solvent, the atactic fraction remains in solution while the isotactic, highly crystalline fraction is recovered by filtration. In other processes, aPP is removed from the isotactic fraction by extraction at high temperature with hydrocarbon solvents. Either way, large quantities of aPP were disposed of by burying in pits. In the late 1950s and early 1960s, roofing technologists in Europe learned that atactic polypropylene provided an inexpensive adhesive for restoring and even improving the lost adhesive qualities of bitumen. However important technological changes were to affect the ready availability of by-product aPP.

Supporting the active species, i.e., the transition metal salt, on inorganic substrates such as silica or magnesium dichloride, and adding some specific internal electron donors during the procatalyst's manufacture and external donors during the catalyst's activation, led to the development of newer generation catalysts, the so-called third- and post-third-generation Ziegler-Natta catalysts, which are highly active and stereospecific [12]. Higher catalyst activity and stereospecificity mean that the concentration of catalyst residues and the amount of atactic polymer produced are substantially reduced. Therefore, the polymer product generally does not require additional purification to remove the atactic or low crystalline fraction.

As existing and new commercial plants increasingly use these high-activity catalysts, the aPP supply from isotactic polypropylene (iPP) manufacture using first-generation Ziegler-Natta catalysts steadily declines. Very little by-product aPP is produced with the use of these new catalysts in iPP manufacture.

D. Amorphous Polyalphaolefins

The recovered aPP product exhibits inconsistent and inferior lot-to-lot reproducibility in such properties as melt viscosity and needle penetration and, more often than not, it is contaminated with other waste products, which requires a product purification step.

In addition, the limited supply of aPP is being exhausted by its use in some products made using old technology and/or equipment. In the early 1980s, a few companies successfully developed processes that produce in a pressurized reactor, on-purpose amorphous polymer products with tight product specifications that can be used directly in hot melt applications, in bitumen formulations for roofing membranes, in pavement modification, in reinforced tapes, etc.

To distinguish the atactic polypropylene by-product from the amorphous polymers made on purpose, these are conventionally called amorphous polyalphaolefins (APAOs).

The APAOs encompass a group of usually low-molecular-weight polyolefins—number average molecular weights from about 7000 to about 20,000.

There are several distinct product types of *on-purpose, reactor-made* amorphous polyalphaolefins:

1. Homopolymers of propylene
2. Copolymers of propylene and ethylene
3. Copolymers of propylene and 1-butene

Also included are:

4. Copolymers of propylene and 1-hexene
5. Terpolymers of ethylene, propylene, and 1-butene

The versatility of a controlled polymerization process is such that by varying the concentrations of comonomer in the reactor, the polymer producer may obtain products with a broad range of properties that depend on the polymer's composition [13]. For example, higher ethylene concentrations result in increasing values in the needle penetration; i.e., the APAOs become softer with higher ethylene concentrations, discussed later.

Proper design of the process of synthesizing the amorphous polyolefin, i.e., choice of catalyst system, of comonomer(s), and of polymerization reaction conditions, results in products that have well-defined physical and mechanical properties such as melt viscosity, softening point, needle penetration, embrittlement temperature, and open time [14].

Given the wide range of physical properties of asphalt, a polyolefinic modifier with controlled and reproducible properties is valuable because it makes the formulation of an asphalt blend with reproducible properties more predictable.

II. APAO CHARACTERIZATION METHODS

The characterization of the polymers discussed in this chapter requires the use of techniques that are widely used in the hot melt adhesives (HMA) industry, in paper/tape lamination or roofing and pavement modification, and that are carried out following standard test methods set forth by the American Society for Testing and Materials (ASTM). Test methods include melt viscosity (ASTM D-3236), needle penetration (ASTM D-1321), and ring and ball softening point (ASTM E-28). Other test methods determine thermal properties such as heats of fusion and crystallization (ASTM D-3417), and glass transition temperature (ASTM D-3418), as well as molecular weight by gel permeation chromatography (ASTM D-3593), tensile properties (ASTM D-638), and rheological properties by dynamical mechanical analysis (ASTM D-4440). These test methods also describe the type of instrument to be used for carrying out the tests.

In the following paragraphs, we will discuss the significance of each one of those characterization methods.

A. Melt Viscosity

The melt viscosity test measures a liquid or molten polymer's internal friction—its resistance to flow under an applied strain. It is a most distinctive property because it determines the flowability and degree of wetting or penetration of a substrate by the molten polymer; it gives an indication of its processability.

Melt viscosity is a measure of a polymer's molecular weight, which is controlled in process by the precise addition of hydrogen as modifier. The higher the concentration of hydrogen in the reactor, the lower the molecular weight and the melt viscosity value of the polymer. The catalyst's response to hydrogen allows the manufacture of products with a wide range of tightly controlled melt viscosity values, from less than 100 cps to as high as 300,000 cps.

Conventionally, the MV is measured and reported at 190°C (375°F) in units of centipoise, cps, or millipascal × seconds.

The melt viscosity value of the APAO depends on the measurement temperature. At different temperatures, the MV values are different, increasing as the temperature decreases, as seen in Figure 1.

B. Needle Penetration

With thermoplastics and elastomers, hardness is often used as a simple measure of stiffness (or softness). Needle penetration (NP) is determined by measuring the depth to which a weighted needle penetrates the polymer over an established period of time, and the value measured indicates the resistance of the polymer to deformation by penetration. In other words, the penetration is the distance, in tenths of a millimeter (dmm), that the needle penetrates vertically into the sample under specific conditions of temperature, load, and time, in seconds. The standard test for APAO is run at 25°C (77°F) with a 100-g needle load for 5 s.

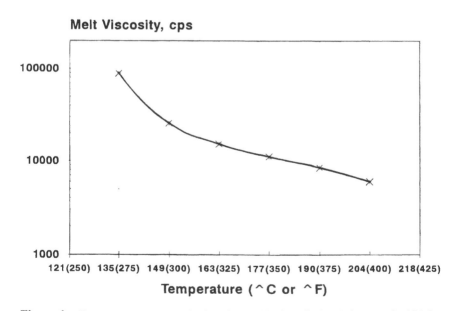

Figure 1   Temperature versus melt viscosity profile for a high ethylene grade APAO.

Copolymer composition determines the softness of an APAO. Increasing concentrations of ethylene in the copolymers result in softer APAOs with increasing needle penetration values, as seen in Figure 2.

C. Ring and Ball Softening Point

Because of the predominantly amorphous nature of these polyolefins, melting does not take place at a sharp temperature but instead these gradually change from very viscous and stiff to softer and less viscous materials as the temperature increases. The ring and ball softening point (RBSP) test measures the precise temperature at which a disc of polymer becomes soft enough to allow a test object such as a steel ball to flow through it. The softening point of a polymer is important because it is a measure of the polymer's heat resistance, application temperature, and solidification point. RBSP and glass transition temperature define the temperature range in which the APAO can be used without softening or becoming brittle.

As is the case for needle penetration, the softening point is determined primarily by the copolymer composition. In particular, for the ethylene copolymers, increasing the comonomer concentration decreases the softening point as evidenced in Figure 2. The APAO homopolymer shows a high RBSP value of

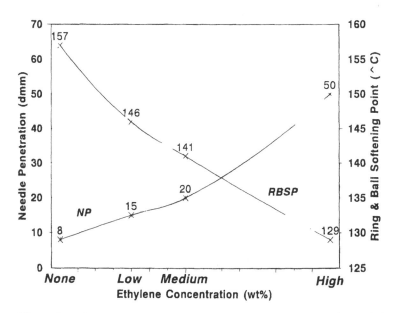

Figure 2   Effect of ethylene concentration on the needle penetration and the ring and ball softening point of APAO.

about 157°C (315°F). As ethylene concentration increases in the polymer, the softening point decreases, with a high-ethylene APAO having a lower RBSP of about 130°C (266°F). For propylene/1-butene copolymers, the RBSP can be as low as 93°C (200°F).

D. Thermal Analysis

Thermal analysis of polyolefins is carried out by differential scanning calorimetry. DSC gives important information on characteristics such as heats of fusion and crystallization, which provide a measure of the degree of crystallinity, or the crystalline content, of a polymer. DSC also allows the determination of the melting point, of the crystalline fraction, which in the case of a polymer is the point at which the rate of absorption of heat reaches a maximum, and of the crystallization point.

The APAOs are predominantly amorphous polymers with some having residual crystallinity, as evidenced from heat of fusion measurements. The homopolymer APAO has the highest crystalline content (which explains why it is the hardest APAO, with the lowest NP values). On the other hand, APAOs with a high ethylene or 1-butene content have little or no measurable crystallinity and therefore are the softest products. The heat of fusion values for APAOs may vary from about 20 J/g for the homopolymers to very low or nonexistent for the propylene/1-butene APAO. In general, as comonomer content in the polymer increases, the crystallinity decreases.

E. Glass Transition Temperature (Low-Temperature Flexibility)

Glass transition temperature. T_g, can be described as the temperature below which there is a virtual cessation of local molecular motion, or alternatively, the temperature below which the polymer is glassy and above which it is rubbery [15].

Most polyolefins possess low glass transition temperatures. This allows the material to endure strains at temperatures well below those at which a typical bitumen is brittle. Polymers with a low glass transition temperature are most effective in increasing the low-temperature fracture toughness of modified bitumen. This is an important property when considering an application for the modified bitumen. For instance, low-temperature flexibility is required in a single-ply roofing membrane or in pavement modification for use in a cold weather climate. Low-temperature flexibility is a characteristic of APAOs.

Several methods of instrumental analysis are used to determine the glass transition of a polymeric material: mechanical/rheological, calorimetric, dilatometric, and dielectric methods are the most common. The calorimetric method has wide acceptance; the instrument routinely used is the differential scanning calorimeter.

Low-temperature flexibility is correlated to the DSC-measured glass transition temperature. As with other physical properties of APAOs, comonomer concentration determines the T_g. The higher the ethylene concentration in the copolymer, the lower the T_g. A higher ethylene content in the copolymer allows greater mobility of the polymer chains because the mobility is determined primarily by the barrier to rotation around backbone carbon-carbon bonds. Replacing a methyl group by a hydrogen atom decreases the barrier to rotation, therefore lowering the T_g, as seen in Figure 3 for different types of APAO. For a 1-butene copolymer, there is again a higher energy barrier to rotation but the longer alkyl substituent (by one methylene unit) actually plasticizes the copolymer [16], with the result that the 1-butene copolymer has a lower T_g than the homopolymer.

F. Molecular Weight and Molecular Weight Distribution

Gel permeation chromatography (GPC) is a separation method ideally suited for polymers. It is one of the most powerful and versatile analytical techniques available for understanding and predicting polymer properties and performance, and it is the only single technique for characterizing the molecular weight distribution (MWD) of a polymer [15].

For a 400 cps melt viscosity APAO, the number average molecular weight, $\overline{M}_n$, and the weight average molecular weight, $\overline{M}_w$, are about 7,000 and 35,000,

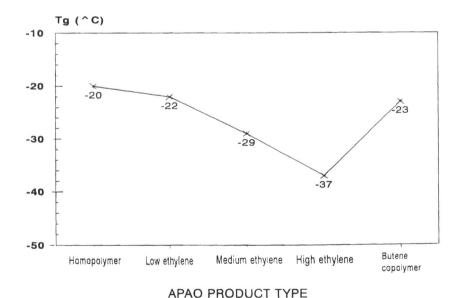

APAO PRODUCT TYPE

Figure 3   Glass transition temperatures for APAOs determined by DSC.

respectively, whereas for an 8000 cps APAO, they are about 14,000 and 86,000, respectively. For all Apaos, the molecular weight distribution, calculated from the $\overline{M}_w/\overline{M}_n$ ratio, is rather narrow, varying from about 5 to about 6.2.

G. Tensile Strength

The determination of the tensile strength of a polymer is an important way to measure its cohesive strength. Figure 4 shows the tensile strength values for various APAOs. The homopolymer has the highest tensile strength value. As the ethylene content in the copolymers increases, the degree of crystallinity and, to a lesser extent, the degree of chain entanglement decreases and, as a whole, the tensile strength decreases. This correlates well with increasing softness (as measured by softening point and needle penetration). Within a homologous series, molecular weight has a slight effect on tensile strength, as chain entanglements as well as intrachain interactions increase with increasing molecular weight. The propylene/1-butene copolymer has elongation values > 700%.

H. Rheology

Polymeric materials are inherently viscoelastic, with the simultaneous existence of viscous and elastic properties. Thus, a great deal of information can be obtained from studying both the viscous (or loss) modulus, G'', and the elastic (or storage) modulus, G', responses to deformation [17]. These responses can be measured

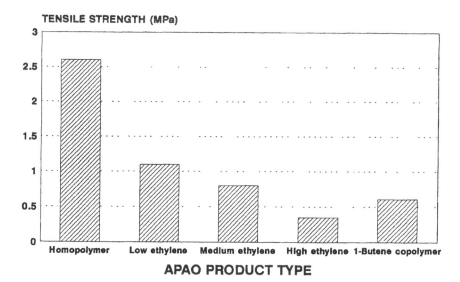

Figure 4 Tensile strength for APAOs.

by dynamic mechanical analysis (DMA) in oscillation. Dynamic oscillation over a frequency range is used to determine the storage modulus as a measure of elasticity (or the energy stored by a polymer). In addition, the glass transition temperature can easily be determined by the peak in tan σ (the ratio between G'' and G') during a temperature sweep at constant frequency.

For a medium ethylene grade APAO, Figure 5 shows a typical curve obtained in dynamic oscillation at a constant temperature of 25°C. The complex viscosity (η^*), a mathematical representation of viscosity, shows a steady decline with frequency, consistent with a narrow molecular weight distribution for APAO. The decrease in viscosity as frequency is increased from 0.01 to 10 Hz indicates that the polymer's viscosity is shear sensitive, a typical characteristic of non-Newtonian fluids. The storage modulus G' shows an increase with frequency. McLaughlin and Latham [18] reported that G' reaches a plateau at higher frequencies for higher molecular weight aPPs. However, no such behavior is noted for the low-molecular-weight APAO.

Not surprisingly, the storage modulus depends on the polymer's structure. The homopolymer shows the highest G' value of all APAOs. As ethylene is incorporated into the chain backbone, the polymers become less stiff, with the high-ethylene copolymers being the softest products. The trend is reversed with the 1-butene copolymer, because of more efficient chain entanglements compared

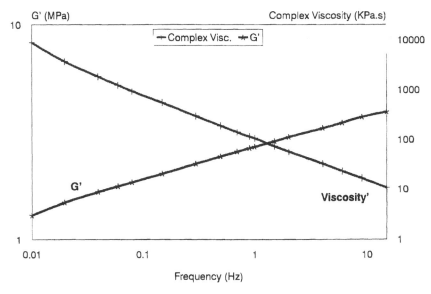

Figure 5 Storage modulus and complex viscosity versus frequency rheogram determined at 25°C for a medium ethylene grade APAO.

with the ethylene copolymer. This trend is observed in Figure 6. The G' was determined at room temperature at an oscillation frequency of 1 Hz.

III. INTRA-APAO BLENDS

Because of the limited number of APAO product types that can be produced in any on-purpose process, a formulator might need specific properties of melt viscosity, needle penetration, or softening point, not available with APAO as a neat product [19]. Intra-APAO blending allows the formulation of blends with a continuum of properties that lie in between the properties of the individual components.

The most commonly used method for establishing *miscibility* (like dissolves like) in polymer-polymer blends is the determination of the glass transition temperatures of the blends versus those of the pure, unblended components. A miscible polymer blend will exhibit a single glass transition temperature value between the T_g values of the pure components [20]. On the other hand, in a compatible blend, the glass transition temperatures of the individual components are present but there is no observable gross phase separation in the blend.

As an illustration of the mutual miscibility of the APAOs, T_g values were determined by DSC for blends of a homopolymer, with a glass transition tempera-

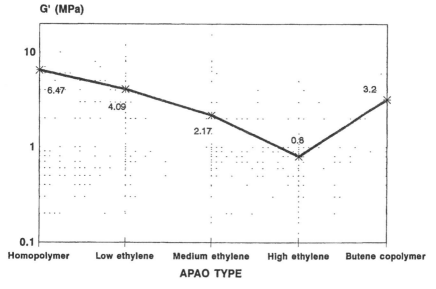

Figure 6 Storage modulus values for homopolymer and for ethylene and 1-butene APAOs.

ture of −20°C, and a high-ethylene copolymer, with a T_g of −37°C. The thermograms of each blend showed only a single T_g that was composition dependent, as shown in Figure 7. These blends were chosen because the difference in the T_g values between the pure components is quite wide, about 17°C. This is important because the glass transitions appear as somewhat broadened inflection points. The larger the difference in the T_g values, the better the resolution. The results indicate that the blends of homopolymer APAO with the high-ethylene APAO are miscible in all proportions. This should be expected given that both components are hydrocarbon polymers for which the difference in the solubility parameters, δ, is less than about 1.0 $(J/cm^3)^{1/2}$ [15]. In general, the glass transitions of intra-APAO blends determined by DSC show single T_g transitions that are composition dependent and indicate their mutual miscibility [20].

When blending is done by mixing a low-needle-penetration homopolymer with a high-ethylene, very soft copolymer, the initial addition of the homopolymer has a significant effect on the needle penetration value of the blend, which decreases rapidly in a semilogarithmic trend until it reaches the low NP value characteristic of the pure homopolymer component.

Similarly, the softening point of the blend increases very rapidly with the addition of small quantities of the homopolymer, reaching a plateau at a blend

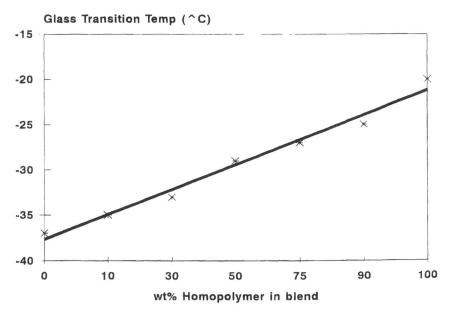

Figure 7 Composition versus glass transition temperature for a high ethylene grade/ homopolymer APAO blend.

composition of about 50:50. Figure 8 shows plots for both the NP and the RBSP as a function of the blend compositions. Again, addition of small quantities of homopolymer causes large initial changes in both the NP and the softening point. On the other hand, the melt viscosity experiences a more linear change with concentration of modifier. Similar behavior has been observed for several other intra-APAO blends [19].

There are obvious benefits of intra-APAO blending when a product with specific characteristics and properties is not available. The desired composition can be obtained directly from the plots of composition versus property. This broadening of the properties can be successfully achieved because, as noted previously, reactor-made APAO can be produced with tightly controlled physical properties.

IV. BLENDS OF APAO AND STYRENE BLOCK COPOLYMERS

Figure 9 corresponds to the storage modulus scans, obtained in dynamic oscillation at 1 Hz as a function of temperature, for two blends of a homopolymer APAO with a styrene-ethylene-butylene-styrene styrene block copolymer (SEBS SBC) and for the neat blend components. The scans were terminated at about room

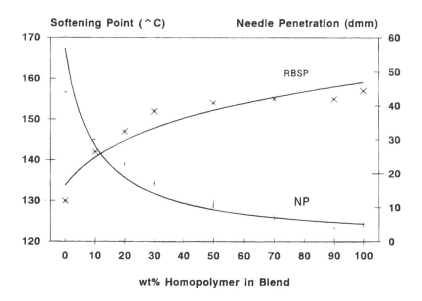

Figure 8   Effect of homopolymer content on the ring and ball softening point and needle penetration of a high ethylene grade/homopolymer APAO blend.

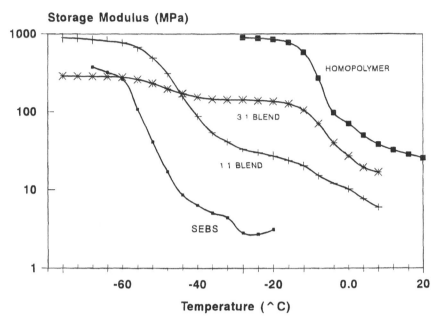

Figure 9 Storage moduli and tan Delta for a homopolymer/SEBS 1:1 blend and for SEBS.

temperature. In the styrene-based block copolymers, past the T_g of the elastomer fraction (EB), G' reaches a plateau. As the temperature continues to increase, the glass transition of the polystyrene phase is reached (about 105°C), past which G' declines further (not shown). The polystyrene phase provides the rigidity above the glass transition temperature of the elastomer fraction. For the two homopolymer APAO/SEBS blends, G' past the glass transition of SEBS (about −45°C) reaches a plateau that is delimited by the T_g of the ethylene-butylene midblock component of the SEBS and the APAO. As shown in the figure in this type of blend, it is the APAO that provides rigidity above the elastomer midblock's glass transition temperature. This rigidity, in terms of G'_1 is higher than what is found for SEBS. Past the T_g of the APAO (about −10°C), G' experiences a steady decline [21].

As shown in Figure 10, the determination of the T_g of blends of APAO with SEBS indicates that the components are not miscible because the glass transition scans of the blends show the glass transition temperatures of the blends' components. Even though the optical appearance of the blends was transparent, the presence of the two well-defined, unshifted glass transition temperatures of the components indicates that the components of the blends are immiscible. A

tan delta

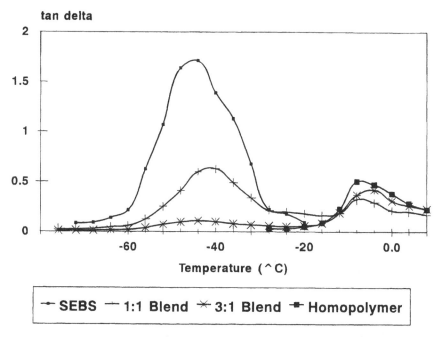

Temperature (^C)

⎯ SEBS ⎯ 1:1 Blend ✳ 3:1 Blend ⎯ Homopolymer

Figure 10 Tan delta for SEBS, homopolymer APAO, and homopolymer/SEBS = 3:1 and 1:1 blends.

similar conclusion can be drawn about the blends of APAO with the styrene-isoprene-styrene (SIS) rubber [21]. These findings do not necessarily imply that the blend's components are incompatible, as judged by the fact that they are homogeneous looking, with no evidence of phase separation.

An advantage of modifying asphalt with blends of APAOs with styrene block copolymers is that the APAO/SBC blends will show improved heat resistance, particularly when using a blend formulated with homopolymer APAO, as the homopolymer shows a substantially high ring and ball softening point of about 150°C. In addition, at temperatures well below 0°C, the APAO provides rigidity above the T_g of the SEBS's midblock, particularly when using homopolymer APAO.

V. APAO-MODIFIED ASPHALT

A comprehensive binder study was carried out to quantify the effects of APAO on three different types of asphalts, two sol types and a gel type. The study was designed to determine the comparative physical character of the modified and unmodified asphalts [22]. Four APAOs of different compositions and melt viscosities were used as asphalt modifiers.

The following paragraphs describe and discuss the most relevant results obtained from the study:

Normally, cracking resistance is sacrificed when rutting resistance is improved and vice versa. Based on the findings that needle penetration at 4°C (39.2°F) is affected only slightly by incorporation of the APAO, which may indicate increased resistance to pavement cracking at low temperatures (both components are hard at this low temperature but APAO has a lower glass transition temperature), and increased viscosity at 60°C (140°F) may indicate increased resistance to pavement rutting at high service temperature, APAOs may improve resistance to rutting *and* cracking simultaneously.

APAO products have less effect than SBS, PE, and SBR on viscosity at 135°C (275°F); this result is a positive finding because it indicates that, by comparison with other polymer-modified asphalts, lower temperatures may be used to bring the APAO-modified binders to the appropriate mixing and compaction viscosities and thus less fuel may be required. APAO polymers are more easily blended with asphalt than polyethylene or neat crumb rubbers. APAOs are quite soft and even fluid at temperatures around 135°C (275°F), which facilitates blending with relatively low energy requirements.

APAOs, like most commercially available polymers, lower the temperature susceptibility of asphalt. They increase the viscosity of asphalt at mixing, compaction, and high pavement service temperatures and have little effect on asphalt consistency at low pavement service temperatures. This result is indicative of increased resistance to rutting in the summer without loss of resistance to cracking in the winter.

APAOs may permit the use of softer than usual asphalts in pavements to reduce cracking in cold weather without diminishing resistance to rutting in hot weather.

We also found that 8% APAO in a given asphalt affects the viscosity at 60°C (140°F) and penetration at 25°C (77°F) about as much as 5% of a higher molecular weight SBS or PE and has significantly less effect than 5% SBR. APAOs also have less effect than SBS, PE, and SBR on viscosity at 135°C (275°F).

In conclusion, APAO as a low-molecular-weight saturated aliphatic polymer lends distinct advantages when blended with asphalt, compared with other higher molecular weight polyolefinic and rubber-type polymers. It enhances the low-temperature flexibility as well as the high-temperature resistance of the asphalt cement, lowering its temperature susceptibility. Because of its low molecular weight and range of softening points, under proper blending conditions, lower temperatures and less energy will be required to achieve a well-dispersed system.

Finally, the advantage of a product tailor made in a polymerization reactor is that the resulting products have well-defined, predictable physical and mechanical properties.

REFERENCES

1. Terms Relating to Materials for Roads and Pavements, ASTM Designation D8-91.
2. E. J. Barth, in *Asphalt Science and Technology*, Gordon & Breach, New York, chapter 2 (1962).
3. D. N. Little, J. W. Button, Y. Kim, and S. J. Ahmed, *Proc. Assoc. Asphalt Paving Technol.*, **55**: (1986).
4. D. N. Little, J. W. Button, R. M. White, E. K. Ensley, Y. Kim, and S. J. Ahmed, Report. No. FHWA/RD-87/001, November 1986.
5. D. W. Gilmore, J. B. Darland, L. M. Girdler, L. W. Wilson, and J. A. Scherocman, American Society for Testing and Materials, Special Technical Publication 899, 1985.
6. J. H. Collins and W. J. Mikols, *Proc. Assoc. Asphalt Paving Technol.*, **54**: (1985).
7. R. L. Terrel and J. L. Walter, *Proc. Assoc. Asphalt Paving Technol.*, **55**: (1986).
8. J. W. Button, Research Report 187-18, Texas Transportation Institute, Texas A & M University, January 1991.
9. D. W. Van Krevelen, *Properties of Polymers*, Elsevier, Amsterdam, chapter 7, (1976).
10. S. A. Hesp and R. T. Woodhams, in *Polymer Modified Asphalt Binders* (K. R. Wardlaw and S. Shuler, eds.), ASTM Publication STP 1108, p. 7 (1992).
11. J. Boor, *Ziegler-Natta Catalysts and Polymerizations*, p. 1, Academic Press, New York (1979).
12. J. C. W. Chien, in *Transition Metal Catalyzed Polymerizations*, (Roderic Quirk, ed.), Cambridge University Press, Cambridge, p. 55 (1988).
13. A. Sustic and B. Pellon, *TAPPI Hot Melt Symposium*, p. 193, 1991.
14. A. Sustic and B. Pellon, *J. Adh. Seal. Council*, **20**: 41 (1991).
15. D. W. Van Krevelen, *Properties of Polymers*, Elsevier, Amsterdam, chapter 6, (1976).
16. A. Eisenberg, *Physical Properties of Polymers*, American Chemical Society, Washington, DC, p. 55 (1984).
17. H. A. Barnes, J. F. Hutton, and K. Walters, *An Introduction to Rheology*, Elsevier, Amsterdam, chapter 1 (1989).
18. K. W. McLaughlin and D. D. Latham, *Polym. Prepr. ACS Div. Polym. Chem.*, **29**(2), 406 (1988).
19. A. Sustic and B. Pellon, *Adhesives Age*, **34**(12), p. 17 (1991).
20. O. Olabisi, L. M. Robeson, and M. T. Shaw, *Polymer-Polymer Miscibility*, Academic Press, New York (1979).
21. A. Sustic and V. Krishnamurthy, in *TAPPI Hot Melt Symposium*, p. 77 (1992).
22. M. A. R. Rodriguez, J. W. Button, and D. N. Little, Report RF7139-1F, Texas Transportation Institute, September 1991.

13

Chemistry and Technology of SBS Modifiers

Elio Diani, Mauro Da Via, and Maria Grazia Cavaliere
EniChem, Milan, Italy

I. INTRODUCTION

Even though bitumen has been in use since ancient times, its easy availability, low cost, and unique properties make it a widely used material in the modern, high-technology world.

Today, the use of bitumen is well established in the waterproofing industry and in the road construction sector. Waterproofing is characterized by a continuous improvement in terms of performance and durability, and at the same time the modern heavy traffic requires roads with enhanced strength and stability.

Bitumen itself is viscoelastic in nature and suffers from some important limitations due to the high sensitivity to temperature of its properties; for these reasons it would hardly match the novel application requirements if not properly modified.

In this respect polymeric materials seem to be the most promising bitumen modifiers for obtaining improved technological properties. First attempts to improve the behavior of bitumen by the addition of small quantities of polymers were not completely satisfactory for highly demanding applications because of processing problems and insufficient property improvement.

Recently, the introduction of styrene-butadiene-styrene (SBS) thermoplastic rubber has provided a sound method for improving bitumen properties. In fact, the thermoplastic nature of SBS copolymers makes the mixing with bitumen extremely easy and effective, and the viscoelastic behavior of bitumen versus temperature can be modified to a great extent and carefully tuned by proper selection of the type and quantity of thermoplastic rubber to match practical requirements.

II. BITUMEN MODIFICATION

A. Theoretical Background

Bitumen can be described as a colloidal substance, in which the dispersed phase, consisting of asphaltenes, is covered by a protective layer of polar resins; this complex, called a micelle, is dispersed in a continuous maltenic phase that consists of a mixture of aromatic and saturate oils.

Depending on the equilibrium between components, colloids can be classified as sol (mainly liquid consistency) or as gel (mainly solid consistency).

In a bitumen, the thermodynamic equilibrium between components is strongly affected by temperature, and for this reason the sol, gel, or intermediate condition can be observed in the same bitumen simply by varying the temperature.

As a consequence, bitumen shows a complex viscoelastic behavior resulting in a combination of all the elemental rheological bodies:

Hooke solid (elastic response)
Maxwell liquid (creep phenomena)
Kelvin solid (delayed elasticity)
Prandtl solid (plastic deformation)

In spite of this complex behavior, standard tests currently in use for bitumen characterization such as ring and ball softening point, needle penetration, and Fraass point are very simple; therefore it is not surprising that they are largely inadequate for a satisfactory and useful description of the rheological behavior of this material.

However, the results obtained by the standard tests can be properly arranged in the well-known Heukelom's abacus [1], which furnishes a simplified but impressive description of the relationship between material consistency and temperature.

According to Bingham [2], consistency is the property of a material that opposes a permanent change of its shape, and it can be defined by the complete flow-force relation.

Figure 1 shows schematically the consistency versus temperature for some models of bituminous materials. The temperature range has been divided into two parts, taking into account service and processing conditions, respectively. The general shape of the consistency curve of an "ideal" material is shown by the dashed line; consistency is constant over the whole service range and suddenly drops in the range of processing temperatures.

The behavior of a real, neat bitumen is quite different; its consistency decreases almost linearly when the temperature increases. Such a material is easy

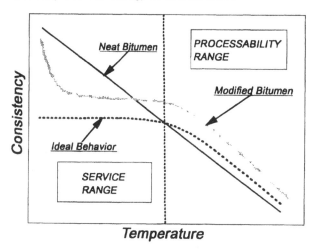

Figure 1   Schematical rheological behavior of bituminous materials.

to process but shows too high a thermal sensitivity at service conditions. But for a polymer-modified bitumen, the consistency-temperature curve closely approaches the ideal material behavior, combining satisfactory processability with quite constant performance at service conditions.

In fact, the rheological behavior of the modified bitumen is completely different from that of the neat bitumen and closely reflects the behavior of the material used for the modification. In this connection, thermoplastic elastomers seem to be the preferred bitumen modifiers for the end use requirements, resulting in a rubber-like behavior in most applications.

Thermoplastic rubbers are block copolymers, generally of the $(SB)_nX$ type, where S represents the polystyrene block, B the polybutadiene (or polyisoprene) block, and X the coupling agent. The result can be a radial- or linear-type macromolecule with terminal polystyrene blocks bound to the central polydiene blocks.

Because of the mutual incompatibility of polydiene and polystyrene, a separation takes place between the "soft" (polydiene) and the "hard" (polystyrene) blocks, resulting in a morphological arrangement in which the polystyrene domains are dispersed in a continuous elastomeric matrix. Such a structure is the reason for the peculiar behavior of these copolymers, which combine processing characteristics typical of thermoplastics with mechanical properties similar to those of a filled and vulcanized elastomer.

A composite material with tailored performance, based on bitumen and SBS thermoplastic rubber, can be obtained provided the blending is properly carried out in terms of both the thermodynamic and the kinetic aspects [3]. The thermodynamic aspect is related to the type and quantity of polymer and bitumen to be mixed, and the kinetic aspect depends on the mixing conditions adopted; they are, of course, closely interdependent.

Interaction between bitumen and SBS thermoplastic rubber can be explained by taking into account (a) the nature of the bitumen, consisting, as previously mentioned, of a mix of hydrocarbons with a very wide chemical composition, and (b) the two-phase character of the thermoplastic rubber.

Although they are the minor component (at most 10–15%), SBS copolymers are able to build a continuous phase: the polybutadiene midsegments absorb a great portion of the malthene fraction of bitumen by swelling, while the styrenic end blocks show poor compatibility with bitumen and segregate in hard-phase domains, acting as physical cross-links for the soft polybutadiene segments. The resulting morphology gives the composite a pseudorubbery plateau in the viscoelastic curve, resembling the behavior of a highly swollen rubber network. The elastic properties are enhanced in the range of temperature relevant for the application, while the ease of processing is not affected, because in the temperature range involved in manufacturing SBS copolymers show their thermoplastic behavior.

The continuous elastomeric phase can be easily identified by examining the mixtures by fluorescence microscopy; very effective phase contrast is obtainable as the SBS copolymer emits a yellow-green fluorescence while the neat bitumen appears dark [4].

Figure 2 shows a microphotograph of a 100/13 bitumen/SBS blend; it was obtained with a Leitz Laborlux 12 instrument at ×400 magnification.

Considering that the SBS weight fraction in the blend is extremely low, it can be reasonably assumed that the continuous phase is composed of the thermoplastic elastomer swollen by the miscible bitumen fractions. Moreover, indirect proof of the extent of bitumen/SBS interaction is given by the variation of bitumen viscoelastic behavior, analyzed by dynamic mechanical spectrometry.

Figure 3 shows curves of the storage modulus (G') versus temperature for both the neat bitumen and a 100/13 bitumen/SBS blend.

Modification with SBS copolymer results in a remarkable improvement in flexibility in the low-temperature range; on the contrary, at high temperatures the neat bitumen begins to flow as a viscous liquid whereas the modified bitumen still retains an appreciable elastic component. This proves the primary role of the continuous elastomeric matrix, whose behavior resembles that of a highly swollen rubber network.

The main effect of the modification with SBS copolymers is a dramatic reduction of the bitumen thermal sensitivity, which is the main technological drawback of neat bitumen.

The results plotted in Figure 3 were obtained by dynamic mechanical analysis. As previously mentioned, the building of a continuous rubbery matrix

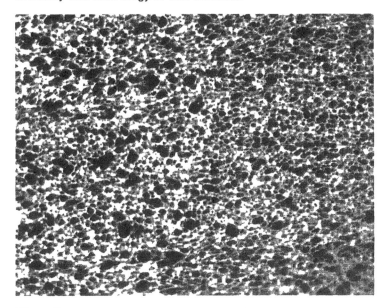

Figure 2 Fluorescence photomicrograph (×400) of a 100/13 bitumen/SBS blend.

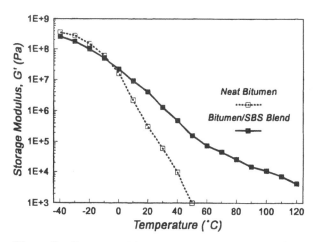

Figure 3 Storage modulus versus temperature for neat bitumen and a 100/13 bitumen/SBS blend.

greatly enhances the bitumen elasticity; for this reason, the storage modulus was chosen as an effective indicator of viscoelastic behavior.

B. Industrial Mixing

Today, the industrial mixing of bitumen with thermoplastic rubber is done using a wide variety of machines, including very slowly rotating paddle mixers as well as very high shear mixers. The mixing time can vary from 1 to 12 h or more and high-shear mixers are the preferred choice in terms of effectiveness.

A block diagram of a plant for the production of polymer-modified bitumen is shown in Figure 4. Hot bitumen and polymer, in the form of powder or crumbs, are batched in a low-shear premixer where the bitumen "wets" the polymer. Premixing time and temperature are adjusted in order to guarantee adequate polymer swelling by the bitumen.

After the premixing step, the bitumen/polymer mixture goes through a high-shear mixer to get the proper dispersion and then is discharged into the storage tank. When necessary, a valve system lets the modified bitumen go through the high-shear mixer again to obtain improved dispersion.

Thermal conditions greatly influence the dispersion of SBS copolymers, and the recommended range of temperature is between 180 and 200°C. At lower temperatures, dispersion becomes difficult, and temperatures exceeding 200°C can promote excessive and uncontrollable cross-linking of polymer.

Concerning the mixer, the speed and the nip between rotor and stator must be carefully adjusted in order to optimize dispersion and minimize mixing time.

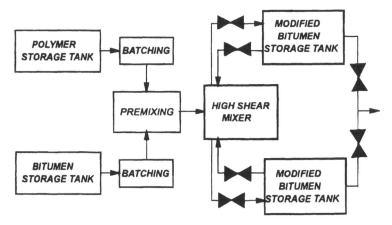

Figure 4   Block diagram of a plant for the production of polymer-modified bitumen.

C. Testing

In spite of the substantial modification of bitumen properties due to the addition of SBS copolymers, it is difficult to make objective predictions about the performance of the final product on the basis of traditional testing. Traditional tests are, in fact, conceived in order to establish a set of specifications for the bitumen and hardly match the actual service conditions.

In this respect, as previously mentioned, dynamic mechanical analysis was found to be a powerful technique for performance assessment, allowing the prediction of viscoelastic behavior under the final service conditions. The complete assessment of a material's viscoelastic behavior usually requires a number of dynamic mechanical experiments performed over a wide range of temperatures and frequencies [5].

Unfortunately, even up-to-date experimental instruments are not able to cover the whole range of interest; both very low and very high test frequencies are outside the range of practically available experimental conditions. In this situation, the Williams, Landel, Ferry (WLF) time-temperature superposition principle proves to be an attractive and valuable technique, particularly when polymeric materials are examined. Results obtained in confined intervals of temperature and frequency can easily be extrapolated to a wider range of conditions, allowing the prediction of performances.

The time-temperature superposition principle for neat bitumen and bitumen/polymer blends was tested and validated using an experimental approach, as outlined in Figure 5 in the case of the storage modulus (G') for a 100/8 bitumen/SBS blend [6.7]. Dynamic measurements were carried out on a Rheometrics RDS II machine, operating in torsion mode and using a parallel plate geometry with cylindrical test specimens (8 mm in diameter). The gap height ranged between 5 and 6 mm and, in order to prevent slippage, test samples were glued.

A series of isothermal runs were obtained in the range -30 to $+80°C$, every $10°C$; the frequency range investigated was 10^{-2} to 10^2 rad/s, and four experimental points were taken for each frequency decade. Owing to the progressive sample softening that takes place when the test temperature is raised, strain was properly adjusted, within the limits of linear viscoelasticity, to get torque values in the right range for the instrument.

According to the WLF time-temperature superposition principle, G' experimental data were reduced to the reference temperature $T_0 = 0°C$, and in this way a master curve was constructed that covers a remarkably wide range of frequencies.

By using the same procedure at different temperatures, it is possible to derive a comprehensive description of the viscoelastic behavior in the frequency and temperature ranges involved in the final use conditions.

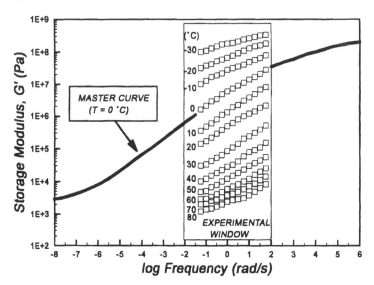

Figure 5   Example of WLF time-temperature superposition.

III.   BITUMINOUS ROAD BINDERS

A.   Binders

Asphalt paving mixes are generally designed to withstand the stress induced by local traffic loads and environmental conditions. However, when subjected to traffic and to climatic elements, the road pavement undergoes both physical and chemical aging and may develop, often in a very short time, rutting (permanent deformation) and cracking.

According to our extensive experience, fatigue cracking and thermal cracking are strictly connected to the quality of the binder, which also plays, in combination with the mineral grading, a primary role in avoiding rutting. This is particularly true in the case of porous asphalts or thin and very thin asphalt concretes.

Methods actually in use for characterization of bituminous binders are mainly concerned with workability, handling, and safety aspects while performance assessment is almost completely neglected. In this respect, the following example, concerning the characterization of some bituminous binders, proves the key role of dynamic mechanical spectrometry in predicting performance.

A comparison was made between a commercial grade of bitumen and two polymer-modified bitumens, obtained by the addition of 8 parts per hundred parts of bitumen (phb) of an SBS copolymer and an ethylene vinyl acetate (EVA) copolymer, respectively. The properties of the materials used are listed in Table 1.

Table 1 Materials Description

Material	Description
Bitumen	Commercial grade of bitumen SOLEA (180/200 pen) supplied by IP (Italiana Petroli)
SBS copolymer	EUROPRENE SOL T 161/B, produced by EniChem Elastomeri (styrene content = 30%, molecular weight = 230×10^3, radial structure)
EVA copolymer	Commercial grade (vinyl acetate content = 19%, melt flow index = 150 dg/min)

Mixtures of bitumen and 8 phb of polymers were prepared using a Silverson L4R high-shear mixer operating at 6000 rpm; temperature was set at 190°C and mixing was done for 20 min. Hot final mixtures were directly poured into the molds, designed to obtain the proper test specimens for characterization. The results of the tests, commonly used in the road sector, are reported in Table 2.

The adhesion test consists of measuring the shear force necessary to separate two small lithic prisms, bonded by a thin layer of bituminous binder [8]. Sample preparation is accomplished according to the "isoviscosity" principle, commonly used in concrete preparation as well. Spreading temperature must be adjusted in order to have a binder melt viscosity ranging between 0.4 and 0.6 Pa s; lithic prisms are then bonded and cooled at room temperature before testing.

Polymer addition always improves bitumen properties, but SBS modification offers the best results, particularly with regard to adhesion to inerts, which is dramatically improved, independently of the chemical nature, basic or acidic, of the stone.

Ring and ball and Fraass values show the reduced thermal susceptibility of polymer-modified bitumens in comparison with the neat bitumen. However,

Table 2 Properties of Neat Bitumen (NB) and Polymer-Modified Bitumens

Properties	NB	EVA	SBS
Ring and ball (°C)	40	67	110
Penetration (dmm)	200	97	61
Fraass breaking point (°C)	−18	−19	−28
Viscosity at 160°C (Pa s)	0.10	0.33	0.80
Viscosity at 180°C (Pa s)	/	0.17	0.50
Viscosity at 200°C (Pa s)	/	0.10	0.35
Adhesion to marble (kg/cm²)	2.95	3.50	7.80
Adhesion to granite (kg/cm²)	3.40	3.95	8.45

dynamic mechanical analysis proved to be a powerful tool for an in-depth examination, allowing better prediction of binder performance.

According to the procedure previously described, a series of master curves of storage modulus versus frequency was obtained at different reference temperatures in the range -30 to $+80°C$, every $15°C$. Data from these master curves were then used to plot storage modulus versus temperature, as shown in Figure 6.

The range of temperature considered covers most of the service conditions for a road. In turn, reference frequencies were selected in an effort to reproduce traffic conditions corresponding to normal travel (10 Hz) and to braking (1000 Hz). The lowest frequency (10^{-7} Hz) accounts for the mechanical stress induced by the natural (yearly) thermal cycle related to the succession of the seasons.

It is important to point out that for the mixture containing EVA copolymer, it was impossible to get reliable measurements of G' at temperatures higher than $60°C$ because of the excessive softening of the samples.

As expected, increasing frequency or decreasing temperature results in increased binder stiffness, but at high frequency or low temperature polymer-modified bitumen shows lower moduli than neat bitumen, which means enhanced flexibility. On the contrary, at the lower frequencies or at the higher temperatures the situation is reversed and modified bitumen shows higher moduli, which means better temperature resistance. As a consequence, in all the combinations of frequency and temperature examined, SBS-modified bitumen clearly offers the best overall behavior.

Regarding the prediction of performance, flexibility at low temperature and strength at high temperature can be related to the ability of binders to withstand both thermal and fatigue cracking.

Although storage modulus, i.e., the elastic response component of a material, can account to a certain extent for rutting resistance, it is possible to use a more specific experimental test, such as creep measurement. This test consists of compressing the sample with a constant load for a fixed time and measuring the deformation induced during the loading and the recovery steps.

Creep measurements were carried out on a Rheometrics RSA II solid analyzer, working in compression mode and using a parallel plate geometry (15 mm diameter) under a constant load of 50 g. Creep curves for modified bitumens are shown in Figure 7; as expected on the basis of G' values, EVA-modified bitumen is stiffer and undergoes lower deformation during the loading step than the SBS-modified bitumen.

When the load is removed, both modified bitumens show a prompt elastic response with a partial recovery of deformation. But for longer relaxation times the behavior is quite different: EVA-modified bitumen is affected by a remarkable permanent deformation whereas the SBS-modified bitumen shows delayed but almost complete recovery in a relatively short time.

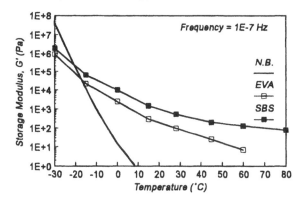

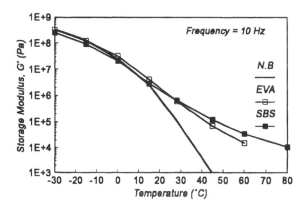

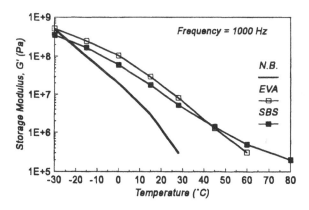

Figure 6 *G'* versus temperature at the indicated frequencies.

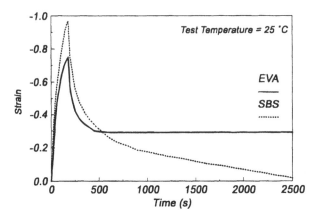

Figure 7   Creep test under constant load.

B. Asphalt Concrete Mixes

In fact, the final product for road paving is a mixture of stones and binder; therefore it was interesting to check whether the good performance of modified binder was transferred to the final concrete mix.

For this purpose, a comparison was made of two asphalt concrete mixes containing, respectively, a conventional road binder (60 pen) and a modified binder (100/8 bitumen/SBS blend) with comparable penetration. A conventional dense-graded concrete was selected, and all testing of aggregate-binder mixtures was done using the same lithic material and grading.

Binder content for all the mixes was 5% by mass. According to the isoviscosity principle, previously mentioned, temperatures of mixing and compaction were carefully selected in order to have comparable melt viscosities for the two binders. In this way the interparticle interlock, resulting in a load-bearing framework, was reasonably comparable in all the test specimens.

Compaction was done according to the Marshall [9] procedure (75 blows) and densities of compacted samples typically ranged between 92 and 94% of the theoretical maximum density.

Materials were prepared in the laboratory to yield cylindrical samples 60 mm in diameter by approximately 120 mm in height. Dynamic measurements were carried out on an MTS 831 machine operating in compression and in stress-controlled conditions. Frequency sweeps ranging from 2 to 20 Hz were performed every 5°C in the range of temperature from −30 to +70°C.

The region of linear behavior was determined, prior to the complete characterization, by performing loading sweeps at each temperature. Experimental data were then used to construct master curves according to the shifting procedure provided by the time-temperature superposition principle.

Final results are shown in Figure 8, where the storage modulus E' is plotted against temperature for the asphalt concrete mixes at the reference frequencies selected when the binders were examined (Figure 6). Binder largely affects the performance of the final asphalt concrete mix, and the mix containing SBS-modified binder performs much better than the conventional one, particularly at the extreme temperature and frequency. As expected, overall behaviors of pure binders are in very good agreement with the performance of the corresponding asphalt concrete mixes.

IV. ROOFING SHEETS

Prefabricated bituminous membranes are composite materials with a multilayer structure consisting of nonwoven glass or synthetic material impregnated and held toghether by a suitable bitumen-based binder. The binder plays a leading role, giving to the overall composite structure waterproofing and physical-mechanical characteristics necessary to fulfill performance and durability requirements. The membrane performance can be predicted by examining the viscoelastic behavior of the basic binder, namely the bitumen.

Table 3 shows the properties obtained in the traditional tests generally used in the waterproofing industry for both a neat bitumen and a 100/13 bitumen/SBS blend. Materials and mixing conditions were the same as previously described in the section concerning road binders.

In spite of the substantial modification of bitumen properties by the addition of SBS copolymer, it is difficult, on the basis of these data, to make objective predictions about the performance of the final product.

Most prefabricated membranes end up on building roofs that are naturally exposed to climatic effects. Thermal cycles that take place according to the natural daily, seasonal, or yearly frequencies play a fundamental role.

From a practical point of view, thermal cycling turns into mechanical cycling because the roof structure consists of different materials with different thermal expansion coefficients. Therefore, the waterproofing membrane, in particular the bitumen-based binder, must be able to withstand stresses induced by the alternate stretching and shrinking of the overall roof structure.

According to the season, extreme environmental conditions can produce winter embrittlement, leading to crack formation, or summer softening, which impairs dimensional stability.

The primary performance of bitumen, when used as membrane binder, was analyzed on the basis of these considerations, employing the same methods previously described. Also in this case, according to the time-temperature superposition principle, a set of master curves was derived, covering a very wide range of temperature and frequency.

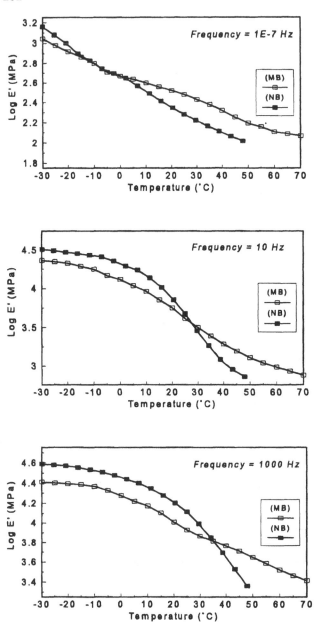

Figure 8 E' versus temperature for the asphalt concrete mixes.

Table 3 Traditional Bitumen Testing

Properties	Neat bitumen	100/13 bitumen/SBS
Penetration (dmm)	195	46
Ring and ball (°C)	43	126
Bending on mandrell (°C)	−16	−30
Viscosity at 180°C (cP)	45	1690
Viscosity at 200°C (cP)	25	1100

Final results are shown in Figure 9, where curves of storage modulus (G') versus temperature are shown. The range of temperature covers most of the real service conditions, and reference frequencies were selected according to the previously mentioned natural cycling of the roof.

The highest frequency (1 Hz) may be representative of the thermal shock induced by a sudden summer storm or the stresses that the roof undergoes during construction and maintenance.

Addition of SBS copolymer improves bitumen performance over the entire temperature range. In comparison with the behavior of neat bitumen, represented by the dashed line, modified bitumen exhibits, at the same time, higher moduli at high temperatures and lower moduli at low temperatures. Under the experimental conditions, it was impossible to test neat bitumen at frequencies lower than 1 Hz, because of insufficient consistency of the test specimens. Moreover, it is

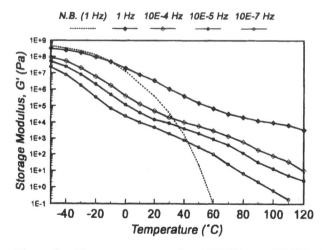

Figure 9 G' versus temperature for a 100/13 bitumen/SBS blend. Dashed line represents neat bitumen at 1 Hz.

important to note that the performance of modified bitumen seems to be satisfactory even under the extreme conditions.

For neat bitumen, the combination of low temperatures and high frequencies may cause embrittlement and failure, whereas high temperatures and low frequencies promote creep and flow. Neat bitumen behaves only in this way, while modified bitumen displays flexibility or consistency enough to withstand both conditions.

Regarding the other waterproofing requirements, such as durability, dynamic-mechanical measurements, combined with suitable aging procedures, could represent a promising technique.

V. CONCLUSIONS

The viscoelastic behavior of bitumen can be modified to a great extent and carefully tuned, in order to match practical requirements, by the addition of styrenic thermoplastic elastomers.

Hot mixing of bitumen and SBS copolymers leads to the formation of a rubber-like continuous phase, which can be directly detected by fluorescence microscopy.

Dynamic-mechanical analysis was found to be a powerful tool for performance assessment, allowing the prediction of viscoelastic behavior under actual service conditions. The storage modulus was chosen as an effective indicator of viscoelastic behavior.

SBS-modified bitumen performs much better than neat bitumen under all experimental conditions investigated. This is particularly evident under boundary conditions, i.e., at extreme temperatures and frequencies.

REFERENCES

1. W. Heukelom, A Bitumen Test Data Chart for Showing the Effect of Temperature on the Mechanical Behaviour of Asphaltic Bitumens, *J. Inst. Petrol.,* **55,** 404, (1969).
2. E. C. Bingham, *Fluidity and Plasticity,* Committee for the Study of Viscosity, New York, pp. 227–286 (1922).
3. E. Diani and L. Gargani, Bitumen Modification with SBS Thermoplastic Elastomers, *4th Eurobitume Symposium,* Madrid, Vol. 1, pp. 276–279 (1989).
4. M. Bouldin, J. M. Collins, and A. Berker, Rheology and Microstructure of Polymer/Asphalt Blends, ACS Rubber Division, Las Vegas, NV, May 29–June 1, 1990.
5. J. D. Ferry, *Viscoelastic Properties of Polymers,* 3rd ed., Wiley, New York (1980).
6. E. Diani, L. Gargani, and L. Vitalini Sacconi, Styrenic Block Copolymers as Bitumen Modifiers for Improved Roofing Sheets, ACS Rubber Division, Detroit, October 8–11, 1991.

7. M. G. Cavaliere, E. Diani, and L. Vitalini Sacconi, Polymer Modified Bitumens for Improved Road Application, 5th Eurobitume Symposium, Stockolm, Vol. 1A, pp. 138–142 (1993).
8. Italian Standard, CNR (Consiglio Nazionale Ricerche) Procedure 2/1951.
9. ASTM Standard D 1559, Standard Test Method for Resistance to Plastic Flow of Bituminous Mixtures Using Marshall Apparatus.

14

Polychloroprene-Modified Aqueous Asphalt Emulsion: Use in Roofs and Roads

S. S. Newaz
Bayer Corporation, Houston, Texas

M. Matner
Bayer AG, Dormagen, Germany

I. INTRODUCTION

Bituminous materials are used around the world in road construction and roofing [1,2]. In the United States, bituminous-type materials are by far the most commonly used covering for flat roofs and built-up roofs (BURs). However, BURs have a history of failure, as do asphalt roads. Various materials, such as asphalt-impregnated glass mats, bitumen-saturated felts, and saturated and coated felts, have been used to improve the life of the asphalt surface. Synthetic rubber roofing offers energy savings, life extension, and resistance to cracks and chips [3].

Elasticized asphalt is another alternative to extend the life of the roof. Here the asphalt is treated with a common commercial polymer in order to improve the overall life of the roof or the road, to cut down the repair frequency, and to save energy. The polymer can be added to hot asphalt or to an asphalt emulsion as an aqueous polymer dispersion.

Our work relates only to the modification of aqueous asphalt emulsions with aqueous polychloroprene (CR) latex. When a mixture of CR latex and asphalt emulsion is coagulated, an elastic matrix is formed that can be used for roofing, deck waterproofing, or road repairs.

CR latex is normally a water-based anionic emulsion [4]. However, a cationic version of this latex is also available. Similarly, an asphalt emulsion can be anionic or cationic depending on the emulsifiers used during the emulsion preparation. We would broadly divide our discussion into two parts: use of an anionic system for roofing and use of a cationic system in road repairs.

II. USE OF AN ANIONIC SYSTEM FOR ROOFING

A dried elasticized CR-asphalt matrix is suitable for repairing a flat roof. Its desirable characteristics include elasticity, high temperature-resistance, cold weather flexibility, resistance to oxidation, low water absorption, and some puncture resistance.

For application, the asphalt emulsion is mixed with CR-latex and an organic thickener is added to attain the desired viscosity. The polymer-modified emulsion can then be coagulated by direct contact with an aqueous solution of an inorganic salt (two-component system). Alternatively, a one-component system can be applied directly. This leads to the combination of choices for the latex type, asphalt type, coagulant, thickener, and coagulant application system.

The typical CR-latex used has a pH of 11.5–12.5, a solids content of 58–59% (see Table 1), and a viscosity of 100 ± 20 cps. An anionic emulsion of asphalt typically contains 65% solids with a pH between 8 and 10 and a viscosity of about 500 cps. When the asphalt and polymeric latex are mixed, an organic thickener is required to achieve phase stability. The addition of a surfactant may be needed for a longer period of stability. The asphalt-latex mixture can be easily coagulated by treatment with a 10% aqueous solution of calcium chloride, calcium nitrate, or aluminum chloride. Table 2 gives a typical mixture of latex and asphalt used for coagulation. Latex and asphalt can be mixed in a wide range of proportions. Table 2 shows a typical cost-effective formulation.

We discuss the subject in terms of (a) choice of the asphalt emulsion and latex, (b) choice of thickener and surfactant, (c) choice of coagulant, (d) choice

Table 1 Physical Properties of Anionic CR-Latex

Property	Parameter
Total solids, %	58–59
pH	11.5–12.5
Viscosity, cps	60–120
Surface tension, dyne/cm	30–40
Mean particle size, nm	180–250
Crystallization rate	Fast or medium
Comonomer	None or sulfur

Table 2 Latex-Asphalt Mixture for Roofing

Ingredient	Weight (%)	Solids content (%)	Dry weight (%)
Asphalt	82–83	65–67	53.6
Latex	16–17	58–60	9.7[a]
Thickener	1–1.2	—	—
Surfactant	0–0.5	—	—

[a] Amounts to 18% rubber per 100 g of dry asphalt.

of application technique, and (e) high- and low-temperature behavior of the asphalt-polymer matrix.

A. Choice of Asphalt and Latex

Anionic asphalt emulsions are graded as rapid setting (RS), medium setting (MS), or slow setting (SS). In order to obtain a desirable elastic rubberized asphalt matrix, the choice of asphalt grade is more critical than the choice of the CR-latex. We have examined several grades of latex with various crystallization rates and with or without copolymerized sulfur in the presence of different grades of asphalt. In each case, as Table 3 shows, the choice of latex makes little difference in how the matrix coagulates.

With SS or MS asphalt emulsions, the coagulation is incomplete or they fail to give any elastic coagulum. The RS type is the asphalt emulsion of choice for making CR-modified elasticized asphalt film. An asphalt with needle penetration of 80–100 (0.1 mm) is preferred. Choice of latex is not critical as far as the coagulation characteristics are concerned, but for high-temperature stability or

Table 3 Choice of Latex and Asphalt for Coagulation[a]

Asphalt type	CR-latex type	Coagulation behavior
RS	Slow crystallization	Coagulates well
	Fast crystallization	Coagulates well
	Sulfur modified	Coagulates well
MS	Slow crystallization	Matrix is stringy
	Fast crystallization	Matrix crumbles
	Sulfur modified	Slow coagulation
SS	Slow crystallization	Does not coagulate
	Fast crystallization	Does coagulate
	Sulfur modified	Does not coagulate

[a] 10% aqueous $Ca(NO_3)_2$ is used as coagulant.

low-temperature flexibility, sulfur-containing CR or rapidly crystallizing types are not suitable (vide infra).

B. Choice of Thickener and Surfactant

Even after the asphalt and latex are mixed by stirring they may slowly separate out into two layers. Phase stability is increased by the use of polymeric thickeners. Usually the thickeners are added to the asphalt-latex mixture as a diluted water solution when the viscosity of the mixture starts to rise. Table 4 shows the effect of various thickeners on the resulting viscosity and phase stability. We have found that the use of 0.75–1.0 wt % polyacrylate is sufficient to eliminate phase separation.

Some nonylphenoxy-type nonionic surfactants, when used in the 0.5 to 1.0 wt % range in the absence of polyacrylate, can give phase stability to the latex-asphalt mixture without changing the viscosity. Phase stability is more pronounced with SS grades of asphalt. With RS grades the use of a nonionic surfactant with or without acrylates gives pronounced phase stability but adversely affects the coagulation.

Cellulosic thickeners, such as methyl cellulose or methylhydroxyethyl cellulose, are suitable for thickening the asphalt-latex matrix. In fact, the cellulosic thickeners are preferred over acrylate types because of lower water absorption. Acrylates give poor water absorption to the asphalt polymer coagulum. Fillers such as kaolin, chalk, and slate clay should be used.

C. Choice of Coagulant

Aqueous solutions of calcium salts work best to destabilize the emulsifier systems of both the asphalt and latex. A 10–20% solution of calcium nitrate or calcium

Table 4 Effect of Thickeners on Latex-Asphalt Mixture

Thickener type	Amount (wt %)	Viscosity (cps)	Phasing characteristics
None	—	240	Almost complete in 1 h
Polyacrylate	0.1	270	Moderate after 3 h
	0.25	300	Moderate after 3 h
	0.5	330	Slight after 3 days
	0.75	450	Trace after 3 weeks
	1.0	580	Trace after 3 weeks
Methocel	1.0	340	Moderate after 3 days
	2.0	540	Slight after 3 weeks
Hydroxypropyl cellulose	1.0	280	Moderate after 3 h

chloride works well. Aluminum chloride or ferric chloride solutions are also known to give the same coagulation characteristics.

D. Choice of Application Technique

For commercial applications, the method of choice is spraying. A dual-nozzle power spray is used to pump an asphalt-latex mixture from one nozzle and an aqueous coagulant from the other. The asphalt-latex destabilizes as soon as it comes into contact with the coagulant and is then spread out as a thin film on the application area. Ambient air drying removes the water and the dried film remains as an elastic covering for the roof.

The spray method of application may be hindered by logistic considerations such as wind. Brushing or troweling can be used as an alternative. These methods are also preferred for most do-it-yourself (DIY) jobs. Phase-stabilized latex-asphalt mixtures can be mixed with dilute coagulant before application, which allows application by manual spray, brush, or broom. This still permits sufficient time for coagulation to occur after application. Popular brushing and troweling are done as either **thin-layer** or **thick-layer** applications. Both application methods can be carried out with a one- or two-component system. With the thin-layer one-component system the asphalt-latex emulsion is simply left to dry under ambient conditions. With the two-component system, a calcium salt–based coagulant is used. In the thick-layer application, the asphalt-latex emulsion is thickened to a very high viscosity (8,000–10,000 cps) and then left to dry as is a one-component system. In the thick-layer two-component application, cement is used as the dehydrating agent for film formation.

E. High- and Low-Temperature Behavior

Elasticized asphalt roofing must survive extreme temperature variations and weathering over the years. A simple ball drop test was performed to check the impact resistance at 0 and 5°C. Table 5 shows the results for asphalt with 20% of different latex types subjected to this test. The samples were attached to an insulated board and conditioned at 5°C for 1.5 h. A steel ball, weighing 534 g,

Table 5 Latex-Asphalt Matrix Crack Under Impact

CR-latex type	Tests at 5°C	Tests at 0°C
S type, slow crystallization	Crack	—
Medium crystallization	No crack	Slight crack
Medium crystallization, no S	No crack	Slight crack
Slow crystallization, no S	No crack	Slight crack

was then allowed to drop through a 7.5 cm × 2.5 m tube. The samples were also treated at 0°C and treated similarly with the 534-g ball. Shattering of the sample was considered to be failure. The latex without sulfur as comonomer fared better overall.

The same ball drop test was performed with different amounts of latex (20–24 wt %) in the asphalt. Only homopolymers were used, and the same ball test was used at successively lower temperatures. Table 6 summarizes the results, along with the results of mandrel bend and elevated temperature (84°C) tests.

For the mandrel bend test, 3 × 30 cm sections were cut and placed in a cold box at −14°C, along with a 20-mm-diameter mandrel for at least 1 h. Each sample was then bent 180° in 3 s or until it broke.

For the elevated temperature test, the panels used for ball drop tests were placed in an 84°C oven at an angle of about 80° for 3 days. No sagging or flowing indicated passing.

Tests show that all the asphalt-latex mixtures passed the elevated temperature test, and use of harder asphalt made the matrix more brittle. A latex content of 20–24% passed the mandrel tests. Increasing the amount of latex improved crack resistance at lower temperatures. However, even a mixture with a 24% latex content did not pass the −14°C crack tests.

We have examined a number of plasticizers in an attempt to improve the low-temperature impact resistance. Table 7 shows this plasticizer study. All the latex-asphalt mixtures contained 20 wt % latex. The cured plaques were subjected to the ball drop test. To get crack resistance at −10°C, 3 or 5 wt % of a plasticizer may be required. However, this has a tendency to make the matrix oily and tacky. The severe ball drop cracking test may not be a true reflection of the low-temperature crack resistance. In the standard Fraass test done on thin films, the crack resistance is observed to −20 or −25°C.

The asphalt-latex roof offers oxidation resistance, lower repair frequency, elastic film to cover uneven cracks, and excellent water and weather resistance.

Table 6   Ball Drop, Mandrel Bend, and Elevated Temperature Test

Tests/conditions	Samples, latex wt %			
	20[a]	20	22	24
Ball drop, 20°C	Pass	Fail	Pass	Pass
Ball drop, 0°C	Fail	Pass	Fail	Pass
Ball drop, −4°C				Fail
Mandrel, −14°C	Fail	Pass	Pass	Pass
Elevated temperature	Pass	Pass	Pass	Pass

[a] Done with harder asphalt RS-1H.

Table 7 Ball Drop Test for Plasticizer Study

Plasticizer	Level (wt %)	Temperature (°C)	Result
None	—	0	Fail
Butyoleate	1	0	Pass
	1	−10	Fail
	3	−10	Pass
	3	−15	Fail
	5	−10	Pass
	5	−15	Fail
Dioctylsebacate	1	0	Fail
	3	0	Pass
	3	−10	Pass
	3	−15	Fail
	5	−15	Fail
Dioctylphthalate	1	0	Fail
	3	0	Fail
	5	0	Pass
	5	−10	Pass

III. USE OF A CATIONIC SYSTEM IN ROAD REPAIRS (SURFACE DRESSING)

Cationic CR-latex has a low pH of 1–2.0, a high solids content, and a moderately high viscosity. Table 8 summarizes the typical physical properties of a cationic CR-latex and cationic asphalt emulsion used in this application. To prepare the cationic asphalt emulsion, hot asphalt is treated in a colloidal mill with a cationic aqueous surfactant. Asphalt with needle penetration of about 200 (0.1 mm) is preferred for cationic emulsion use. The cationic latex can be added in the

Table 8 Properties of Cationic Products

CR-latex	
Total solids, %	58
pH	1.5
Viscosity, cps	270
Surface tension, dyne/cm	43
Particle size, nm	170–200
Asphalt Emulsion	
Total solids, %	63
Viscosity, cps	300
Particle size, nm	1000–15,000

colloidal mill together with the cationic aqueous surfactant or can be added as a postadditive. The usual amount used is about 3–5% CR on dry weight basis. But even lower amounts of 2–3% can be effective.

To make a CR-modified asphalt emulsion by postaddition, the latex is heated to 60°C, then added to the asphalt emulsion at 100°C. The usual storage temperature is 80°C and application on the road is done at 60°C. The cationic latex-asphalt mixture is applied to the road at this temperature. Then the gravel is spread into the emulsion phase, followed by treatment of the spread gravel with a heavy roller. Table 9 summarizes the properties of cationic asphalt mixture.

This combination of high mechanical shear, higher applied temperature, and contact with anionic gravel surface (charge transfer) brings about the required destabilization of the latex-asphalt surfactant system and makes it possible for the cationic system to be used for surface dressing of roads.

The polychloroprene-modified asphalt improves the characteristics of bitumen material in several ways: tenacity and toughness, low- and high-temperature flexibility, adhesion of the gravel, aging, and brittle and softening points. The bitumen used in the unmodified state has 200 mm penetration, a softening point of 37–40°C, and a brittle point between −7 and −3°C. The use of 3–5% CR-latex improves the softening point to 55°C and the brittle point to −20 to −25°C (Tables 10 and 11). Vialit cold or hot gravel tests show the temperature for 10% gravel loss from the matrix. The temperature range studied is −25° to −5°C for the cold and 70–72°C for the hot gravel test. The difference in temperature in the hot and cold gravel tests gives the range of temperature tolerance before

Table 9   Asphalt Properties with Cationic Latex

Property	Untreated asphalt	Cationic CR-latex-modified asphalt
Softening point, °C	37–40	55
Break point, °C	1–4	−15 to −20
Particle size, nm	170–200	Bimodal, 200 and about 12,000
Difference, hot and cold gravel test (Vialit test),[a] °C	50	70

[a] Vialit test. A stainless steel plate (200 × 200 × 2 mm) is coated with 1.4 g of a modified asphalt emulsion (the final film thickness after evaporating the water should be 1.4 mm). One hundred pieces of standard gravel (quartz, diameter 8/10) are implanted into the liquid emulsion film. The water is evaporated and the plate is conditioned stepwise at different temperatures in 5° steps. Every time the desired temperature is reached, the back side of the plate, which is in a horizontal position, is hit three times by a steel ball (weight 510 g) falling from a height of 500 mm. This test is repeated at different temperatures (going up or down) until 10% of the original amount of gravel is lost. This temperature is noted as the limit temperature.

Table 10 Properties of Anionic Latex-Modified Asphalt: Bitumen 80/100E

Property	Without polymer	Modified by 15% polychloroprene (anionic)
Needle penetration, DIN 52010		
At 0°C (0.1 mm)	8	11
At 25°C (0.1 mm)	58	44
Softening point (DIN 52011), °C	70	>140
Flowing (AIB DIN 52123), part 2		
At 70°C, mm	70	0
At 80°C, mm	>140	0
At 100°C, mm	>140	0
Mandrel test at low temperature (DIN 52123), break, °C	3	−13
Swelling properties in demineralized water at room temperature, weight increase %		
After 25 days	5	5
After 50 days	7	NA

Table 11 Properties of Cationic Latex-Modified Asphalt

	Baypren latex K content (% latex dry related on asphalt dry)		
	0	3	4
Needle penetration, DIN 52010			
At 0°C (0.1 mm)	14	18	21
At 25°C (0.1 mm)	134	128	129
Softening point (DIN 52011), °C	46.5	54	55
Brittle point, Fraass (DIN 52012), °C	−13	−22	−22.5
Ductility at 4°C (DIN 52013), cm	13	42	41.5
Elastic recovery at 4°C (DIN 52103), cm	0	67.5	66.5
Tenacity, Nm	0.4	6.6	7
Toughness, Nm	2.7	8.9	8.8
Impact test at low temperature (Vialit), °C (10% loss of gravel)	7	−20	−23
Impact test at high temperature (Vialit), °C (10% loss of gravel)	71	108	114
Impact test; difference of temperature, °C	64	128	137

significant (10%) gravel loss is experienced. The CR-modified asphalt shows a much wider range of this tolerance temperature.

Other polymers such as Natural Rubber (NR), styrene-butadiene-styrene block co-polymer (SBS), styrene-butadiene rubber (SBR), and ethylene-propylene diene monomer (EDPM) are also used in asphalt modification [5]. These polymers were tested in a road trial as part of a comparative study undertaken by the Bundesanstalt für Strassen und Verkehrswesen, Germany (Federal Department for Roads and Transportation) with the Technical University of Munich. The test road, Bundenstrasse B-17, an important north-south connection, is located in the southern part of Germany. Test strips with different polymer modifications were built in 1989. A committee of 14 experts checked the condition of the road strips for 5 years. In 1994 the trial was complete and the results were summarized. The unanimous opinion of the experts was that the part of the road constructed using polychloroprene-modified asphalt was in the best condition.

REFERENCES

1. W. R. Gorman and A. M. Usmani, American Chemical Society, Rubber Division Philadelphia Meeting, April 1995.
2. A. M. Usmani, *Polym. News*, **20**, 242 (1995).
3. A. M. Usmani. Rubber Polymers in Roofing Applications, *Polym. News* (1995).
4. A. G. Bayer. Technical Information Bulletin. *Lattices Polymer Dispersion.*
5. Annual reports, 1989–1994. *FORSCHUNGSGESELLSCHAFT* für Strassen und Verkehrs wessen. Comittee: BAK 7.1.5.1. (Research Institute for Road and Traffic, Cologne, Germany)

15

Polymer Science and Technology

Arthur M. Usmani
Usmani Development Company Indianapolis, Indiana

I. INTRODUCTION

A. Polymers

Polymers are a valuable and integral part of modern life and are examples of useful technology. They are one of the most versatile innovations of chemistry, tailored for hundreds of uses and constantly evolving into even more beneficial products.

Polymers are truly giant molecules whose molecular weights can range from 10,000 up to 1,000,000 or even higher. Thus, it is even possible to see one of these giant molecules, in the form of a vulcanized automobile tire or a molding of phenol/formaldehyde resin, with the unaided eye. Some of the larger molecules, such as enzymes and nucleic acids, can be seen in the electron and atomic force microscopes.

The physical properties of polymers can vary markedly (rubbery, rigid, tough, water resistant, water soluble, etc.) depending on their chemical composition and the lengths of the chains that make up these giant molecules. Most of the solid things that we see around us (fibers, cloth, rubber adhesives, floor tiles, surface coatings, etc.) are polymers, as are all plants and animal tissues. People are polymers, albeit very complex ones. Obviously, then, these giant molecules, or polymers, can and do occupy an important place in our daily lives. Without them, life would be very difficult or impossible. The polymer industry is the world's second largest industry and the principal employer of chemists and chemical engineers.

Polymers are made by chemical reactions that link together small molecules, called monomers, to form large molecules with constantly repeating patterns, called polymers. Depending on the monomers, the way the molecules are linked, and the chemical process used in their manufacture, polymers can possess a wide variety of properties. Some are much stronger than steel, others can be made into very lightweight foams. More than ever, high-performance plastics are becoming the materials of choice for use in construction, automotive design, satellites, packaging medical applications, and many other important areas.

Polymer science is a relatively young discipline, whereas polymer technology has served mankind since early times but more specifically during the past 120 years. There are more than 40 different families of polymers, which are used for everything from the nose cones of space vehicles to artificial heart valves to plastic turf for sports. Polymers are used in homes, automobiles, offices, space cabins, and hospitals—in fact, almost everywhere.

B. Polymer Technology

During the past 50 years, many polymer industries, based on thermoplastics and thermosets, films, fibers, elastomers (rubbers), cross-linked polymers, coatings, adhesives, sealants, composites, and polymeric foams, grew and mushroomed into very large scale production. For example, about 26 million metric tons of plastic resins are processed annually in the United States in about 10,000 plastic processing plants.

C. Development of Polymer Science

During the 1920s and the 1940s, there were increasingly successful efforts to provide far more and less expensive basic building units via the synthesis of new monomers, to determine efficient reactions to link repeating units in chains (i.e., mechanism of polymerization), to establish quantitatively the molecular weight and molecular structure of polymers (i.e., characterization), and to explore structure-property relationships (i.e., molecular engineering). By 1950 polymer science was a well-established and worthwhile discipline, but this new science was not accepted by all organic chemists.

The synthesis, characterization, processing, and applications of polymers are described in a very large number of articles and books. Readers may wish to refer to some important books that are listed at the end of the chapter.

II. DEVELOPMENT OF POLYMERS

In this chapter, we will systematically outline the major chemical types of polymers that are now available. This discussion is largely historical and starts with natural polymers, modified natural polymers, and synthetic polymers. Each of these types is then broken down to show the many different types of materials that have been developed on a "polymer tree" basis. These materials include plastics, fibers, coatings, and rubbers. These products may be thermosetting or thermoplastic materials that are produced by condensation or addition polymerization. They may be plasticized or filled and blended with other polymers, or they may be reinforced by strong fibers, such as glass or carbon fibers.

A polymer is composed of molecules that are characterized by repetition of one or more types of chemical units. In polymer chemistry, "polymer" is usually considered to mean "high polymer," i.e. macromolecule.

A. Early Historical

Early man used a variety of naturally occurring polymers as shown in Figure 1. These natural materials included stone, wood, bark, animal skins, cotton, wool, silk, natural rubber, bitumens, and glass.

Keratin, the principal constituent of wool, contains sulfur atoms as crosslinks in the main polypeptide chains. Silk, like wool, is composed of α-amino acids joined together to form long polypeptide chains, but it does not contain sulfur. Animal skin is a reinforced or stabilized proteinaceous form of fibrous collagen.

During prehistoric time, man began to modify the naturally occurring polymeric materials to make them more suitable for his use. As shown in Figure 2, these early modifications were mechanical. More complex mechanical modifi-

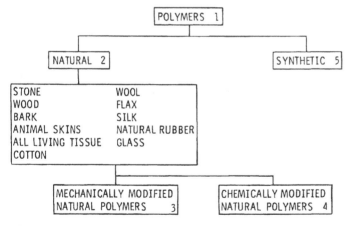

Figure 1 Classifications of polymers, natural and synthetic. (The numbers indicate the descending fixed generation of polymers by type.)

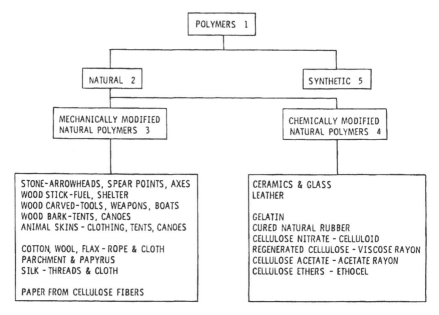

Figure 2 Mechanical and chemical modification of natural polymers. (The numbers have the same significance as in Figure 1.)

cations, such as twisting of cotton, wool, and flax to form threads and then weaving the threads into cloth, came later. Thin-skinned parchment and papyrus vegetable tissues were developed for writing. The mechanical modification of cellulose fibers to make paper was developed by the Arabs in Spain in the 12th century.

Whereas mechanical modification of natural polymers has been practiced for centuries, the use of chemical modification to make entirely new materials is comparatively new; certainly, almost all of these modifications have taken place during the past century. Examples of chemically modified polymers include regenerated cellulose (rayon), cellulose nitrate (celluloid), and cellulose acetate (acetate rayon).

B. Modern Historical

The greatest progress occurred, however, when man started using the raw chemical building blocks found in nature, such as petroleum, to make synthetic materials. It then became possible to design new products for specific end uses by molecular tailoring and engineering. The industries built around petrochemicals are shown in Figure 3.

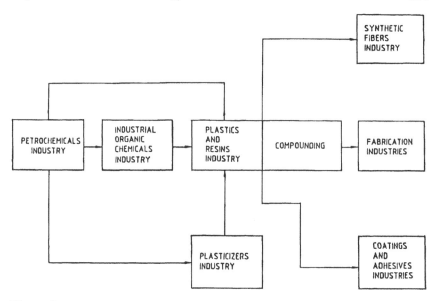

Figure 3 An illustration of industries built around petrochemicals.

Synthetic materials may be organic, semiorganic, or totally inorganic as shown in Figure 4. Semiorganic materials, which contain elements other than carbon, such as silicon (Si), nitrogen (N), phosphorus (P), and boron (B), are comparatively few in number and are recent developments. The most important member in this class with a noncarbon backbone, i.e., $(-SiO-)_n$, is silicone.

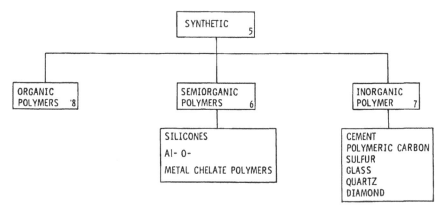

Figure 4 Classification of synthetic polymers.

These silicone polymers may be either rigid or rubbery, depending on the nature of the organic groups attached to the silicon atoms in the polymer chain or backbone. Because they contain major amounts of inorganic elements and usually have good chemical and thermal stability, Teflon (polytetrafluoroethylene), Kel-F (polychlorotrifluoroethylene), and polysulfides are classified as semiorganic polymers.

By far the most widely used synthetic materials are polymers that consist mainly of carbon and hydrogen plus a minimum amount of nitrogen, oxygen, or other inorganic elements. In Figure 5, the synthetic organic materials are divided into two major categories. In the first category are the addition-type polymers, which are based essentially on ethylene and substituted ethylene. The second category is that of condensation polymers, which may be either thermoplastic or thermosetting. The thermoplastics melt reversibly on heating and can be thermoformed to make moldings, fibers, films, etc. Some of the important thermoplastics are nylon, polyethylene terephthalate (PET), and polycarbonates (PCs), as shown in Figure 5.

Thermosetting polymers include familiar materials, such as phenol-formaldehyde, urea-formaldehyde, and melamine-formaldehyde polymers. Thermosetting polymers (thermosets) are three-dimensional (cross-linked) insoluble macromolecules that cannot be softened or processed by heating. More recently

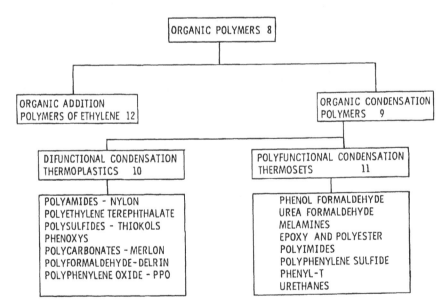

Figure 5 Branching of organic polymers into addition and condensation; subbranching of condensation into thermoplastic and thermoset types.

developed thermosetting polymers include the widely used epoxy resins, which are made by reacting a diphenol (e.g., bisphenol A) and epichlorohydrin. This "macromonomer" is then reacted further with either amine or anhydride hardeners (cross-linking agents). The epoxy resins are widely used as adhesives for wood, metal, and other materials and even more extensively in fiberglass laminates and filament-wound composite structures. Epoxy resins continue to be one of the fastest growing thermosetting polymers.

Having outlined schematically the condensation and other nonaddition organic polymers, let us turn to the organic addition "ethylene polymers." In Figure 6, the polyfunctional ethylene polymers, vulcanized rubbers and thermoset plastics, are presented.

In Figure 7, we consider ethylene with its four hydrogen atoms as the basic building block (repeating unit). When none of the four hydrogen atoms is replaced and the monomer is subjected to polymerization, polyethylene is formed. Because of its simplicity and low cost, polyethylene is the most important polymer in this group and is in wide use today. It combines a moderate degree of rigidity and heat resistance with good low-temperature and electrical properties and excellent chemical inertness. Radar, television, and many high-

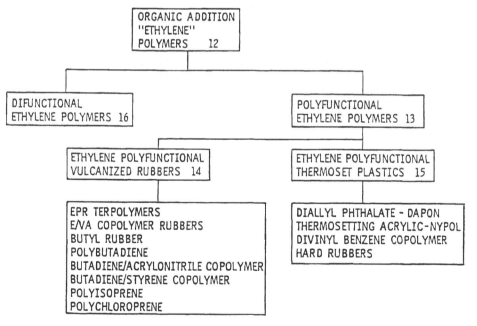

Figure 6　Diagram of family descent for ethylene polymers, specifically polyfunctional types.

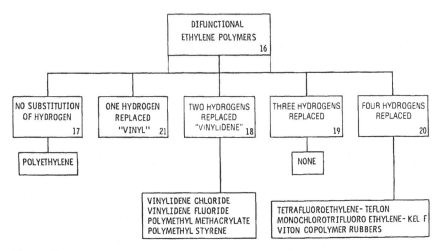

Figure 7   Diagram of family descent from difunctional ethylene polymers.

frequency devices are possible only because of the very low high-frequency dielectric loss of polyethylene. Polyethylene is available in a series of densities and crystallinities.

If we replace one of the hydrogen atoms of ethylene with some other group or element, a host of vinyl polymers is obtained, as depicted in Figure 7 and in detail in Figure 8. The most important members of this group include

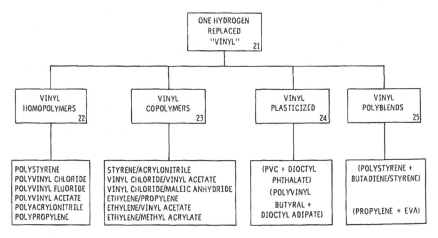

Figure 8   Depiction of vinyls.

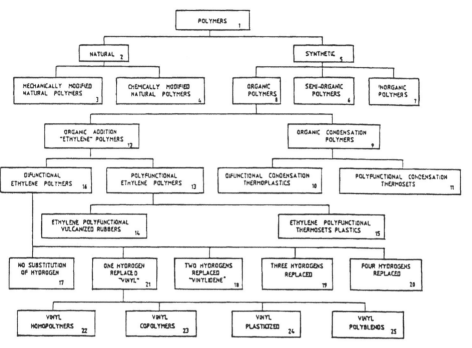

Figure 9   Block diagram showing complete development of the polymer tree.

polymers, copolymers, and flexibilized compositions based on styrene, vinyl chloride and acrylonitrile. Copolymers contain two or more repeating units (mers).

We can also replace two of the hydrogen atoms of ethylene as shown in Figure 7. Examples of such disubstituted ethylenes are vinylidene chloride (saran), in which the hydrogen atoms on one of the two carbon atoms of ethylene are replaced by chlorine atoms, vinylidene fluoride (Kynar), in which two hydrogen atoms are replaced by fluorine atoms, and methyl methacrylate, in which two hydrogen atoms are replaced by a methyl group and a methyl ester group, respectively.

There are no important commercial polymers in which three hydrogen atoms of ethylene have been replaced. The replacement of all four hydrogen atoms by fluorine, chlorine, etc. leads to stable, heat-resistant engineering plastics for specialty applications.

In Figure 9, the complete development of the polymer tree is shown in a block diagram.

III. CLASSIFICATION AND PROPERTIES OF POLYMERS

In general, the preparation of polymers can be described as follows:

$$\text{Monomer(s)} \rightarrow \text{polymer}$$

The small molecules that combine with one another to form polymers are called monomers, and the reactions by which they combine are called polymerizations. Polymerization is perhaps the most important discovery of the past 60 years.

A. Types of Polymers and Polymerizations

Two major types of polymerization are used to convert monomers into polymers. These methods were originally referred to as addition and condensation polymerization by Carothers. Addition polymerization is also called chain, chain-growth, or chain-reaction polymerization. Condensation polymerization is also called step-growth or step-reaction polymerization. The major distinction between addition and condensation polymerization is the difference in the kinetics of the polymerization reactions.

1. Polymer Composition and Structure

In step-growth polymerization, two difunctional monomers are allowed to react with the elimination of some small molecule, such as water. For example, a dicarboxylic acid can react with a diol to produce a polyester and water is eliminated as a small molecule. Step-growth polymerization is initiated by the polymerization of one of the carboxylic groups of the diacid with one of the hydroxyl groups of the diols. The free hydroxyl or carboxyl group of the resultant dimer can react further with an appropriate functional group of another monomer or dimer. This growth process is repeated throughout the polymerization until a high polymer product is obtained.

Typical step-growth polymers are shown in Table 1. The reactions by which they are formed are also indicated. It will become obvious from the table that many different combinations of reactants can be used in the preparation of step-growth polymers.

Some naturally occurring polymers, such as cellulose, starch, wool, and silk, are also step-growth polymers. Thus, cellulose is the polyether formed by the dehydration of D-glucose. The structure of cellulose (polyglucose) is shown in Figure 10. Hydrolysis of cellulose yields D-glucose.

In chain polymerization (addition polymerization), monomers containing carbon-carbon double bonds participate in a chain reaction. The repeating unit of a chain polymer has the same composition as the monomer. Table 2 shows some of the common chain polymers and the monomers from which they are produced.

Table 1 Typical Step-Growth Polymers and Their Formation Reactions

Polymer	Linkage	Polymerization reaction
Polyamide	$-NH-CO-$	$H_2N-R-NH_2 + HO_2C-R'-CO_2H \rightarrow H\,(NH-R\,-CO\text{-}NHCO-R')_nOH + H_2O$ $H_2N-R-NH_2 + ClCO-R'-CoCl \rightarrow H\,(NH-R-NHCO-R'-CO)_nCl + HCl$ $H_2N-R-CO_2H \rightarrow H\,(NHCO-R)_nOH + H_2O$
Protein, wool, silk	$-NH-CO-$	Naturally occuring polypeptide polymers; degradable to mixtures of different amino acids. $H\,(NH-R-CONH-R'-CO)_nOH + H_2O \rightarrow H_2N-R-CO_2H + H_2N-R'-CO_2H$
Polyester	$-CO-O-$	$HO-R-OH + HO_2C-R'-CO_2H \rightarrow H\,(O-R-OCO-R'-CO)_nOH + H_2O$ $HO-R-OH + R''O_2C-R''-CO_2R \rightarrow H\,(O-R-OCO-R'-CO)_nOH + R''OH$ $HO-R-CO_2H \rightarrow H\,(OCO-R)_nCOOH + H_2O$
Polyurethane	$-O-CO-NH-$	$HO-R-OH + OCN-R'-NCO \rightarrow (O-R-OCO-NH-R'-NH-CO)_n$
Polysiloxane	$-Si-O-$	$Cl-SiR_2-Cl \xrightarrow[-HCl]{H_2O} HO-SiR_2-OH \rightarrow H\,(O-SiR_2)_nOH + H_2O$
Phenol-formaldehyde	$-Ar-CH_2-$	
Urea-formaldehyde	$-NH-CH_2-$	$H_2N-CO-NH_2 + CH_2O \rightarrow (HN-CO-NH-CH_2-)_n + H_2O$
Melamine-formaldehyde	$-NH-CH_2-$	
Polyacetal	$-O-CH-O-$ with $\mid R$	$R-CHO + HO-R'-OH \rightarrow (O-R'-OCHR)_n + H_2O$

G-G-G-G-G-G-G,   ,G-G-G-G-G-G-G-G-
              G   G´
  ,G-G-G-G-G-G-G´   G-G-G-G-G-G-G-G-G,
 G´   G
  `G-G-G-G-G-G-G,   G-G-G-G-G-G-G-G-G´
              G   G´
  G-G-G-G-G-G-G´   G-G-G-G-G-G-G-G-G,
 G´   G
  `G-G-G-G-G-G-G-G-G-G-G-G-G-G-G-G-G-G-G´

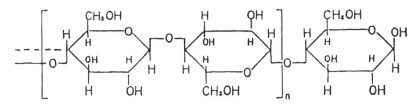

Figure 10   Structural formula for cellulose.

2. Polymerization Mechanism

Polymerizations are classified as step-growth and chain polymerizations depending on the mechanism. Step-growth polymerization can be schematically depicted as proceeding slowly from monomer, to dimer, trimer, and so on to polymer.

$$M + M \rightarrow (-M-)_2$$
$$(-M-)_2 + M \rightarrow (-M-)_3$$
$$(-M-)_2 + (-M-)_2 \rightarrow (-M-)_4$$
$$(-M-)_3 + M \rightarrow (-M-)_4$$
$$(-M-)_3 + M \rightarrow (-M-)_4$$
$$(-M-)_3 + (-M-)_2 \rightarrow (-M-)_5$$
$$(-M-)_5 + M \rightarrow (-M-)_6$$
$$|$$
$$|$$
$$|$$
$$(-M-)_{n-1} + M \rightarrow (-M-)_n$$

Table 2 Typical Chain Polymers

Polymer	Monomer	Repeating unit
Polyethylene	$CH_2=CH_2$	$-CH_2-CH_2-$
Polyisobutylene	$CH_2=C\overset{\diagup CH_3}{\diagdown CH_3}$	$-CH_2-C\overset{\diagup CH_3}{\diagdown CH_3}$
Polyacrylonitrile	$CH_2=CH-CN$	$-CH_2-\underset{\underset{CN}{\mid}}{C}-$
Polyvinyl chloride	$CH_2=CH-Cl$	$-CH_2-\underset{\underset{Cl}{\mid}}{C}-$
Polystyrene	$CH_2=CH-C_6H_5$	$-CH_2-\underset{\underset{C_6H_5}{\mid}}{CH}$
Poly(methyl methacrylate)	$CH_2=C\overset{\diagup CH_3}{\diagdown CO_2-CH_3}$	$-CH_2-C\overset{\diagup CH_3}{\diagdown CO_2-CH_3}$
Polyvinyl acetate	$CH_2=CH-OCOCH_3$	$-CH_2-\underset{\underset{OCOCH_3}{\mid}}{CH}$
Polytetraflouroethylene	$\underset{\underset{F}{\mid}}{\overset{\overset{F}{\mid}}{C}}=\underset{\underset{F}{\mid}}{\overset{\overset{F}{\mid}}{C}}$	$-\underset{\underset{F}{\mid}}{\overset{\overset{F}{\mid}}{C}}-\underset{\underset{F}{\mid}}{\overset{\overset{F}{\mid}}{C}}-$
Polyformaldehyde (acetal)	$CH_2=O$	$-CH_2-O-$
Polyisoprene	$CH_2=\underset{\underset{CH_3}{\mid}}{C}-CH=CH_2$	$-CH_2\diagdown _{\diagup CH_3}^{\diagup CH_2-}C=CH$

B. Linear, Branched, and Cross-Linked Polymers

Polymers can be classified as linear, branched, or cross-linked depending on their structure. Branching can occur in both step-growth and chain polymerizations. It should be pointed out that linear polymers containing pendant groups are not considered as "branched" polymers. When the polymer molecules are linked to each other at points other than their ends, the polymers are called cross-linked, infinite network, or vulcanized polymers. Light cross-linking can be used to impart elastic properties to polymers, whereas a high degree of cross-linking will impart high rigidity and dimensional stability.

Structures of linear, branched, and cross-linked polymers are shown in Figure 11.

C. Chemical Properties of Polymers

In general, the chemical reactivity of the functional groups in a polymer is similar to that of an analogous small organic molecule. Thus, an acid group attached to a polymer will react in the same way as it would if were present in a small molecule or monomer. For example, a hydroxyl group can be esterified, an aromatic ring can be sulfonated, and a methylene group in a polymer can be chlorinated. In this way, new polymers can be synthesized by modifying existing polymers using a variety of chemical reactions.

In polymer reactions, the concept of functional group reactivity is assumed to be independent of molecular size. However, in practice the reaction rate and yield observed with certain polymer functional groups may differ significantly from those for the corresponding low-molecular-weight homologs. The ester group in polymethyl methacrylate is more resistant to hydrolysis because of stearic effects and hydrophobicity of its surroundings than the ester group in methyl propionate. Conversely, the presence of suitable neighboring groups can increase the reactivity of pendant functional groups dramatically. For example, the rate of transesterification of pendant phenyl ester groups surrounded by pendant carboxyl groups is much higher than that of phenyl acetate.

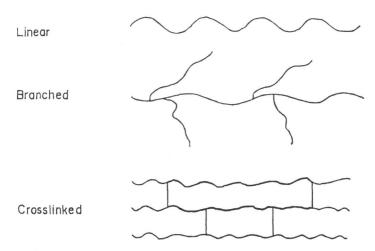

Linear

Branched

Crosslinked

Figure 11 Structure of linear, branched, and cross-linked polymers.

For polymer design, Usmani and Salyer have considered the properties obtained with specific chemical groups in a polymer. For low moisture absorption, a polymer should contain low hydroxyl, amide, and ether linkages and high aromatic and alkyl groups. For good adhesion properties it is important to have some hydroxyl, carboxyl, or nitrogen atoms or groups in the polymer. The chemical groups and associated physical properties of polymers are shown in Table 3.

D. Physical Properties of Polymers

The properties of any material depend on the structure, shape, and size of the component molecules and on the nature and magnitude of intermolecular bonding forces. The unusual physical properties of a polymer are associated with interactions between polymer chains. These interactions consist of molecular bonds and physical entanglements. In general, the magnitude of these interactions depends on the nature of the intermolecular bonding forces, the molecular weight, chain packing, and the flexibility of the polymer chain.

1. Bonding Forces

Intermolecular or van der Waals forces in polymers are similar to those in small organic molecules. Polymers contain many repeating units and thus the total secondary bonding is many thousand times greater for polymers than for monomer molecules.

Intermolecular forces are of three kinds: dipole, induced dipole, and London dispersion forces. The bond energies of these forces are modest, in the range of 2–10 cal/mole, and they exert great influence on physical properties. In the absence of strongly polar substituents, the London dispersion forces predominate.

Table 3 Chemical Groups and the Associated Physical Properties of Polymers

Group	Property Obtained
Aromatic	Thermal and mechanical stability, stiffening effect
Cycloaliphatic	Thermal and mechanical stability, stiffening effect
Alkyl	Moisture and chemical resistance
Alkyl, long, chain	Moisture resistance, plasticization, and toughening effect
Ether linkage	Chemical resistance, increased flexibility, and moisture sensitivity
Amide	Toughening and adhesion
Hydroxyl	Adhesion, moisture sensitivity
Di- and multifunctional vinyl	Curing, three-dimensional network. Increased thermal stability but also reduced toughness (increased brittleness)

Hydrogen bonding, a strong dipole-dipole interaction, is present in many polymer systems. Hydrogen bonding forms only with O, N, F, and Cl, which are strongly electronegative atoms, each having an unshared electron pair. Hydrogen bonding occurs in polyamides, cellulose, and polyvinyl alcohol.

Many commercial polymers contain polar groups that provide strong dipole-dipole interactions between the chains. Ester groups, cyano groups, and halogens are common pendant groups. Linear nonpolar polymers, e.g., polyethylene, have only van der Waals attraction between the chains and hence must have relatively high molecular weight and must be closely packed to have useful mechanical properties.

Ionic interactions between chains are also possible, e.g., in ionomers and electrodeposited polymers. Ionic bond energies are large, of the order of 100 kcal/mole, and hence the amount of interactions between the chains is extremely large. This results in outstanding strength and impact resistance.

Intermolecular bonds found in polymers are shown schematically in Figure 12.

2. Cohesive Energy

Intermolecular forces lead to the aggregation of molecules, and the energy change involved is called cohesive energy. Cohesive energy influences properties, e.g., volatility, viscosity, surface tension, solubility, and compatibility. Cohesive energy is usually expressed in cal/cm^3.

3. Crystallinity and What Determines Polymer Crystallinity

Polymers differ from typical low-molecular-weight organic compounds in their morphology. Most polymers possess the characteristics of both crystalline solids and highly viscous liquids. Crystallinity in polymers indicates ordered regions, whereas the amorphous state indicates disordered regions.

Some polymers are highly crystalline, whereas others are completely amorphous and some may be either amorphous or highly crystalline depending on the crystallization conditions. Totally amorphous polymers, e.g., atactic polystyrene and polymethyl methacrylate, are generally assumed to consist of randomly coiled and entangled chains. Styrene polymerized by Ziegler-Natta catalysis is partially crystalline (40–60% crystallinity). Amorphous polystyrene is clear, whereas crystalline polystyrene is opaque because of the spherulite structure of the crystalline phase.

Polymer crystallinity is a controversial subject. Two models for polymer crystallinity have been proposed. The fringed-micelle theory, developed in the 1930s, considers polymers as consisting of small ordered crystalline regions called crystallites that are embedded in a disordered amorphous matrix. This model is shown in Figure 13.

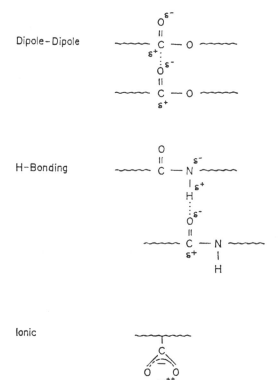

Figure 12 Intermolecular bonds responsible for properties in polymers.

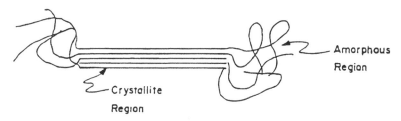

Figure 13 Fringed-micelle model of crystallinity.

The folded-chain lamella theory was formulated in the 1950s, when single polymer crystals, in the form of their platelets (termed lamellae), measuring $\sim 10,000 \times 100$ Å were grown from polymer solutions. Polymer molecules are longer than 100 Å. Thus, polymers are presumed to fold back and forth in an accordion-like manner in crystallization. A schematic two-dimensional representation of the folded-chain lamella model is shown in Figure 14.

An excellent discussion of polymer crystallinity has been provided by Uhlmann and Kolbeck and more recently by Geil. The extent of the tendency for crystallization is important from practical considerations. Crystallinity affects thermal as well as mechanical properties. Crystallinity developed in a polymer sample is a consequence of both thermodynamic and kinetic factors. Some polymers are crystalline because their structure is conducive to chain packing, e.g., high-density polyethylene (HDPE). Others, e.g., nylons, are crystalline because of strong secondary forces. Crystallinity in nylons can be significantly increased by stretching to facilitate ordering and alignment of polymer chains. Polystyrene, polyvinyl chloride, and polymethyl methacrylate show poor crystallization tendencies because the pendant groups stiffen the chains and this stiffening leads to difficulties in chain packing. Excessive rigidity in polymers due to high cross-linking, e.g., in phenol-formaldehyde. urea-formaldehyde, and melamine-formaldehyde, prevents crystallization completely. In contrast, excessive chain flexibility, e.g., polysiloxanes and natural rubber, leads to an inability for chain packing.

4.   Thermal Transitions

Polymer materials are characterized by two major transition temperatures. The crystalline melting temperature (T_m) is the melting temperature of the crystalline domains in a polymer sample. The glass transition temperature (T_g) is the temperature at which the amorphous domains of a polymer assume characteristic properties of the glassy state. The glass transition temperature (T_g) is the onset of the

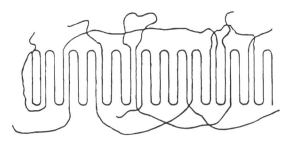

Figure 14   Schematic of lamella model of polymeric crystals.

rubbery state as the temperature is increased. Amorphous plastics are used below T_g, whereas elastomers must be used above T_g.

The thermal transitions depend on the morphology of the polymer. A completely amorphous polymer will have only a T_g, whereas a completely crystalline polymer will theoretically have only a T_m. Semicrystalline polymers exhibit both T_g and T_m. Generally, thermal transitions are measured by changes in properties, such as specific volume, heat capacity, stiffness (modulus), and refractive index. Popular techniques for determining thermal properties are differential scanning calorimetry (DSC), differential thermal analysis (DTA), thermomechanical analysis (TMA), and dynamic mechanical analysis (DMA).

DSC, DTA, TMA, and DMA analyses are all related to thermal behavior as a function of heating rate or time. In Figure 15, TMA scans of fluorenone polyesters with inscribed thermal transitions are shown. The modulus loss tangent data obtained for a fluorenone polyester are shown in Figure 16. The glass transition temperature, as indicated by loss modulus maxima, is 280°C. There are also various apparent T_g transitions. The peak damping value (tan $\delta > 1.0$) observed is comparable to that for polymers that exhibit a sharp transition from glassy to rubbery behavior.

5. Molecular Weights

Step-growth and chain polymerizations provide chains with many different lengths; therefore, average chain lengths and average molecular weight terms

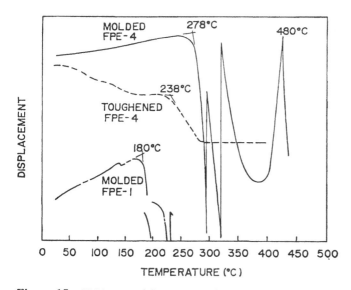

Figure 15   TMA scan of fluorenone polyesters.

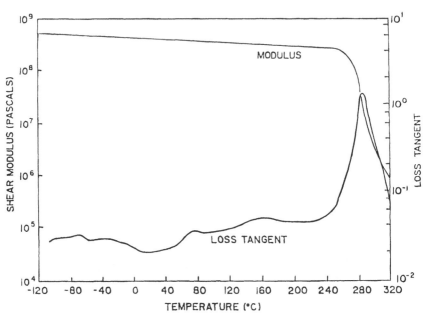

Figure 16   DMA of a fluorenone polyester sample.

should be used. The number average degree of polymerization (X_n) is defined as the average number of repeating units in the polymer chains. The number average molecular weight (M_n) equals X_n multiplied by the molecular weight of the repeating unit.

The molecular weights of polymers influence polymer properties in synthesis, processing, and end use applications. Low oligomers (low-molecular-weight polymers) have relatively low molecular properties below a critical X_n. The molecule begins to develop mechanical properties. e.g., tensile strength, at molecular weight greater than X_n as shown in Figure 17. In nylons that are highly hydrogen bonded, X_n may be as low as 50. Hydrocarbon polymers, e.g., polyethylene, must have X_n greater than 100 for this process to start. Above the threshold X_n, the mechanical properties increase rapidly with increasing molecular weight until a second higher critical molecular weight X_n is reached at which further increase in molecular weight produces very little change in mechanical properties. As would be expected, the value of the second critical X_n also depends upon the intermolecular bonding forces present. This X_n is roughly 200 in nylons and greater than 700 for polyethylenes.

Currently, much of the plastics processing is performed by extrusion and injection molding at relatively high temperatures, e.g., 260°C and very high pressures, e.g., 2900 kPa. Extrusion and injection molding involve high capital

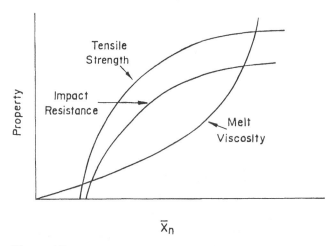

Figure 17 Dependence of properties on molecular weight in a hypothetical polymer.

costs and place practical limits on the size of the finished parts that can be produced.

A new process that has emerged into significant commercial utilization is reaction injection molding (RIM), wherein articles are molded directly from low-molecular-weight monomeric precursors at low temperature and low pressure. RIM is now being used to make automobile fascia and house window frames and doors. At present, RIM processing is limited almost entirely to polyurethanes, but other rapidly polymerizing thermosetting resins are being developed.

The concept of average molecular weight values can be easily understood by visualizing that there are 4 balls each weighing 40 kg and 100 balls each weighing 0.01 kg. Thus:

		Total kg
4	40.0 kg ball	160
100	0.01 kg ball	1
104		161

$$\text{Number average, } M_n = \frac{\text{total weight}}{\text{total number}} = \frac{161}{104} = 1.55 \text{ kg/ball}$$

Therefore,

$$M_n = \frac{\sum N_i M_i}{\sum N_i} = 1.55 \text{ kg/ball}$$

Also

$$\text{Weight average, } M_w = \frac{\text{mass of each species weighted by mass per ball}}{\text{total mass}}$$

$$= \frac{\sum N_i W_i}{\sum N_i M_i}$$

$$= \frac{\sum N_i M_i^2}{\sum N_i M_i}$$

$$= \frac{4 \times (40)^2 + 100 \times (0.01)^2}{161}$$

$$= 39 \text{ kg/ball}$$

Consider another example with three molecules having molecular weights of 1.00×10^5, 2.00×10^5, and 3.00×10^5:

$$M_n = \frac{1.00 \times 10^5 + 2.00 \times 10^5 + 3.00 \times 10^5}{1 + 1 + 1} = 2.00 \times 10^5$$

$$M_w = \frac{(1.00 \times 10^{10}) + (4.00 \times 10^{10}) + (9.00 \times 10^{10})}{6.00 \times 10^5} = 2.33 \times 10^5$$

Most thermodynamic properties are related to the number of particles present and are thus dependent on X_n. Bulk properties, e.g., viscosity and toughness, are particularly affected by M_w. However, melt elasticity is more closely dependent on M_z. The M_z is the third power average and is shown mathematically as follows:

$$M_z = \frac{\sum N_i M_i^3}{\sum N_i M_i^2}$$

Thus, the M_z for the preceding example will be

$$\frac{(1.00 \times 10^{15}) + (8.00 \times 10^{15}) + 27.00 \times 10^{15}}{(1.00 \times 10^{10}) + (4.00 \times 10^{10}) + (9.00 \times 10^{10})} = 2.57 \times 10^5$$

In a similar way, $z + 1$ and higher average molecular weights can be calculated but they are not of major interest. Also, for a polymer mixture it is clear that $M_z > M_w > M_n$. As the heterogeneity decreases the condition $M_z \geq M_w \geq M_n$ approaches. For a completely homogeneous polymer $M_z = M_w = M_n$. The ratio of M_w and M_n is often used to describe the molecular weight heterogeneity and is called the polydispersity index.

Various techniques are available to determine molecular weights. Number average molecular weight, M_n, can be determined by Raoult's techniques, which

are dependent on colligative properties, e.g., ebulliometry (boiling point eleva-
tion), cryometry (freezing point depression), osmometry, and end-group analysis.
Gel permeation chromatography (GPC) is the most popular method at present.
Typical M_w determination methods are light scattering (Zimm plot) and GPC.
The M_z values are usually determined by ultracentrifugation.

6. Polymer Solubility

The mixing process for a solvent (A) and a polymer (B) involves separation of
polymer molecules and their distribution in the entire solution. This solubility
and heat of solution will be closely dependent on cohesive energy density (CED).
For mixing to occur, the change in Gibbs free energy (ΔF) must be negative.

$$\Delta F = \Delta H - T\Delta S$$

where ΔH, ΔS, and T are change in enthalpy, change in entropy, and Kelvin
temperature, respectively.

In a thermodynamically ideal solution, ΔH is zero and ΔS is a function of
the composition of the mixture. Real solutions deviate from ideality in having
nonzero heats of mixing, or nonideal entropies of mixing, or both. A regular
solution, like the ideal solution, is an imaginary concept rarely seen in practice.
In the regular solution concept, the entropy of mixing follows the ideal law but
the heat of mixing is finite. Such solutions are easy to treat theoretically as
demonstrated by Hildebrand and Scott. For the regular solution it is assumed
that the internal latent heat of vaporization per unit volume may be identified
with cohesive energy density. The presence of intermolecular forces consisting
only of London dispersion forces is also assumed. The function used in calculations is the square root of the cohesive energy density commonly called the
solubility parameter (δ). The heat of mixing ΔH is

$$\Delta H = u_M \phi_A \phi_B (\delta_A - \delta_B)$$

where u_M is the volume of the mixture, ϕ_A and ϕ_B are volume fractions of A
and B, and δ_A and δ_B are solubility parameters of A and B.

For solubility, ΔF must be negative (zero is the limiting case). For the
limiting case, $\Delta H = T\Delta S$. Since $T\Delta S$ is negative, it is sufficient that $\Delta H = 0$,
that is, $\delta_A = \delta_B$ for solubility. Experimental methods are available for determining
the solubility parameter of solvents and other low-molecular-weight compounds.
However, δ can be calculated from the repeating units of a polymer or estimated
as the midpoint of the solubility parameter range of solvents in which the polymer
is found to be soluble.

Highly hydrogen-bonded materials violate the assumptions of regular solution theory and therefore the solubility predictions are less reliable than for
nonpolar materials. Hansen and Burrell have accounted for this deficiency by
taking into account the hydrogen bonding; classifying solvents as strongly, moder-

ately, or poorly hydrogen bonded; and considering the ranges of solubility parameter consistent with solubility in a specific class. For polar systems, δ is replaced by the parameter β, which also takes into account the dipole interaction parameter ϵ.

$$\beta = (\delta_A - \delta_B)^2 + (\epsilon_A - \epsilon_B)/T$$

Further refinements have been made by introducing a quantitative measure of hydrogen bonding that, when used along with dielectric constants and solubility parameters, can be used to predict solubility by a three-dimensional plot method.

Flory and Huggins independently developed a more rigorous treatment of polymer solutions. In this theory, polymer molecules are considered to consist of a large number of identical repeating units, such as a, b, c. Then the distribution of the molecule over a lattice is used in the statistical thermodynamic calculation of the entropy of mixing of small molecules. If unit a is placed at a point in the lattice, then unit b can be placed only at points that are at a unit distance from a and so on. If the solution is concentrated enough, the occupied lattice sites are distributed at random. Thus, the entropy of mixing of polymer solutions must be smaller for simple molecules because there will be fewer ways in which the same amount of lattice sites can be occupied. Using this model, the Flory-Huggins treatment provides an expression for the partial molar Gibbs free energy of dilution, which includes the Flory-Huggins interaction parameter, x_1. The Flory-Huggins theory has limitations but may be used to predict the equilibrium behavior between liquid phases containing an amorphous phase. The theory may be also used to predict the cloud point in polymer solutions.

The shortcomings of the Flory-Huggins lattice theory were overcome by Flory and Krigbaum by assuming the presence of an excluded volume from which long-range interactions can begin. A polymer molecule is in the theta state in a θ solvent at a given temperature and behaves like an ideal statistical coil. The θ temperature, also called Flory temperature, can be defined as a critical miscibility temperature at the limit of infinite molecular weight.

7. Polymer Compatibility

Generally, compatibility between polymers is poor and phase separation in solution is common. Like solubility, compatibility is treated theoretically by considering the free energy of mixing of the system. A typical compatible blend is polypropylene and the copolymer of ethylene and vinyl acetate, which can be used as a superior liner film for disposable diapers.

8. Polymer Viscosity in Dilute Solutions

The ratio of the viscosity of a polymer solution to that of the solvent is known as relative viscosity (η_r). Specific viscosity (η_{sp}) is $\eta_r - 1$. The intrinsic viscosity $[\eta]$ is η_{sp}/c, as c (concentration) approaches a zero value.

The relationship between the intrinsic viscosity and the molecular weight of a dissolved polymer for a specific polymer-solvent system at a given temperature is given by:

$$[\eta] = KM^\alpha$$

where K and α are empirical parameters. The parameter α equals 0.5 at the θ temperature and may approach 0.8 for very good solvents. The determination of molecular weights by viscometry has been discussed by Meyerhoff.

9. Viscoelasticity and Mechanical Properties

Polymers show a combination of solid and liquid properties, e.g., elasticity and viscosity, and are therefore viscoelastic in nature.

Let us now consider the rapid application of a stress to such a material. Elastic deformation will occur instantaneously, but this will gradually give way to flow and thus the stress will be relieved. The stress decays exponentially with time. The rate of this stress relaxation is characterized by the relaxation time, which is defined as the time in which the stress decays to $1/e$ of its initial value. Therefore, the higher the stress applied, the greater will be the force inducing flow and the faster the decay. Stress relaxation is mathematically expressed as follows:

$$T = T_0 e^{-t/\lambda}$$

where T is the residual stress T_0 the initial stress, t the time of stress decay, and λ the relaxation time of the material.

In a similar way, the process of stress application leads to the concept of a retardation time regulating the approach of an equilibrium rate of shear in the material after the stress is applied.

Flow can be seen as a process of continued stress application and stress relaxation. Thus, viscosity is related to the relaxation time and the retardation time in a rather complex way. General rheological behavior can be depicted in terms of a visco-elastic spectrum, which is associated with a spectrum of relaxation and retardation times. Rheological studies involve analysis of the retardation-relaxation-time pattern. Excellent treatments of the subject matter have been provided by Tobolsky, Nielsen, Ferry, and Meares.

BIBLIOGRAPHY

Arthur, J. C., Jr. (ed.), *Polymers for Fibers and Elastomers*, ACS Symp. Ser. No. 260, American Chemical Society: Washington, DC (1984).

Ash, M., and Ash, I., *Encyclopedia of Plastics, Polymers and Resins*, Chemical Publishing Co., New York (1982).

Batzer, H., and Lohse, F., *Introduction to Molecular Chemistry*, Wiley, New York (1979).

Bawn, C. E. H., *Macromolecular Science*, Buttersworth, Woburn, MA (1975).

Billmeyer, F. W., Jr., *Textbook of Polymer Science*, 3rd ed., Wiley, New York (1984).

Bolker, H. I., *Natural and Synthetic Polymers: An Introduction*, Academic Press, New York (1974).

Bovey, F. A., and Winslow, F. H. (eds.), *Macromolecules, an Introduction to Polymer Science*, Academic Press, New York (1979).

Brandup, J., and Immergut, E. H. (eds.), *Polymer Handbook*, 3rd ed., Wiley, New York (1989).

Brydson, J. A., *Plastics Materials*, 3rd ed., Buttersworth, Woburn, MA (1975).

Burrell, H., Solubility Parameters, *Off. Dig. Fed. Paint Soc.*, **27**, 726 (1955).

Carothers, W. H., Types of Polymerization, *J. Am. Chem. Soc.*, **51**, 2548 (1929).

Carraher, C. E., and Tsuda, M. (eds.), *Modification of Polymers*, ACS Symp. Ser. No. 121, American Chemical Society, Washington, DC (1980).

Carraher, C. E., and Moore, J. A. (eds.), *Modification of Polymers, Polymer Science and Technology*, Vol. 21, Plenum, New York (1983).

Cassidy, P. E., *Thermally Stable Polymers*, Marcel Dekker, New York (1980).

Chanda, M., and Roy, S. K., *Plastics Technology Handbook*, Marcel Dekker, New York (1987).

Chompff, A. J., and Newman, S. (eds.), *Polymer Networks: Structure and Mechanical Properties*, Plenum, New York (1971).

Cowie, J. M. G., *Polymers: Chemistry and Physics of Modern Materials*, Intext, New York (1974).

Craver, C. D. (ed.), *Polymer Characterization*, Plenum, New York (1971).

Crowley, J. D., Teague, G. S., and Lowe, J. W., Solubility Parameters, *J. Paint Technol.*, **38**, 296 (1966).

Davidson, T. (ed.), *Polymers in Electronics*, ACS Symp. Ser. No. 242, American Chemical Society, Washington, DC (1984).

Driver, W. E., *Plastics Chemistry and Technology*, Van Nostrand-Reinhold, Cincinnati (1974).

Elias, H. G., *Macromolecules*, Vols. 1 and 2, Plenum, New York (1976).

Erickson, J. L. (ed.), *Orienting Polymers*, Lecture Notes in Mathematics 1063, Springer-Verlag, Berlin (1984).

Ferry, J. D., *Viscoelastic Properties of Polymers*, Wiley, New York (1961).

Fettes, E. M. (ed.), *Chemical Reactions of Polymers*, Wiley, New York (1964).

Flory, P. J., Polymer Solutions, *J. Phys. Chem.*, **46**, 151 (1942).

Flory, P. J., *Principles of Polymer Chemistry*, Cornell University Press, Ithaca, NY (1953).

Geil, P., Polymer Crystals, *JEMMSE*, **3**, 1 (1981).

Gibson, G. W., and Granito, C. E., *Witwesser Chemical Line Notation, an Introduction*, Merck Index, Merck, Rahway, NJ (1976).

Goddard, E. D., and Vincent, B. (eds.), *Polymer Adsorption and Dispersion Stability*, ACS Symp. Ser. No. 240, American Chemical Society, Washington, DC (1984).

Gordon, M., and Platé, N. A. (eds.), *Liquid Crystal Polymers I*, Advances in Polymer Science 59, Springer-Verlag, Berlin (1984).

Gordon, M., *High Polymers, Structure and Physical Properties*, Addison-Wesley, New York (1964).

Hall, I. H. (ed.), *Structure of Crystalline Polymers*, Elsevier, London (1984).

Ham, G. E. (ed.), *Copolymerization*, Wiley, New York (1964).

Han, C. D. (ed.), *Polymer Blends and Composites in Multiphase Systems*, Advances in Chemistry Series 206, American Chemical Society, Washington, DC (1984).

Hansen, C. M., Solubility Parameters, *J. Paint Technol.*, **39**, 104 (1967).

Harper, C. A. (ed.), *Handbook of Plastics and Elastomers*, McGraw-Hill, New York (1975).

Hienmenz, P. C., *Polymer Chemistry: The Basic Concepts*, Marcel Dekker, New York (1984).

Hildebrand, J. H., and Scott, R. L., *Regular Solutions*, Prentice-Hall, Englewood Cliffs, NJ (1962).

Hildebrand, J. H., and Scott, R. L., *The Solubility of Nonelectrolytes*, Reinhold, New York (1950).

Huggins, M. L., Polymer Solutions, *J. Phys. Chem.*, **46**, 151 (1942).

Jenkins, A. D., *Polymer Science*, Vols. 1 and 2, Elsevier, New York (1972).

Kaufman, H. S., and Falcetta, J. J. (eds.), *Introduction to Polymer Science and Technology*, Wiley, New York, 1977.

Kaufman, M., *Giant Molecules*, Doubleday, New York (1968).

Keller, A., Polymer Crystals, *J. Polym. Sci. Symp.*, **59**, 1 (1977).

Kinloch, A. J. (ed.), *Durability of Structural Adhesives*, Applied Science Publishers, London (1983).

Kinsey, R. H., Ionic Polymer, *Modern Plastics Encyclopedia*, Vol. 57, p. 34 (1980).

Kirshenbaum, G., *Polymer Science Study Guide*, Gordon & Breach, New York (1973).

Kirshenbaum, G. S., *Polymer Nomenclature Committee ACS, Polymer News*, **9**, 101 (1983); **10**, 169 (1983).

Klempner, D., and Frisch, K. C. (eds.), *Polymer Alloys, Polymer Science and Technology*, Vol. 10, Plenum, New York (1977).

Kumar, A., and Gupta, S. K., *Fundamentals of Polymer Science and Engineering*, Tata McGraw-Hill, New Delhi (1978).

Labana, S. S. (ed.), *Chemistry and Properties of Cross-Linked Polymers*, Academic Press, New York (1977).

Labana, S. S., and Dickie, R. A. (eds.), *Characterization of Highly Cross-Linked Polymers*, ACS Symp. Ser. No. 243, American Chemical Society, Washington, DC (1984).

Lee, L. H. (ed.), *Adhesion and Absorption of Polymers*, Polymer Science and Technology, Vol. 12A, Plenum, New York (1980).

Lenz, R. W., *Organic Chemistry of Synthetic High Polymers*, Wiley, New York (1976).

Lenz, R. W., and Stein, R. S. (eds.), *Structure and Properties of Polymer Films*, Polymer Science and Technology, Vol. 1, Plenum, New York (1973).

London, F., Intermolecular Forces, *Trans. Faraday Soc.*, **33**, 8 (1937).

Mandelkern, L., *An Introduction to Macromolecules*, Springer-Verlag, New York (1973).

Mandelkern, L., Polymer Crystals, *Acct. Chem. Rev.*, **9**, 81 (1976).

Martuscelli, E., Palumbo, R., and Kryszewski, M. (eds.), *Polymer Blends: Processing, Morphology and Properties*, Plenum, New York (1980).

Mathias, L. J., and Carraher, C. E. (eds.), *Crown Ethers and Phase Transfer Catalysis in Polymer Science*, Plenum, New York (1984).

Meares, P., Rheology, in *Polymers, Structures and Bulk Properties*, Princeton University Press, Princeton, NJ (1965).

Meyerhoff, G., Viscometry, *Fortschr. Hochpolym. Forsch.*, **3**, 59 (1961).

Mittal, K. L. (ed.), *Physicochemical Aspects of Polymer Surfaces*, Vols. 1 and 2, Plenum, New York (1983).

Mittal, K. L. (ed.), *Adhesive Joints: Formation, Characterization and Testing*, Plenum, New York (1984).

Moore, J. A. (ed.), *Reactions of Polymers*, Reidel, Dordrecht. The Netherlands (1973).

Morawetz, H., *Macromolecules in Solution*, Wiley, New York (1965).

Morawetz, H., and Zimmering, P. E., Rates of Polymeric Reactions, *J. Phys. Chem.*, **58**, 753 (1954).

Munk, P., *Introduction to Macromolecular Science*, Wiley, New York (1989).

Nielsen, L. E., *Mechanical Properties of Polymers*, Reinhold, New York (1962).

Odion, G., *Principles of Polymerization*, Wiley, New York (1981).

Parker, D. B. V., *Polymer Chemistry*, Applied Science, Barking, UK (1974).

Pasika, W. M. (ed.), *Advances in Macromolecular Chemistry*, Vols. 1 and 2, Academic Press, London (1968 and 1970).

Pethrick, R. A., and Zaikov, G. E., *Polymer Yearbook*, Vol. 3, Horwood Academic Publishers, London (1986).

Pizzi, A. (ed.), *Wood Adhesives: Chemistry and Technology*, Marcel Dekker, New York (1973).

Rabek, J. F., *Experimental Methods in Polymer Chemistry*, Wiley, New York (1980).

Ravve, A., *Organic Chemistry of Macromolecules: An Introductory Textbook*, Marcel Dekker, New York (1967).

Rodriquez, F., *Principles of Polymer Systems*, 3rd ed., Hemisphere, New York (1989).

Roff, W. J., and Scott, J. R., *Fibers, Films, Plastics, and Rubbers: A Handbook of Common Polymers*, Buttersworth, Woburn, MA (1971).

Rosen, S. L., *Fundamental Principles of Polymeric Materials for Practicing Engineers*, Cahners, Boston (1971).

Rowell, R. (ed.), *The Chemistry of Solid Woods*, Advances in Chemistry Series 207, American Chemical Society, Washington, DC (1984).

Saegusa, T., and Dall'Asta, G. (eds.), *Polymerization Reactions*, Advances in Polymer Science 58, Springer-Verlag, Berlin (1984).

Saunders, K. J., *Organic Polymer Chemistry*, Wiley, New York (1973).

Schultz, J., *Polymer Material Science*, Prentice-Hall, Englewood Cliffs, NJ (1974).

Seymour, R. B., *Introduction to Polymer Chemistry*, McGraw-Hill, New York (1971).

Seymour, R. B., *Modern Plastics Technology*, Reston, VA (1975).

Seymour, R. B., Solubility Parameters of Organic Compounds, in *Handbook of Chemistry and Physics*, C-720, 56th ed., CRC Press. Orlando, FL (1975).

Seymour, R. B., *Conductive Polymers*, Plenum, New York (1981).

Seymour, R. B., *Introduction to Polymer Chemistry and Technology*, Audio Triple Action Impact Course, American Chemical Society, Washington, DC (1981).

Seymour, R. B. (ed.), *History of Polymer Science and Technology*, Marcel Dekker, New York (1982).

Seymour, R. B., Introduction to Polymer Chemistry and Technology (an Audio-Course), American Chemical Society, Washington, DC (1982).

Seymour, R. B., *Plastics vs. Corrosives*, SPE Monograph Series, Wiley, New York (1982).

Seymour, R. B., *Engineering Polymers Sourcebook*, McGraw-Hill, New York (1989).

Seymour, R. B., *Polymer Composites*, USP, Zeist, The Netherlands (1989).

Seymour, R. B., *Modern Plastic Composite Technology*, ASM International, Materials Park, OH (1991).

Seymour, R. B., and Carraher, C. E., *Structure Property Relationships in Polymers*, Plenum, New York (1984).

Seymour, R. B., and Carraher, C. E., *Polymer Chemistry: An Introduction*, 2nd ed., Marcel Dekker, New York (1988).

Seymour, R. B., and Carraher, C. E., *Giant Molecules*, Wiley, New York (1990).

Seymour, R. B., and Cheng, *Advances in Polyolefins*, Plenum, New York (1987).

Seymour, R. B., and Mark, H. F., *Applications of Polymers*, Plenum, New York (1987).

Seymour, R. B., and Mark, H. F., *Handbook of Organic Coatings*, Elsevier Science Press, New York (1990).

Seymour, R. B., and Mark, H. F., *Handbook of Manmade Fibers*, Elsevier Science Press, New York (1991).

Seymour, R: B., and Steiner, R. H., *Plastics for Corrosion Resistant Applications*, Reinhold, New York (1955).

Sharples, A., Crystallinity in Polymers, in *Polymer Science*, (A. D. Jenkins, ed.), North-Holland, Amsterdam (1972).

Small, P. A., Solubility Parameters, *J. Appl. Chem.*, **3**, 71 (1953).

Sorenson, W. R., and Campbell, T. W., *Preparative Methods of Polymer Chemistry*, 2nd ed., Wiley, New York (1968).

Sperling, L. H. (ed.), *Recent Advances in Polymer Blends, Grafts, and Blocks*, Polymer Science and Technology, Vol. 4, Plenum, New York (1974).

Stahl, G. A. (ed.), *Polymer Science Overview*, ACS Symp. Ser. No. 175, American Chemical Society, Washington, DC (1981).

Stevens, M. P., *Polymer Chemistry: An Introduction*, 2nd ed., Oxford University Press, New York (1990).

Strepikheyev, A., Derevitskaya, V., and Slonimsky, G., *A First Course in Polymer Chemistry*, Mir Publications, Moscow (1971).

Tager, A., *Physical Chemistry of Polymers*, Mir Publications, Moscow (1972).

Tobolsky, A. V., *Properties and Structures of Polymers*, Wiley, New York (1960).

Treloar, L. R. G., *Introduction to Polymer Science*, Springer-Verlag, New York (1970).

Uhlmann, D. R., and Kolbeck, A. G., Polymer Crystals, *Sci. Am.*, **233**, 96 (1975).

Usmani, A. M., Fabrication of Polymers, *J. Macromol. Sci. Chem.*, **A18**(2), 251 (1982).

Usmani, A. M., Polymer blends, *J. Macromol. Sci. Chem.*, **A22**, 293 (1985).

Usmani, A. M., and Rheineck, A. E., Electrodeposition of Polymers, *Adv. Chem. Ser.*, **119**, 130 (1973).

Usmani, A. M., and Salyer, I. O., Chemical Properties of Polymers, *J. Macromol. Sci. Chem.*, **A13**(7), 937 (1979).

Usmani, A. M., and Salyer, I. O., *J. Sci. Ind. Res.*, **38**, 555 (1979).

Usmani, A. M., and Salyer, I. O., Effect of Pendant Group on Polymer Properties, *J. Mater. Sci.*, **16**, 915 (1981).

Usmani, A. M., and Salyer, I. O., in *Use of Renewable Resources for Polymer Applications* (C. E. Carraher and L. H. Sperling, eds.), Plenum, New York (1982).

Usmani, A. M., and Salyer, I. O., Spherulite Structure of Crystalline Polymers, *J. Elastomers Plastics*, **12**, 90 (1990).

Usmani, A. M., Salyer, I. O., Butler, J. M., Chartoff, R. P., and Burkett, J. L., *ACS Organic Coatings and Plastics Chemistry Proceedings*, **44**, 434 (1981).

Vollmert, B., *Polymer Chemistry*, Springer-Verlag, New York (1973).

Walton, A. G. (ed.), *Structure and Properties of Amorphous Polymers*, Elsevier, Amsterdam (1980).

Williams, D. J., *Polymer Science and Engineering*, Prentice-Hall, Englewood Cliffs, NJ (1971).

Williams, H. L., *Polymer Engineering*, Elsevier, New York (1975).

Williams, J. G., *Fracture Mechanics of Polymers*, Ellis Horwood, Chichester, UK (1984).

Young, R. J., *Introduction to Polymers*, Chapman & Hall, London (1981).

16

Factors Influencing Compounding of Constituents in Bitumen–Polymer Compositions

Janusz Zielinski
Warsaw University of Technology, Płock, Poland

I. INTRODUCTION

Petroleum asphalts are used as the fundamental raw materials to secure various types of end products. Especially significant are those having insulating-seal and coating properties. These types of asphalts are intended primarily for the road and construction industries. Growing requirements with respect to the qualities of these products have forced the modification of certain properties of asphalts, namely the elimination of undesirable characteristics such as narrow range of application temperatures and appreciable brittleness at low temperatures, as well as unsatisfactory thixotropic and adhesion properties. Especially useful changes in the functional properties of asphalts can be obtained through the addition of polymers.

The puzzle of asphalt–polymer compositions concerns problems at the boundary of two scientific disciplines: chemistry and polymer and petroleum technology. The complexity of this puzzle is the cause of as yet unexplained phenomena, such as mixing of bitumens with polymers, the effect of interactions of individual components on the properties of compositions, or the suitability of applied experimental methods. The patent literature is the primary source of information in the area of asphalt–polymer compositions. Most of the recent research work concerning these compositions describes formulations, methods of preparation, and practical applications. The resulting compositions have been the subject of a large number of investigations.

A.   Methods of Modification

The modification of asphalts by various additives, such as high-molecular-weight compounds, requires the recognition of critical phenomena like the miscibility of the composition's components. One must also recognize how to obtain specific properties based on such modifications. Methods whose purpose is to contribute suitably beneficial properties are defined as either chemical or physical methods.

1.   Chemical Methods

Chemical methods rely upon the action of molecules or chemical compounds with the constituents of an asphalt, which bring about changes in the asphalt's chemical composition. These include reaction with oxygen, sulfur, sulfur dioxide, and chlorine. Various properties are thus obtained from such a treatment. These methods, especially oxidation, have been applied in industry for many years to obtain oxidized asphalts from vacuous petroleum residues. A separate problem is the chemical modification of asphalt based on the direct acquisition of polymers in an asphalt environment. Several types of polymer reaction are listed below.

> Radical polymerization: Gundermann [1], among others, used experiments dedicated to the polymerization of styrene in an asphalt environment as a basis. He promoted the idea of inhibiting the reaction using asphaltenes.
> Polymerization via an ionic approach: obtaining an asphalt–polybutadiene composition in the presence of an alkali metal catalyst [2].
> Polycondensation: reactions with phenol-type compounds found in tars produced from the alkylation of propylene with benzene; direct reactions with formalin in asphalt medium [3].
> Polyaddition: experiments performed on an asphalt–polyurethane composition, obtained by an addition reaction in a molten asphalt environment. It was found that the reaction for the formation of polyurethane in asphalt occurs without difficulty and with high yields when run at 145°C for 2–6 h [4].
> Copolymerization: experiments performed by irradiating an aqueous dispersion of vinyl monomer in direct contact with road asphalt [5], or methyl methacrylate [6], with visible light of a chosen wavelength.

2.   Physical Methods

Physical methods are based on mixing asphalt with inert chemical additives but changing the composition's properties. For example, bitumen can be modified with inorganic substances, primarily of mineral origin, such as talc or chalk (used as fillers), as well as organic materials like oil-based plasticizers, antioxidants, and polymers.

Among the various physical methods, two different paths to asphalt modification by polymers can be distinguished. In the first, no further modification can be made when the polymer content is the range of 10–20%. Most of the applications of these materials are as adhesives in highway engineering. The second path involves obtaining mixtures of composite materials, which contain asphalt, polymers, fillers, plasticizers, and other components. Materials of this type are not dominated by the contents of one of the components and find applications in the construction industry and in motorization, as well as in other specific disciplines. Another related path is modifying polymers through the addition of petroleum asphalts (e.g., filling, plasticizing). The range of investigations in this last area is not large. Applications are similar to those that have already been discussed.

II. MISCIBILITY OF THE COMPOSITION'S COMPONENTS

The inability of polymers to mix with asphalts in an unlimited fashion is a rather general phenomenon. The behavior of bitumens under the influence of polymer additives has not yet been completely explained. It is generally accepted that bitumen–polymer systems are heterogeneous mixtures of varying character and degree of miscibility of their components. Choosing the proper components for these mixtures is not simple, because of the difficulties related to their miscibility. A properly prepared composition should be at least microheterogeneous. Stabnikov [7], using the concept of miscibility, found a way to disperse polymen uniformly ensuring in bitumen a homogeneous material, as verified by its physicochemical properties.

Rozental et al. [8] distinguished the following fundamental elements, which determine the likelihood of obtaining homogeneous asphalt–polymer mixtures:

1. The polymer should swell or be soluble in aliphatic or aromatic solvents and should distinguish itself by having a structure similar to that of oils and asphalt resins. If the polymer does not display such characteristics, then a third component that mixes well with bitumen needs to be introduced.
2. It is most beneficial to introduce the polymer into the asphalt in the form of a latex or oligomer. It is usually necessary to break up hard and brittle polymers as a first step.
3. In the case of applications involving highly oxidized asphalts containing relatively small amounts of oil-based components, the polymer should be plastified by the addition of aromatic or naphthenic oils.

4. It is beneficial to mix polymers with asphalts at temperatures that will ensure the elimination of processes that might degrade or destroy the polymer. As a rule, the mixing temperature should not be below 150°C.

The miscibility of components of asphalt–polymer compositions is influenced by two factors. The first is the method by which the composition is obtained. It is especially important to determine the proper choice of parameters, such as temperature, mixing time, and speed of mixing. The second factor is related to the properties and structure of the asphalt and the polymer, as well as the source and amount of the added polymer.

Work performed by Zielinski has shown that the comiscibility of components of asphalt–polymer compositions is dependent on both internal and external factors. Of the external factors, temperature is the most important. In the case of application of multicomponent structures, the proper order of addition of components is as important as temperature. Decisive internal factors include amount of oils and asphaltenes in the bitumen, the bitumen's degree of crystallinity, molecular weight, macromolecular structure, and amount of polymer.

III. METHODS

The simplest method for obtaining asphalt–polymer compositions is to introduce a polymer solution or dispersion into the asphalt, followed by mixing (e.g., in the production of paints, lacquers, and road surfaces [9,10]). Obtaining kinetically stable asphalt–polymer solutions in organic solvents, however, is dependent on several external and internal factors, as demonstrated by Zielinski and Milczarska [11].

Another method for obtaining compositions is to use a molten-state scenario. The asphalt is mixed with a polymeric thermoplastic component in an appropriate mass ratio. A modification of this method is the homogenization of both components by milling. This process is generally applied to high-molecular-weight polymers or to those that are not prone to swelling.

Both ready-to-use products and concentrates can be prepared by the methods described above. The products can be used in subsequent technical operations. Besides asphalt–polymer concentrates, polymer-filler concentrates also find use here [12]. A microscopic method for examining asphalt–polymer compositions (e.g., polypropylene and others) has been developed [13]. Using this method, one can assess the influence of composition preparation technology on the degree of mixing of the components.

IV. PARAMETERS

One of the fundamental operations that determines the miscibility of constituents in asphalt–polymer compositions is the optimization of the parameters used to prepare these systems. The choice of temperatures or mixing times of the mixture (with respect to homogeneity of the components) is dependent on several factors, such as the nature of the components, the degree of oxidation of the asphalt, and the structure, molecular weight, and properties of the polymer.

As a rule, the temperatures used for preparing the compositions are in the range 130–250°C, and the time needed to achieve homogeneity does not exceed 3 h. It is especially important to establish a temperature for a particular type of asphalt with a group of polymers (e.g., polypropylene or polyethylene), so as not to initiate degradation or destruction of the polymer. The physical modification of asphalts (e.g., with olefinic polymers) is dependent on mixing the components at elevated temperatures (135–250°C). The actual temperature is determined by the type of asphalt. This process occurs relatively easily and generally does not require the introduction of additional components, like solvents or plasticizers [14].

Rubbers are most susceptible to being destroyed at elevated temperatures [15]. Natural and synthetic rubbers were the first materials to be applied on a large scale as polymer additives for asphalts. There are four common ways to mix rubbers with asphalts. The first uses a latex admixture with the softened asphalt and homogeneity is achieved with temperatures under 80°C. The second way is to use a latex admixture with an asphalt emulsion. The third is to use a rubber admixture added to a fluidized asphalt. This can be applied at room temperature or warmed up to 40–80°C. The final technique uses a rubber or powdered rubber admixture with asphalt. The mixture is heated until a homogeneous binder is obtained. Rozental et al. [16] have found that rubbers cannot be exposed to temperatures of 180–200°C for more than 2 h without having their fundamental properties adversely affected. It is significant to note, however, that asphalts are commonly heated to such temperatures.

An important element in the production of asphalt–polymer compositions is the preparation of raw materials. As a rule, asphalts are introduced into such mixtures in the one of the following forms: solid or semisolid, powder, granules, or as an emulsion. Most often, the asphalt is introduced into the mixture without any preliminary preparation and is simply heated above its softening point. Polymers that are added to asphalts are typically in the form of tablets, granules, irregular chips (generally thermoplastic polymers), latexes, powders or liquid hydrocarbon solutions (refers primarily to rubbers), prepolymers (e.g., polyurethane), and concentrates with asphalt (generally polyolefins).

V. CONSTITUENTS OF THE COMPOSITIONS

The miscibility of components of asphalt–polymer compositions is limited by:

1. The structure and properties of the asphalt. Only highly polar polymers are found in this group, having the following preferred characteristics:

a small degree of complexity and a regular spatial structure, moderate van der Waals forces, viscoelastic properties in a wide temperature range, and a reasonably low vitrification temperature. Asphalts are poor solvents for most polymers, despite containing compounds like petroleum and aromatic oils, as well as resins and asphaltenes, which promote precipitation of polymers in an asphalt environment. For every asphalt there exists a limit to the polymer content. If this limit is exceeded, a sudden change in the properties of the components occurs. It may also become impossible for a homogeneous mixture to be formed. Asphalts that contain a large amount of aromatic oils mix well with polymers having aromatic or unsaturated chains, whereas asphalts from paraffin oil mix well with polymers having saturated chains [17]. Kraus and Rollman [18] hypothesized that the influence of polymer on asphalt properties is greater when the amount of low-molecular-weight oils in the asphalt is large.

2. The origin and amount of the applied polymer. Many investigations have been performed regarding the modification of asphalt properties by use of the following groups of polymers: polyolefins (polyethylene, polypropylene, polybutene, and their copolymers), rubbers and powdered rubber, thermoplastic polymers (polyvinyl chloride, polystyrene, polyvinyl octane, polyvinyl alcohol, polybutadiene), hardenable plastics (epoxy resins, polyester, or silicone as auxiliary materials), thermohardenable plastics (phenol-formaldehyde resins).

A. Polyolefins

The amount of added polyolefin depends on the degree of oxidation and the chemical structure of the asphalt and usually does not exceed 15% w/w. Asphalts modified by polyethylene, polypropylene, or polybutene are characterized by and compared to the original asphalts in the following areas [19,20]: an expanded range of plasticity and a concurrent expanded range of utilization (an increase in the softening temperature and a decrease in the fragility temperature), smaller penetration, improved adhesion to crystal mineral surfaces, and improved resistance to aging. Asphalts modified by atactic polypropylene are distinguished by plastoelastic features, which increase when an oil-based plasticizer has been added [21].

Homogeneous and stable compositions are obtained from the permanent external structure of asphalt. These rheological compositions behave as non-Newtonian liquids in terms of their structural viscosity [22]. Compared to unmodified asphalt, asphalt–polymer compositions distinguish themselves by increased thixotropic properties, especially in the case of groups that display a homogeneous structure and contain increased amounts of polymer. These properties grow as the degree of oxidation of the asphalt increases.

B. Rubbers

A large number of investigations concerning rubber–bitumen compositions can be found in the literature. In addition to natural rubbers, many synthetic rubbers have been used as additives to modify asphalt. Among these types are butadiene, butadiene–styrene, chloroprene, and butadiene–acrylonitrile rubbers. Especially advantageous properties can be attained with asphalts modified by styrene–butadiene–styrene copolymers [23,24]. These materials are used primarily in the road construction industry. The most beneficial concentration of added rubber is 3–6% w/w. This ensures that the resulting composition will have the necessary homogeneity.

Several benefits are realized when rubber formulations are added to asphalts: an increase in the softening temperature (natural rubber and neoprene in latex form are most effective), an increase in asphalt ductility (mainly at low temperatures), increased adhesion to aggregates, increased resistance of the asphalt to the effects of motor fuel, improvements in resistance to aging, and an increase in deformative elastic recovery.

C. Thermoplastic Polymers

Materials of this type are typically added to asphalts at concentrations of 6–8% w/w. These are especially used with hard asphalts, oxidized asphalts, and asphalts that contain a small amount of oil fraction. Asphalts modified by thermoplastic polymers are characterized by low penetration and ductility, high softening temperature, good adhesion, and improved resistance to both aging and the effect of diesel oil.

D. Chemically Cured Materials

From this group of materials, epoxy and polyester resins are the most commonly used substances for the improvement of asphalts. Adding an epoxy resin, for example, to road asphalt results in an improvement in its adhesion to aggregates; increases in resistance to aging, abrasion, and slipperiness; and increases in elasticity and in resistance to organic solvents.

E. Thermally Cured Materials

Only uncured phenol-formaldehyde resins can be used as modifiers for asphalts. Asphalts modified in this way are characterized by increased penetration and ductility at 25°C, as well as lower softening and fragility temperatures. Currently preferred are thermoplastic polymers (e.g., divinylthermoelastoplastic). Adding these polymers at 2–3% w/w ensures the asphalt's complete elastic recovery in the temperature range −30 to 60°C [25].

VI. PHYSICAL STRUCTURE OF THE COMPOSITIONS

Because of the complex physicochemical form of asphalt–polymer compositions, their structures are discussed hypothetically. The high molecular weight of many polymers makes it difficult to dissolve them in asphalt. For this reason, a considerable number of approaches dealing with the structure of asphalt–polymer compositions suggest utilizing the polymer either in dissolved form, in a swelled state (partial dissolution), or in the form of a dispersion (macromolecular dissolution). The existence of any of these forms of the polymer is dependent on its properties, as well as those of the asphalt.

Most studies have shown that asphalt–polymer compositions are mixtures that are found most often in the form of a solution or a dispersion. According to Rozental et al. [8], these compositions should be considered to be three-component structures containing solvents (asphalt oils) into which asphaltenes and polymer are diffused. In most of the cases, the polymers are unreactive relative to the asphalt (i.e., they do not undergo chemical reactions). The degree of dispersion is dependent on the mixing conditions, especially temperature. At higher mixing temperatures, the polymer finds itself in a softened state, a condition that reduces the uniformity of its dispersion into the asphalt.

Barth [26] believes that a polymer dissolved in asphalt can be the cause of high elasticity at low temperature. A portion of the polymer, however, remains in a state of macromolecular diffusion. It may then simultaneously influence increases in mechanical strength and reduced plastic flow.

The mechanism of swelling is likely to be the cause of the composition's elastic characteristics. It has been observed that, upon introduction of the polymer into the composition, the structure takes on elastomeric qualities rather than viscoelastic ones [27]. The effect of swelling of the polymer in asphalt (in a 1:1 ratio of dissolved phase to swelled phase) is dependent on the parameters of preparation of the composition, the chemical and molecular structure of the components, and their properties.

Beem and Brasser [28] established that polymer adsorbed by the paraffin oil contained in asphalt causes the asphalt to harden and, at the same time, alters its properties. When rubber flakes are mixed with asphalt, it has been observed that the granules absorb hydrocarbons contained in the oils, effecting the swelling of the rubber granules. Following the loss of part of the oil component, asphaltenes become more active in this arrangement, resulting in hardening of the asphalt.

The preceding hypotheses partially elucidate the mechanism governing the ability of polymers to mix with asphalts and do not describe all of the discussed structures.

According to Zelezko and Pecenyj [29], polyethylene and isotactic polypropylene form a coarse-grained suspension when mixed with asphalt. In this suspension, polymer molecules absorb the oil-based hydrocarbons contained in the asphalt. At the same time, they also function as fillers. With respect to widening the range of plasticity of oxidized or partially oxidized asphalts, the best additives are polymers with a linear structure or rubbers of regular structure, which contain 80–90% cis groups. During mixing, mutual diffusion of the constituent molecules occurs.

According to Walther [30], one result of introducing polymers into asphalt is that the asphalt can form a homogeneous diffuse phase or can be diffused into a homogeneous polymer phase. A reversal of phases can occur at low polymer concentrations. For example, polyisobutylene at only 5–7% w/w leads to a structure in which the asphalt can act as a filler. The first alternative structure responds to the modification, the second to the asphalt–polymer compositions. A similar effect of the phase reversal in the dispersion has been observed in asphalts modified by the addition of polyurethane at 50% w/w [31].

In assessing the structure of asphalt–polymer compositions, Zielinski examined many aspects, including those related to polymer mixtures. These mixtures are treated as heterogeneous, dispersed, colloidal systems, for which reversal of the phase dispersion is a characteristic. It was found that, for bitumen–nonpolar thermoplast compositions (polyolefin), the appearance of the phase dispersion reversal is a characteristic effect. Another essential phenomenon is the presence of regions suitable for inhomogeneous and unstable structures. For the most part, such a structure is dependent on the bitumen's degree of aromaticity and grows along with its size. For compositions comprising asphalt and atactic or isotactic polypropylene, a nonlinear change in their physicochemical properties has been observed.

Factors that determine the magnitude of the changes in the composition's properties can be separated into three distinct groups. In the first group, small amounts of polymer are utilized. In these compositions, dispergation of the dispersed molecules' dimensions occurs. The compositions are dominated by interactions between components of the bitumen's diffuse phase and polymer macromolecules. The second group deals with inhomogeneous and stable compositions. In this region, a change in valence of asphalt–polymer interactions occurs and dispersed molecules that are found in the first group prevail, along with swelled polymer molecules. The third group contains large quantities of polymer, which is dominated by mutual interactions between polymer macromolecules, resulting in gel, colloid, and associated micelle structures. A three-dimensional configuration is formed in developed agglomerations built from ordered polymer fragments of high molecular weight and dispersed asphalt molecules. The composition is filled with components of the bitumen environment that have been additionally enriched by dissolved polymer fragments of low molecular weight.

Transitions between the various groups of compositions occur continuously and are dependent on the amount and origin of the polymer as well as the structure of the bitumen group. At the same time, interactions are occurring between polymer macromolecules and the medium, the solvation layer, and the nucleus of the dispersed bitumen. These interactions determine the characteristics of the composition's structure and its stability.

Asphalt–polymer compositions should be treated as complex polydisperse structures of varying character. At the same time, one must also consider variations in the degrees of miscibility of a composition's constituents. These are mixtures, relative to the amount and origin of polymer and asphalt, that can be regarded as either colloidal arrangements with sol-gel structure (molecules in the diffuse phase undergo small changes with respect to unmodified bitumen molecules) or as associated micelle colloids with gel structures.

VII. APPLICATIONS

Polymers can induce various changes in bitumen properties, as determined by the possibilities of their versatile applications. Asphalt–polymer compositions are found in many applications, which are listed as follows.

1. In the building industry, they have been used in the production of insulating-seal materials. Examples include: roof shingles, dilatant putties, hydro-insulating films and coatings, pourable sealing compounds, and cements.
2. In highway engineering, for pavement and materials to fill crevices.
3. In the machine industry, as anticorrosive coatings and as materials to suppress vibrations and absorb sounds.
4. Compositions are used to stabilize soils, for impregnations on cartons, as glues and binders to connect steel to wood.
5. Replacements for specific types of polymers for roads and industry. They are distinguished by markedly improved thermorheological and strength qualities when compared to unmodified asphalts or to petroleum by-products.

REFERENCES

1. E. Gundermann, *Strasse*, **5**, 198 (1975).
2. U. S. Patent 3,547,850.
3. H. Lopienska and M. Smarzynska, *Nafta*, **30**. 484 (1974).
4. A. Bukowski and J. Gretkiewicz, *Przemysl Chem.*, **60**, 467 (1981).
5. U. S. Patent 3,068,764.
6. B. Liszynska, *Investigations of the Products of Polymerization of Methyl Methacrylate Conducted in the Presence of Petroleum Asphalt*, Wyd. Politechniki Warszawskiej, Warsaw (1983).
7. N. W. Stabnikov, *Asfaltopolimerbetonye Oblicovki Severnykh Gidrotekhnitheskikh Sooruzeni*, Stroizdat, Petersburg (1980).

8. D. A. Rozental, L. S. Tabolina, and W. A. Fiedosova, *Nefteperabatywayscaja i Neftekhimiceskaja Promyslennost*, Izd. Centr. Naucono-Issl. Inst., Moscow (1988).
9. J. Zielinski and A. Bukowski, Lakier asfaltowy, Patent RP 152262 (1991).
10. S. S. Newaz and M. Matner, Polychlorophene-Modified Aqueous Asphalt Emulsion: Use in Roofs and Roads, ACS, Division of Industrial and Engineering Chemistry, Chicago, August 1995.
11. J. Zielinski and T. Milczarska, *Ropa Uhlie*, **31**, 641 (1989).
12. Patent RFN 244348 (1975).
13. J. Zielinski, *Investigations of the Structures and Properties of Compositions from Polypropylene and Petroleum By-products*, Wyd. Politechniki Warszawskiej, Warsaw (1982).
14. A. Bukowski and K. Piotrowska, *Polim. Tworzywa Wielkoczasteczkowe*, **23**, 258 (1978).
15. A. Bukowski, J. Zielinski, and J. Gretkiewicz, *Polim. Tworzywa Wielkoczasteczkowe*, **31**, 305 (1986).
16. D. A. Rozental, W. A. Fiedosova, and T. S. Chudiakova, *Chim. Techn. Topliw. Masel.*, **23**, 13 (1979).
17. E. Gundermann and A. Ulrich, *Plaste Kautschuk*, **10**, 472 (1963).
18. G. Kraus and W. Rollman, *Kautschuk Gummi Kunststoffe*, **34**, 645 (1981).
19. A. Rjna, M. Masarykova, and M. Rewus, *Ropa Uhlie*, **27**, 51 (1985).
20. A. W. Bratcikov, G. S. Szyfris, and H. F. Szarafijev, *Stroit. Mater.*, **3**, 23 (1983).
21. J. Zielinski, *Erdol Kohle Erdgas Petrochem.*, **42**, 456 (1989).
22. J. Zielinski, *Erdol Kohle Erdgas Petrochem.*, **41**, 258 (1988).
23. M. G. Cavaliere, M. Da Via, and E. Diani, Bitumen Modification with Rubber for Road Binders and Roofing Sheets, ACS, Division of Industrial and Engineering Chemistry, Chicago, August 1995.
24. B. Brule, Polymer Modified Asphalt Cement Used in the Road Construction Industry: Basic Principles, ACS, Division of Industrial and Engineering Chemistry, Chicago, August 1995.
25. A. S. Kolbanowskaja, *Trudy Soyuz. Dor. Neft. Issl. Inst.*, **46**, 10 (1970).
26. E. J. Barth, *Asphalt—Science and Technology*, Gordon & Breach, New York (1962).
27. G. Hameau and M. Duron, *Bull. Liaison Lab. Petr. Chem.*, **81**, 121 (1976).
28. F. J. van Beem and P. Brasser, *J. Inst. Petrol.*, **59**, 91 (1973).
29. E. P. Zelezko and B. G. Pecenyj, *Nefteper. Nefiechim.*, **7**, 13 (1972).
30. H. Walther, *Chem. Rundsch.*, **45**, 849 (1969).
31. J. Gretkiewicz, *Investigations on the Preparation, Structure and Properties of Asphalt-Polyurethane Compostions*, Wyd. Politechniki Warszawskiej, Warsaw (1979).

17

Development and Characterization of Asphalt Modifiers from Agrobased Resin

Sangita and I. R. Arya
Central Road Research Institute, New Delhi, India

R. Chandra and Sandeep
Delhi College of Engineering, Delhi, India

I. INTRODUCTION

The soaring price of imported petroleum crude may pose difficulties in maintaining the Indian road network because indigenous crudes yield poor quality asphalts. In view of nonavailability and increased price of imported crude as well as to improve the quality of asphalt obtained from indigenous crudes, it is necessary and desirable to find ways and means to conserve the available resources of paving materials, e.g., binder and aggregate. This aim can be achieved by (a) improving the quality of available asphalt, (b) proper mix design, (c) quality control during construction of pavements, and (d) rejuvenation and recycling of old pavements.

The use of additives is one of the ways to conserve the road materials by improving the quality (performance and durability) of asphalt surfacings. These additives range from antioxidant, antistripping chemicals to polymers, resin, recycling agents, fiber, sulfur, carbon black, and lime. Durability of asphalt surfacings depends on resistance to change in flow properties during construction and in-service aging. As these surfacings age, asphalt gradually becomes hard and brittle, leading eventually to failure due to cracking and raveling. Oxidation is one of the prime causes of age hardening. During this process, asphaltenes are formed as a result of molecular interaction between polar molecules and other asphalt components, resulting in increased viscosity. The use of metal complexes and soluble organomanganese, organocobalt, and organocopper compounds has been found to be effective in reducing age hardening and hence increasing the durability of pavements [1,2]. The product, asphalt antioxidant blend, is better known under the trade name CHEMCRETE. Use of manganese modifier in asphalt and macadam is claimed to improve the temperature susceptibility of the mixes, thereby improving their physicomechanical properties, i.e., Marshall stability, resistance to permanent deformation, and dynamic stiffness. To enable the organometallic compound to be dispersed rapidly in asphalt the material is blended with a carrier. The antioxidant modifiers in the range currently differ from each other with respect to the type and concentration of the organometallic compounds and the viscosity of the carrier oil used.

Among the antioxidants used to retard the age hardening of asphalt, metal chelates are not economical. Therefore some researchers have paid attention to the use of lime, carbon, and carbon products as cheaper antioxidants for paving asphalt.

Agrobased resin is a by-product that is obtained by processing the Anacardium Occidentale linn tree that grows at altitudes below 1000 feet in the southern region of India. This resin is one of the versatile raw materials that could be used for the manufacture of adhesives, epoxy resins and laminates, paints and varnishes, foundry core oil chemically resistant cements, rubber-reinforced resins, platicizers, antioxidants, and a variety of other industrial products [3–7]. Antioxidants developed from one of the distillates of agrobased resins have proved to be successful in the rubber industry [8–10]. The present chapter deals with the separation and characterization of the main constituents of agrobased resin. Development, characterization, and evaluation of antioxidant activity of the dithiocarbamates synthesized from the phenolic constituents of agrobased resin/*m*-cresol and a diethylamine are discussed in the following.

Rejuvenation of age-hardened asphalt surfacings is another way of conserving the available road construction materials. During the process of aging, the consistency of an asphalt becomes imbalanced as a result of atmospheric weathering, making the asphalt hard and brittle. This results in raveling and development of fatigue cracks. The addition of plasticising agents to bring back the binder to its original consistency is termed rejuvenation.

Recycling is a process of considerable economic and technological importance as the available conventional materials can be utilized over a longer period of time. Second, the maintenance of India's existing road network is costly, time consuming, and material intensive. Recycling of existing pavement materials has emerged as a viable rehabilitation and maintenance alternative as it offers several advantages: low construction cost; conservation of aggregates, energy, and transportation cost of materials; and prevention of environmental pollution, besides keeping the road level constant.

Asphalt pavement recycling may be done in three different ways: (a) surface recycling, (b) in-place recycling, and (c) off-site central plant recycling. In off-site central plant recycling the distressed pavement layers are scarified and removed from the road. The materials are then crushed and mix design analysis is conducted on the resulting blend. Asphalt and rejuvenating agents are added to hot recycled aggregates through batch or continious drum dryer mixer. The resulting hot mix is then placed on the road using conventional methods. A rejuvenating agent is required to restore the physicochemical properties of a hardened asphalt in a mix to be recycled before its laying. Considerable research work has been carried out at the global level for the development of rejuvenating agents [11–19]. Satisfactory rejuvenating agents should have (a) compatibility with oxidized asphalt, (b) ability to restore the physicochemical properties of deteriorated asphalt, (c) good wetting, and (d) resistance to age hardening during construction and in-service aging.

Materials used as rejuvenating agents are usually high-penetration asphalts, flux oils, mobil oils, lube stocks, furnace oils, highly aromatic extract oils like phenol extracts, furfural extracts, etc. Liquid hydrocarbons in the boiling range 205 to 399°C are superior additives. Agrobased resin [20] has also been used to develop rejuvenating agents. Recently, recycling agent RA250 was developed [21] in India and used for rejuvenation of various grades of asphalts.

Because of the enormous inherent advantages of pavement recycling and mounting importance of recycling agents and their nonavailability in India, a stimulus has arisen for the development of indigeneous recycling agents. In the present study an attempt has been made to study the possibility of using black residue (R) of agrobased resins (left after distilling the constituents, A) as an additive for restoration of physical properties of age-hardened asphalt.

II. EXPERIMENTAL

A. Materials

Agrobased resin obtained by processing the Anacardium Occidentale tree is a dark brown viscous liquid with a specific gravity of 0.952. The viscosity determined with a Brookfield viscometer using spindle LV-1 was 9 cps at 30°C. Diethylamine (Glaxo), carbon disulfide (Glaxo), formalin solution (Ranbaxy), and benzene (AR, BDH) were used as received. *m*-Cresol (BDH) was distilled before use. An 80/100 grade bitumen having a penetration value of 96 (at 25°C) and softening point of 43°C was used in the present study. Aged asphalt with penetration value of 12 (at 25°C) and softening point of 90°C was recovered by the Abson method from samples of an asphalt aggregate mix, which was collected from distressed pavement.

B. Separation of Constituents A and R from Agrobased Resin

Agrobased resin was first decarboxylated by heating it slowly to 200°C in an open vessel while stirring intermittently. After cooling to ambient temperature, decarboxylated product was then transferred to a round-bottom flask fitted with a thermometer, condenser, and vacuum inlet tube to receive the vacuum distillate. The temperature of the liquid was raised to 220°C and maintained for some time under reduced pressure to obtain the desired fraction. The pale yellow liquid (product A) was distilled off and collected for synthesis of mannich base and antioxidant. The black residue (R) left in the flask was preserved for rejuvenation of hardened bitumen.

C. Synthesis of Mannich Base

Equimolar quantities of A (molecular weight 304, distilled from resin under vacuum) and diethylamine were placed in a round-bottom flask. Formaldehyde solution (40%) was added dropwise, stirring the mixture at room temperature. After addition of formaldehyde solution, the reaction mixture was stirred further for a period of approximately 3 h. At the end of the reaction a tan viscous liquid was obtained that was dried in a desiccator. Another sample of Mannich base was prepared by carrying out the reaction between *m*-cresol, diethylamine, and formalin under similar conditions. This was used as a standard Mannich base. These Mannich bases were designated MB-1 and MB-2, respectively.

D. Preparation of Antioxidants (Dithiocarbamates)

Mannich bases synthesized by the procedure described above were refluxed at 125°C using a slight molar excess of carbon disulfide for 2–3 h. The Mannich bases were converted into corresponding dithiocarbamates. The products were separated from unreacted materials using a separating funnel. A yield of about 50% was obtained. Thus two antioxidants, A-1 and A-2, were prepared and studied for their efficacy (antioxidant activity).

E. Characterization Techniques

Agrobased resin and its constituent (A) were characterized using Fourier transform infrared (FTIR) and nuclear magnetic resonance (NMR) spectroscopy. These two techniques were also used for structural characterization of Mannich bases and antioxidants (dithiocarbamates) prepared from Mannich bases. The FTIR spectra were recorded with liquid films of product using a Nicolet DX-5 FTIR spectrophotometer. The NMR spectra of products were recorded on a Jeol JNM-FX-100 FT-NMR spectrophotometer in CDCl$_3$.

Penetration grade asphalt, residue (product R left after distillation of agrobased resin), and hardened bitumen, which were recovered from distressed pavements, were characterized by routine tests, e.g., penetration (at 25°C) and softening point according to ASTM procedures D-5-86 and D-36-26, respectively. Flash points of 80/100 grade asphalt and product R were determined according to ASTM method D-92.

F. Evaluation of Antioxidant Activity (Age-Hardening Test) of Products A-1 and A-2

Antioxidant-asphalt mixtures were prepared by adding 1 and 2% of the antioxidant (A-1 and A-2) to the 80/100 grade asphalt. These mixtures and neat asphalt were then subjected to the thin-film oven test (TFOT) and ultraviolet (UV) oxidation test. For TFOT, 25 g of material was spread in a dish (55 mm diameter) and heated in a circulatory air oven for $5\frac{1}{4}$ hours at 163°C. Percent weight loss during the test, softening point, and penetration before and after TFOT as well as the UV oxidation test were determined to evaluate the efficacy of these antioxidants. To carry out the UV oxidation, weighed quantities of neat asphalts as well as asphalts containing 1 and 2% of antioxidant (A-1) were placed in metallic Petri dishes in the form of thin films. Samples thus prepared were kept at 50 ± 2°C in an air oven equipped with UV light sources for a maximum of 21 days. Changes in weight were observed after 7, 15, and 21 days of exposure, whereas softening point and penetration were measured and compared with each other only for the antioxidant-asphalt mixtures obtained after the maximum exposure time, 21 days.

G. Asphalt from Distressed Pavements

A weathered (oxidized) asphalt mix sample was collected from a badly distressed old pavement by a core cutting method. Approximately 2 kg of weathered mix was heated at 140–150°C for 3 h and then placed in an extractor. A sufficient amount of benzene was added to the extractor to dissolve and separate the binder from aggregate. The mix was extracted with benzene until colorless washings were observed. All the washings were collected and benzene-soluble asphalt was then recovered by the Abson method (ASTM D-1856-79, 1984).

H. Rejuvenation of Weathered Asphalt Using Product R

Product A was used to synthesize an antioxidant, whereas black residue (product R) left in the flask after distillation was preserved and used to rejuvenate aged asphalt. For this purpose, blends of residue (product R) and aged asphalt recovered from distressed bituminous pavements were prepared by adding varying amounts (10–50% w/w) of R to hardened asphalt in a molten stage. The mixtures thus prepared were thoroughly mixed and tested for their physical properties, e.g., softening point and penetration, by standard test procedures.

III. RESULTS AND DISCUSSION

A. Agrobased Resin and Its Constituents

Because of the heterogeneous nature of agrobased resin it was necessary to characterize the resin before using it in any formulation. Structural characterization of the resin was done by analyzing its FTIR and proton NMR spectra, which are shown in Figures 1 and 2, respectively. In the FTIR spectrum (Figure 1) the

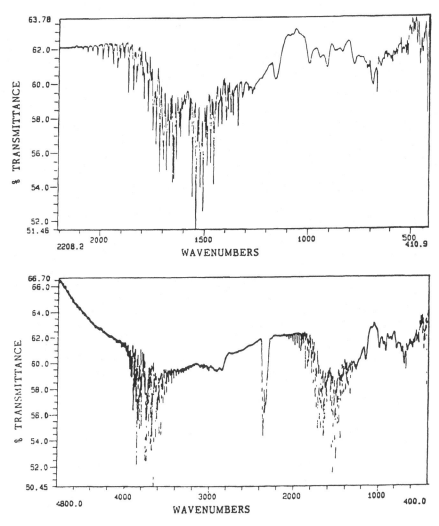

Figure 1 FTIR spectrum of agrobased resin.

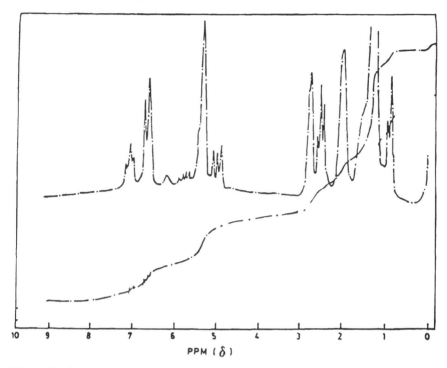

PPM (δ)

Figure 2 ¹H-NMR spectrum of Agrobased Resin.

stretching vibrations due to the phenolic hydroxyl group appear at 3352 cm⁻¹. A sharp absorption band in the region 3000–2870 cm⁻¹ characteristic of C−H stretching is also present. The strong absorption band at 1592 cm⁻¹ is due to skeltal vibrations of the aromatic ring structure. A medium intensity band at 920 cm⁻¹ is indicative of the presence of a trans double bond in an aliphatic side chain. In the ¹H-NMR spectrum (Figure 2) of agrobased resin, aromatic protons appear as a quadret and multiplet at 7.0 ppm and 6.4–6.8 ppm, respectively. The olefinic protons of the side chain appear as a multiplet in the ranges 4.6–5.8 ppm and 1.28–2.80 ppm. The methylene protons in the side chain appear as a distorted triplet centered around 0.8–0.9 ppm.

The FTIR and NMR spectra of the main constituent (product A) of agrobased resin are shown in Figures 3 and 4. Typical absorption bands due to an aromatic ring, a hydroxyl group, and an unsaturated double bond present in a side chain are observed in FTIR spectrum (Figure 3). In the NMR spectrum (Figure 4) of product A, signals due to aromatic protons appear as multiplets, as observed in the spectrum of agrobased resin. The signals observed around 0.87 are due to the methylene protons of the side chain. The hydroxyl proton of

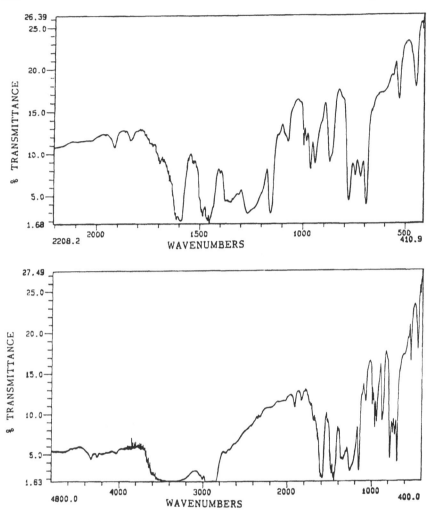

Figure 3 FTIR spectrum of product A.

product A could not be distinguished from the olefinic protons in the NMR spectrum. In order to investigate the −OH signal in samples of this constituent, the ¹H-NMR spectrum of product A was recorded in dimethyl sulfoxide (DMSO)-d₆ solvent. An additional singlet of strong intensity appears at 9.1 ppm, as well as other signals obtained in CDCl₃ when DMSO-d₆ was used as a solvent (Figure 5). This clearly indicates the presence of an −OH group in product A. Thus, the presence of a hydroxyl (phenolic) group was confirmed in product A as indicated

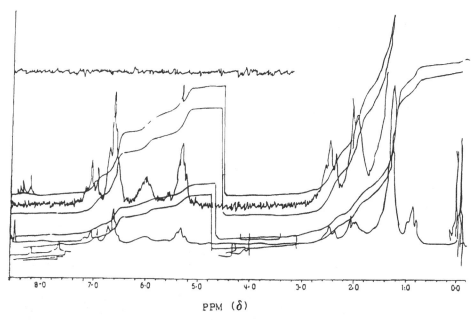

Figure 4 ¹H-NMR spectrum of product A in CDCl₃.

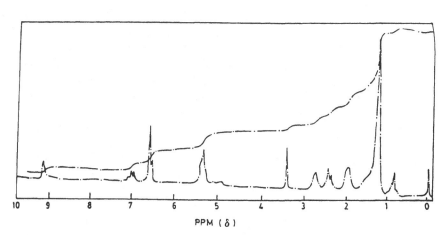

Figure 5 ¹H-NMR spectrum of product A in DMSO-d₆.

in its FTIR spectrum. The general structure of product A as confirmed by spectroscopic methods discussed above can be depicted as follows.

OH

$C_{15}H_{31}$ - *n*

n = 0, 2, 4, 6, . . .

B. Characterization of Mannich Base and Dithiocarbamates (Antioxidants)

Mannich bases MB-1 and MB-2 were prepared by reacting diethylamine with
product A and *m*-cresol, respectively, according to the procedure described in
Section II. These Mannich bases were then converted into their respective dithiocarbamates, which were designated antioxidants A-1 and A-2. The results of
spectroscopic techniques that were used to characterize Mannich bases and antioxidants are discussed below. In the FTIR spectrum of Mannich base MB-1 (Figure
6), characteristic absorption bands at 1632 cm^{-1} due to the presence of a tertiary
amine functionality were observed in addition to other expected absorption bands.
Similar bands were observed in the spectrum of MB-2.

The NMR spectrum of Mannich base (MB-1, Figure 7) shows a multiplet
around 1.0 ppm due to overlapping of a methylene group–diethyl moiety with
the terminal methyl of the side chain. A methylene signal is located at 1.4 ppm,
and allylic methylene gives a broad signal around 2.0 ppm. The apparant quadret
at 2.6 ppm comprises methylene of any ethyl as well as the benzylic methylene
protons. The sharp singlet at 3.7 ppm can be assigned to methylene protons
flanked by an aromatic ring at one side and hydrogen on the other side. The
formation of Mannich base takes place because before its formation signals of
olefinic protons as a triplet at 5.3 ppm and signals of aromatic protons as a
multiplet between 6.5 and 7.2 ppm were obtained. A singlet at 9.0 ppm obviously
due to phenolic hydroxyl protons was present in the spectrum of MB-1.

The FTIR spectra of dithiocarbamates (A-1 and A-2) obtained from product
A and *m*-cresol–based mannich base are shown in Figures 8 and 9. The absorption
peaks at 691 and 623 cm^{-1} due to the C=S group confirm the formation of
dithiocarbamates, which are designated A-1 and A-2, respectively.

In the NMR spectrum of dithiocarbamate based on *m*-cresol mannich base
(Figure 10), a triplet at 1.12 ppm due to a methyl proton of a diethyl moiety and
a singlet at 2.3 ppm due to methyl protons attached to an aromatic ring are
observed. A singlet at 3.7 ppm accounting for two protons of methylene attached

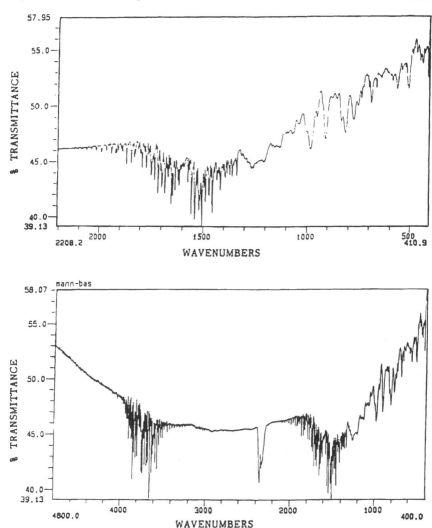

Figure 6 FTIR spectrum of Mannich base (MB-1) prepared from product A.

to a benzene ring and two quadrets at 3 and 4 ppm due to protons of a diethyl moiety are also present in the spectrum; this suggests nonequivalence of the protons of two methylene groups of the diethyl moiety. This may be explained by hindered rotation along the N–C bond of the diethyl group of carbamate. A multiplet due to aromatic protons at 6.0–7.2 ppm is also observed.

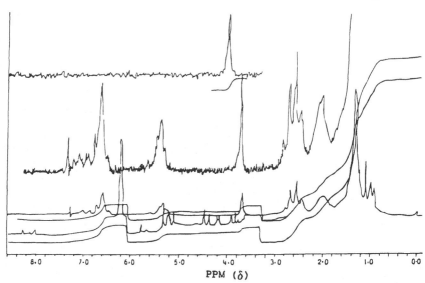

Figure 7   ¹H-NMR spectrum of MB-1.

The NMR spectra of dithiocarbamates obtained from product A–based mannich base are shown in Figure 11. A multiplet around 1.0 ppm due to overlapping of side chain methylene and methylene diethylamine moieties is observed. The methylene hydrogen gives a broad signal at 1.35 ppm, while the allylic methylene protons are observed around 2.0 ppm and benzylic methylene protons at 2.55 ppm. The complex multiplet in the range 3.4 to 4.2 ppm could be analyzed as two overlapping quadrets and is due to methylene group of diethylamine comparing to standard dithiocarbamate nonequivalence in this case. Interestingly, the benzylic methylene flanked by an aromatic ring on one side and sulfur on the other side is considerably deshielded and appeared at 5.46 ppm beside the olefinic proton singlet at 5.35 ppm as a triplet and the aromatic proton at 6.4 to 7.2 ppm as a multiplet.

C.   Properties of Neat Asphalt, Age-Hardened Asphalt, and Residue R (Obtained from Agrobased Resin)

Mainly penetration and softening point tests were performed on neat bitumen (paving grade), aged bitumen, and residue left after distillation of agrobased resin. The properties of these three materials are given in Table 1. Softening point and penetration values of neat asphalt confirm that the binder used in this study is an 80/100 grade asphalt, which is generally used for paving the roads in India.

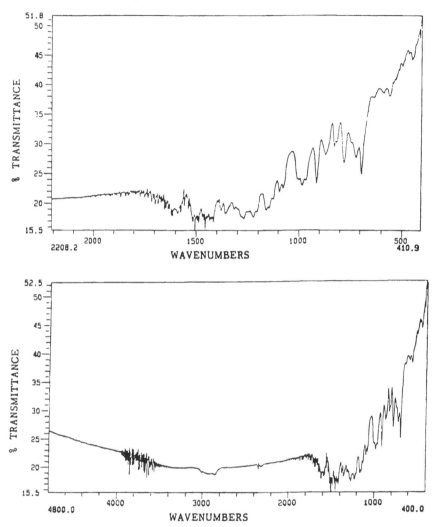

Figure 8 FTIR spectrum of antioxidant A-1.

Aged asphalt recovered from badly cracked pavements showed a penetration value of 12 (at 25°C) and softening point 90°C. These values indicated the extensive age hardening of such pavements. Residue left after distillation of agrobased resin is a black sticky resinous tar. A softening point 28°C and penetration value of 300 were obtained for this residue (R) (Table 1). Flash points of residue R and neat asphalt were 227 and 326°C, respectively.

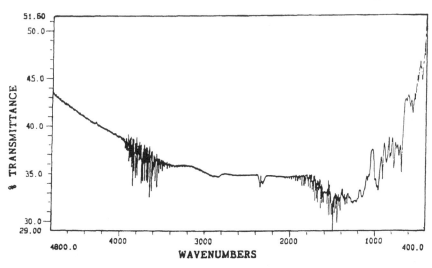

Figure 9   FTIR spectrum of antioxidant A-2.

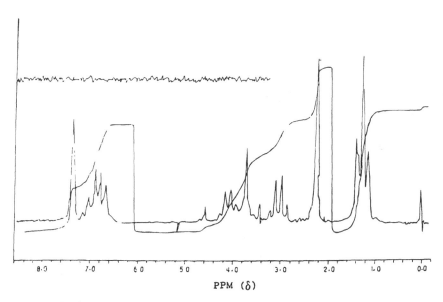

Figure 10   ¹H-NMR spectrum of antioxidant A-2.

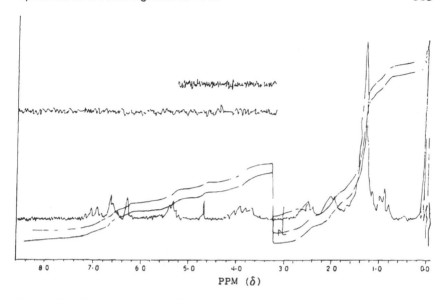

Figure 11 ^{1}H-NMR spectrum of antioxidant A-1.

Table 1 Properties of Asphalt, Black Residue (R), and Age-Hardened Asphalt

Sample	Softening point (0°C)	Penetration (25°C, 100 g, 5 s)	Flash point (°C)
1. Paving grade asphalt	43	95	326
2. Residue (R) obtained after distillation of agrobased resin	28	300	227
3. Age-hardened asphalt (AH) recovered from distressed pavements	90	12	—

D. Evaluation of Antioxidant Activity (Age-Hardening Test)

Mixtures of asphalt and antioxidant (dithiocarbamates) containing 1.0–2.0% A-1 and A-2 were prepared. Samples of approximately 25 g of asphalt-dithiocarbamates mixture and neat asphalt were weighed in separate metallic Petri dishes of equal diameter to obtain thin films of nearly equal thickness. Five films were thus prepared and heated to 145°C in an air oven so that the sample would spread uniformly in the form of a thin film. Such thin films were heat aged at 163°C $\pm$ 5°C for $5\frac{1}{4}$ hours in a circulating air oven. After the completion of the thin-film oven test the samples were cooled down to room temperature.

The plates were weighed again to determine the weight loss during the thin-film oven test. Penetration at 55 at 25°C and softening point after the TFOT were also determined, and results of these studies are given in Table 2. Percent weight loss was higher, i.e., 0.22 to 0.32, in asphalt-dithiocarbamate mixtures containing product A–based antioxidant (A-1) than in those containing *m*-cresol–based antioxidant (A-2) and asphalt having no antioxidant. This may be due to unsaturated aliphatic side chains present in product A, which decomposes and evaporates at the test temperature. In neat asphalt and asphalt containing 1–2% *m*-cresol–based dithiocarbamate only 0.02 to 0.047% weight loss due to evaporation of volatile matter could be observed. From penetration test results as observed after TFOT (Table 2), it is clear that the neat asphalt was hardened much more (penetration 55 at 25°C) than other antioxidant-asphalt mixtures, for which higher penetration values, 60 to 78, were obtained after exposing these samples under similar test conditions. Asphalt treated with 1 to 2% of product A–based dithiocarbamate (i.e., antioxidant A-1) showed less age hardening (higher penetration value, 72–78) than binder treated with *m*-cresol–based dithiocarbamate (penetration 60 and 67), indicating better efficacy of antioxidant A-1 than antioxidant A-2. To evaluate the effect of concentration on extent of age hardening, 1 and 2% of dithiocarbamates A-1 and A-2 were added to asphalt. Penetration values observed after TFOT confirmed less age hardening (higher penetration value) in samples containing 2% antioxidant than in those containing 1% antioxidant. Furthermore, the extent of age hardening as measured by relative penetration values was noticed less in mixtures containing only 1% antioxidant A-1 than in mixtures containing 2% of *m*-cresol–based dithiocarbamate, again confirming the better efficacy of dithiocarbamate (antioxidant) synthesized from product A. The softening points were also determined for all the samples after the thin-film oven test

Table 2   Results of Age Hardening Test of Asphalt and Asphalt-Antioxidant Mixtures After TFOT

Sample[a]	Weight loss during TFOT (%)	Penetration of residue of TFOT (25°C, 100 g, 5 s)	Softening point of residue (°C)
Asphalt 80/100	0.0364	55	51
Asphalt + 1% A-1	0.2263	72	47
Asphalt + 2% A-1	0.3282	78	44
Asphalt + 1% A-2	0.047	60	50
Asphalt + 2% A-2	0.020	67	48

[a] A-1, antioxidant (dithiocarbamate) developed from product A; A-2, antioxidant (dithiocarbamate) developed from *m*-cresol.

and are presented in Table 2. Softening point values observed for various samples were in accordance with penetration values.

The UV oxidation test, which is a complementary test to predict the age hardening of asphalt in service, was performed on neat asphalt thin films as well as asphalt films containing 1 and 2% of dithiocarbamate A-1. These thin films were exposed to UV light at 50 ± 5°C for various intervals of time (7, 15, and 21 days). Percent weight change due to oxidation, penetration at 25°C, and softening point determined for these samples are summarized in Table 3. Percent increase in weight after 7 days of exposure was greater in asphalt (0.17%) than in asphalt containing 2% dithiocarbamate. The extent of oxidation was greater after 15 and 21 days of exposure of these samples to UV light at service temperature; e.g., percent increases in weight were 0.53 and 0.57% after 15 and 21 days, respectively, in asphalt film compared with 0.38 and 0.42% in asphalt-dithiocarbamates mixtures containing only 1% dithiocarbamate. Percent increase in weight was lowest (0.36 and 0.41%) in asphalt mixtures containing 2% dithiocarbamate. The penetration at 25°C and softening point of these samples were determined only after 21 days of exposure of samples to UV light (Table 3). Comparison of the penetration values indicated more hardening of neat asphalt (i.e., lower penetration value) than of the asphalt-dithiocarbamate mixtures containing 2% dithiocarbamate (A-1). Addition of 1% dithiocarbamate marginally checked the oxidative hardening of asphalt.

E. Rejuvenation of Age-Hardened Asphalt

Product A distilled from agrobased resin was used to synthesize antioxidant (dithiocarbamate A-1), and an attempt has been made to utilize the residue (black tar left after distillation) for rejuvenation of hardened asphalt obtained from the blacktop surfacings after in-service aging. For this purpose asphalt mix was collected from old and badly distressed pavements using a core cutter. Asphalt was then extracted and recovered using the Abson method. A low penetration value at 25°C (i.e., 12) and a high softening point (i.e., 90°C) confirmed the

Table 3 Results of UV Oxidation Test

Sample	Percent 7 days	Weight 15 days	Change after 21 days	Penetration of residue of UVOT (25°C, 100 g, 5 s)	Softening point of residue of UVOT (°C)
Asphalt	0.17	0.53	0.57	42	55
Asphalt + 1% A-1	0.16	0.38	0.42	43	55
Asphalt + 2% A-1	0.11	0.36	0.41	51	52

severe age hardening of the recovered asphalt. Varying amounts of black tar residue (R) were added to hardened asphalt to restore it to the nearly original (softer grade) consistency of softer binder. Penetration at 25°C and softening point tests, which are generally used for the consistency of a paving grade asphalt, were carried out on rejuvenated binder. The results indicate (Table 4) that addition of 10% residue had a marginal effect on penetration and softening point values of age-hardened asphalt, whereas 30/40 penetration grade asphalt having a softening point of 63°C could be obtained by adding 35% black tar residue to hardened asphalt. Addition of 50% black residue yielded a 60/70 grade binder with a softening point value of 53°C. To obtain the paving grade 80/100 bitumen above 50% residue must be added to hardened asphalt. Thus a waste from distillation of agrobased resin may be worth using as a rejuvenating agent for recycling of old distressed pavements.

IV. CONCLUSIONS

The use of agrobased resin for development of modifiers is an economical and novel approach. Modifiers A-1 and R were developed and identified from the constituents (A and R) separated by distilling the agrobased resin under reduced pressure. The Mannich reaction has been used successfully for the synthesis of two antioxidants (A-1 and A-2) from the phenolic constituent of agrobased resin (A) and *m*-cresol, respectively, by reacting these with diethylamine (DEA). Agrobased resin, its constituent (A), intermediate products (mannich base MB-1 and MB-2), and antioxidants (dithiocarbamates, A-1 and A-2) were characterized using Fourier transform infrared (FTIR) and nuclear magnetic resonance (^{1}H-NMR) spectroscopy. Asphalt and asphalt-antioxidant mixtures containing 1 and 2% of antioxidants A-1 and A-2 were subjected to the thin-film oven test (TFOT) and UV oxidation test (UVOT) to evaluate the efficacy of antioxidant additives developed from agrobased resin. Penetration and softening point tests on neat asphalt residue and asphalt-antioxidant mixture residue (obtained after TFOT and UVOT) indicated better antioxidant activity of A-1, which was synthesized from the phenolic constituent of agrobased resin.

Table 4 Properties of Hardened Asphalt and Rejuvenated Asphalt

Sample	Softening point (0°C)	Penetration (25°C, 100 g, 5 s)
Age-hardened asphalt (AH)	90	12
AH + 10% R	87	13
AH + 20% R	80	15
AH + 35% R	63	36
AH + 50% R	53	73

Product R, a residual product left over from the distillation of agrobased resin, was characterized for softening point, penetration, and flash point. Age-hardened asphalt (AH, penetration 12 at 25°C) recovered from an asphaltic concrete mix of a deteriorated pavement was then blended with varying amounts of product R. Softening point and penetration data on asphalt and asphalt modified with product R were then collected. Preliminary results confirmed the potential of modifier R to be used for restoration of the consistency of highly oxidized asphalt. On the basis of these studies the following conclusions were drawn:

1. Constituent A of the agrobased resin can be used for synthesis of low-cost antioxidants.
2. Age hardening, i.e., hardening due to oxidation of asphalt binders, can be retarded by adding 1–2% of dithiocarbamates synthesized from product A, which can be easily obtained by distillation (under reduced pressure) of a agrobased resin.
3. The concentration and chemical nature of dithiocarbamates affect the extent of oxidation of asphalt.
4. Dithiocarbamate based on product A acts as a good antioxidant for asphalt exposed to UV radiations and thus may prevent in-service aging of asphalt.
5. The waste (residue R) left after the distillation of agrobased resin could be utilized as a rejuvenating agent by restoring the consistency of oxidized asphalt. The extent of rejuvenation depends on the concentration of residue R used in rejuvenated asphalt.

ACKNOWLEDGMENT

The authors are thankful to the Director, Central Road Research Institute, for his kind permission to publish this research work.

REFERENCES

1. T. W. Kennedy, L. O. Cummings, and T. D. White, Changing Asphalt Through Creation of Metal Complexes, in *Proceedings of the Association of Asphalt Paving Technologists* (AAPT), Vol. 50 (1981).
2. D. Whiteoak, *The Shell Bitumen Handbook*, Shell Bitumen, Surrey, UK, p. 153 (1990).
3. J. H. P. Tyman, *Chem. Ind.*, January , 59 (1980).
4. H. Fries, E. Esch, and T. Kempermann, *Rubber India*, November, 15 (1979).
5. B. Banerjee and C. S. Inamdar, *Eur. Rubber J.*, November, 23 (1976).
6. V. R. Shah and V. Krishnan, Versatile Cardanol, Derivatives in Rubber Compounds, in *Proceedings of IRMRA 10th Rubber Conference*, India, p. 222 (1981).
7. N. D. Ghatge, *Rubber India*, May, 13 (1979).

8. A. M. I. Jayawardene, Synthesis and Evaluation of Antioxidants Based on Mannich Reaction in Natural Rubber Compounds, M.Sc. thesis, University of Sri Jayawardenapura, Sri Lanka (1982).

9. A. Coomarasamy, L. B. K. Silva, and R. Suranimala, Some New Antioxidants for NR, in *Proceedings of the IRMRA*, Cochin, India, February, p. 56 (1977).

10. A. Coomarasamy and A. M. I. Jayawardene, The Use of Cashew Nut Shell Liquid Based Antioxidants and Activator in Natural Rubber Compound, in *Proceedings of IRMRA*, India, p. 93–105 (1983).

11. D. D. Davidson, W. Caneisa, and S. J. Escobou, Recycling of Deteriorated Asphalt Pavements—A Guideline for Design Procedure, in *Proceedings of Association of Asphalt Paving Technologists* (AAPT), USA, Vol. 46, p. 498 (1977).

12. W. J. Kari, Prototype Specifications of Recycling Agents Used in Hot-Mix Recycling, in *Proceedings of Association of Asphalt Paving Technologists*, USA, Vol. 49, p. 178–192 (1980).

13. D. H. Little, R. J. Holmgreen, and J. A. Epps, Effect of Recycling Agent on the Structural Performance of Recycled Asphalt Concrete Materials, in *Proceedings of Association of Asphalt Paving Technologists*, USA, Vol. 50, p. 32–61 (1981).

14. C. R. Ganon, R. H. Wombles, C. A. Ramcy, J. P. Daris, and W. V. Little, Recycling of Conventional and Rubberized Bituminous Concrete Pavements Using Recycling Agents and Virgin Asphalt as Modifiers, in *Proceeding AAPT*, USA, Vol. 49, p. 85 (1980).

15. M. G. Arora and K. A. Rehman, Laboratory Study on Recycling Potential of Deteriorated Asphalt Pavement in Saudi Arabia, in *Proceedings 6th International Asphalt Conference*, AAPA Publication, January, p. 67 (1986).

16. S. P. Dekold and S. N. Amirkhanian, TRR No. 1337, p. 79 (1992).

17. J. S. Caba, F. J. Criado, and L. Valero, The Recycling of Bituminous Mixtures with Rejuvenating Agents, in *Proceedings, 5th Eurobitume Congress*, Stockholm, p. 837 (1993).

18. D. R. Marini, Innovations in Machinery for the Recycling of Asphalt Pavement—Hot and Cold Methods, in *Proceedings, 4th Eurobitume Symposium*, Madrid, p. 888 (1989).

19. J. J. Emeny and M. Terao, TRR No. 1337, p. 18 (1992).

20. P. K. Jain and Sangita, Rejuvenation of Recycled Weathered Road Mixes—A Review and Development of a Rejuvenating Agent from Agrowaste, Symposium on Planning Control and Management of Construction Projects, Institute of Engineering, Roorkee, India, p. V-17 (1991).

21. A. Krishna, M. P. Kala, B. R. Tyagi, S. P. Srivastva, and A. K. Bhatnagar, Rejuvenation of Bituminous Pavements, in *Proceedings New Horizons in Roads and Road Transport*, Vol. 1, ICORT, p. 120 (1995).

18

Polymer Network Formation in Asphalt Modification

Arthur M. Usmani
Usmani Development Company, Indianapolis, Indiana

I. INTRODUCTION

Asphalt is an old and a low-cost thermoplastic material. The main advantages are low cost ($0.06/lb) and ease of availability from petroleum refining. Major disadvantages of asphalt are permanent deformation under load (rutting) and rapid drop in viscosity as temperature increases (high temperature dependence). Modification of asphalt by thermoplastics, thermosets, or rubber polymers mitigates these shortcomings.

Paving and roofing are the principal applications. In paving, asphalt is polyblended with minor amounts of rubber polymers. Higher amounts of rubber or polypropylene modifications produce a polymer network structure useful in roofing. In this chapter we will describe chemical aspects, polymer modification of asphalt, polymer modifiers, secondary structure of asphalt, and mechanisms of inversion and reversion.

II. HISTORICAL

Ancient Egyptians and Iraqis used asphalt for construction, mummification, and waterproofing. Petroleum asphalt became available from petroleum refining about 100 years ago. In the United States alone, usage of asphalt exceeds 70 billion pounds every year. Asphalt finds application in paving, road construction, roofing, coatings, adhesives, and batteries. There are more than 10 million miles of roads and highways and millions of commercial and residential buildings in the United

States that use asphalt. So asphalt is a very important chemical and the most widely used thermoplastic material.

III. AN OVERVIEW OF ASPHALT

A. Chemical Aspects

Despite its wide use, asphalt is not a well-characterized chemical. Asphalt is a complex mixture of organic and inorganic compounds and their complexes.

Asphalt, although viscoelastic in nature and a very low cost thermoplastic, is not a suitable material by itself for use in roofing and paving applications. The main disadvantages of asphalt are rutting and high temperature dependence. To mitigate temperature dependence and other shortcomings, asphalt can be modified by many types of polymers, e.g., thermoplastics, thermosets, and rubber polymers [1–5]. Because of cost considerations, polypropylene and styrene-butadiene-styrene (SBS) modifications find application in roofing and paving. Amorphous polypropylene (APP), a waste by-product in the manufacture of isotactic polypropylene (IPP), was initially used in Europe to modify asphalt for roofing membranes and other applications. Thus, this technology that we know and apply in the United States originated from Europe in the early 1960s. With advances in Ziegler-Natta catalysts, this APP source became limited in the United States. APP is still available in Europe, Mexico, and Asia because of their older polypropylene plants.

Copolymer polypropylene containing 2–10% ethylene has increased clarity, toughness, flexibility, and lowered melting points [5]. Copolymer polypropylene instead of APP is more popular in the United States and to a lesser extent in Europe and Asia to modify asphalt for roofing applications. Ultimately, APP will phase out when the older waste by-product–producing polypropylene plants close in Europe and other parts of the world.

Thermoplastic block copolymers with styrene endblocks and rubbery mid-blocks, e.g., butadiene (SBS), isoprene (SIS), and their hydrogenated versions (SEBS, SEPS) are common modifiers for asphalt [6,7].

B. Composition

The precursor of asphalt is crude oil. Crude oils are the remains of organisms that in the distant past inhabited the inland seas and coastal basins. They became buried in sediments before microorganisms consumed them. The organisms were transformed into crude oil by a process not well understood yet [8]. Because of the origin of crude oil from living organisms, the composition of the crude oil is complex. Furthermore, the composition varies considerably from oil field to oil field. A simple exercise in organic chemistry shows that a C_{10} hydrocarbon has 75 isomers, whereas a C_{20} hydrocarbon has 366,319 possible isomers.

Asphalt contains a complex mixture of molecules and their isomers with about 24 to 150 carbons. This means that there may be over 10^{16} isomers in asphalt. The complexity of the composition is thus comparable to that of life molecules, e.g., protein, enzyme, and DNA. Asphalt contains many classes of compound, e.g., branched aliphatics, cycloalophatics, aromatics, polar aromatics, heterocyclics, and asphaltenes. Each class of compounds exists in asphalt in an immensely large number of isomers. Polar aromatics impart viscoelasticity to asphalt.

The molecular weights of the various species are in the range 300 to 2000. Asphalt has a significant heteroatom content. These include primarily sulfur and nitrogen and to a lesser extent oxygen, vanadium, nickel, and iron. The heteroatoms play an important role in the physical properties of an asphalt. The polar heteroatoms are capable of intermolecular associations. These associations regulate properties, such as emission, boiling range, solubility, viscosity, hardness, adhesion, glass transition temperature, and decomposition temperature. The molecular association makes asphalt behave as a polymer-like material although the molecular weight is low.

The composition as well as rheology and durability of an asphalt is unique to the precursor crude blend. Yet it is note worthy that asphalts from many sources have performed well in roads and on roofs.

C. Analytics

Classification of asphalt is as asphalt cement-2.5, AC-5, AC-10, AC-20, and AC-30. Viscosity, viscosity as a function of temperature, and penetration are commonly employed to characterize asphalt. For complete characterization, a combination of the techniques listed in Table 1 is useful [9–17].

IV. POLYMER MODIFICATION OF ASPHALT

We consume about 70 billion pounds of asphalt annually in the United States in paving and roofing. To mitigate temperature dependence and rutting, asphalt is usually modified by polyblending with rubber polymers, e.g., SBS, SEBS, SIS, and ethylene/vinyl acetate/carbon monoxide (E/VA/CO) at 2–5 wt % polymer level in paving applications. For roofing, both rubber and polypropylene modifications are used but at a much higher polymer loading. Typically 11–15 wt % SBS rubber polymer (linear, radial, or combination) or 20–25 wt % polypropylene (APP/IPP or copolymer PP/IPP) based on asphalt is used in roofing. The higher loading leads to polymer network formation wherein the polymer becomes the continuous phase and asphalt the discontinuous phase. The phase inversion leading to polymer network formation occurs in 15 to 45 min when asphalt and polymer are mixed at 350–390°F under high shear. SBS rubber needs more shear than polypropylene.

Table 1   Asphalt Characterization Methods

1. Fractionation by precipitation/distillation
    1.1 Solvent-nonsolvent precipitation
    1.2 Chemical, eg., acid, base precipitation
    1.3 Vacuum distillation
    1.4 TGA
2. Chromatographic separation
    2.1 Gas chromatography (GC)
    2.2 TLC/FID (Iatroscan)
    2.3 Reverse-phase high-performance liquid chromatography (RPHPLC)
    2.4 Size exclusion chromatography (SEC)
    2.5 Ion chromatography
3. Spectroscopy
    3.1 Fourier transform infrared (FTIR), Near infrared (NIR)
    3.2 UV-visible
    3.3 ^{1}H and ^{13}C NMR
    3.4 2D NMR
    3.5 Electron spectroscopy for chemical analysis (ESCA)
    3.6 XRF
4. Thermal analysis
    4.1 DSC
    4.2 DMA
    4.3 TMA
    4.4 TGA
5. Molecular weight determination
    5.1 SEC
    5.2 Vapor phase osmometry (VPO)
    5.3 Mass spectrometry (MS)
6. Hyphenated techniques

The mechanism of polymer network formation is not fully understood at present [18,19]. At elevated temperatures, sorption of aromatics and cycloaliphatics by polypropylene chains begins, the chains swell and extend, and phase inversion takes place. Polymer chains become the continuous phase and asphalt particles, mostly spherical, the discontinuous phase. The asphalt particle size in the 10 to 150 μm range can be regulated, at will, by composition and processing parameters. Ultraviolet (UV) microscopy is used to follow the inversion and morphology of structures. Electron microscopy is another technique for studying these structures. We show phase inversion schematically below.

P	Aromatics/	P	Asphalt	P
Collapsed	$\longrightarrow$	Swollen,	$\longrightarrow$	Continuous
chains	Cycloaliphatics	extended		phase
	(Asphalt)			

In the case of SBS modification, styrene blocks are solvated by aromatics and swelled by cycloaliphatics. This results in chain extension with the polybutadiene block serving as the elastic band. Phase inversion takes place and SBS polymer becomes the continuous phase. Because of the high molecular weight of SBS, the shape (linear, star shaped), and the longer butadiene chain lengths, considerably lower levels of SBS polymers than of polypropylene are required to achieve phase inversion.

V. POLYMER MODIFIERS

Modification of asphalt only by polypropylene or SBS makes economic sense in roofing applications. Typically, a copolymer polypropylene contains 6–9% ethylene. Although it can be made in a wide range of molecular weights, by Ziegler-Natta catalysis and regulating hydrogen, a number average molecular weight of about 50,000 with a polydispersity in the 4.0 to 5.0 range is suitable for asphalt modification. The glass transition temperature of copolymer polypropylene should be in the -23 to $-29°C$ range. Thus, copolymer polypropylene not only improves the rheological properties but also mitigates the stiffness of asphalt under cold temperatures prevalent in winters. Minor amounts of isotactic polypropylene added to asphalt besides copolymer polypropylene improves thermostability and form stability of roofing membranes. The crystallinity of useful IPP should be around 50% with a high cold crystallization temperature above 120°C.

SBS thermoplastic elastomers have polystyrene endblocks and polybutadiene midblocks with a star (tetrabranched) or linear structure. Styrene and butadiene blocks are mutually insoluble and hence SBS polymer has two glass transition temperatures (100 and $-91°C$). Furthermore, SBS has a lattice structure consisting of polystyrene blocks dispersed in the polybutadiene phase. This structure is destroyed either by dissolving the polymer in a solvent such as tetrahydrofuran or toluene or by heating it above the T_g of the polystyrene block. The lattice reassociates when the solvent evaporates or the polymer cools below 100°C. Generally SBS with a styrene-to-butadiene ratio of 30:70, molecular weight in the 90,000 to 250,000 range, and polydispersity below 1.07 is useful in modifying asphalt for roofing membranes. Useful linear SBS has a molecular weight in the 90,000 to 125,000 range, whereas for radial SBS the molecular weight should be around 250,000. By judicious choice of linear and star-shaped SBS and their amounts, one can obtain desired properties of the compound. Star-shaped SBS will provide high viscosity, a high softening point, and better low ptemperature flexibility to the compound. Linear SBS will impart low viscosity, a low softening point, and better toughness.

To provide strength and reduce cost, SBS- as well as polypropylene-modified asphalts are filled with fillers, e.g., calcium carbonates and clays. Asphalt is a fuel, and to make the roofing membrane fire resistant is a challenge. Alumina trihydrate is not suitable because it decomposes at the polymer modification temperature, resulting in partial loss of water and causing frothing during processing. Calcium borate decomposes above 250°C and releases about 23% water, which can slow the burn rate. Incorporation of about 30% inexpensive calcium borate into either SBS- or polymer-modified asphalt membranes considerably improves the fire resistance. Calcium borate with minor amounts of ammonium polyphosphate, e.g., Monsanto's PhosCheck, considerably enhanced the fire resistance of polypropylene-modified asphalt through high char formation [20].

The unsaturation in SBS produces some cross-linking even during modification of asphalt. The processing temperature during modification should not exceed 200°C to mitigate this problem. Use of an antioxidant, e.g., a hindered phenol, even at a very low level, helps mitigate the undesired cross-linking during modification. During service, the SBS-modified asphalt roofing membrane begins cross-linking as it ages and this is the main cause for eventual failure of the membrane. The UV and outdoor stability of polypropylene is poor due to the methyl groups. Polypropylene degrades during the service of polypropylene-modified asphalt membrane. The degradation eventually results in the failure of the membrane. We took advantage of the opposing types of mechanism to develop a cross-linking resistant SBS-modified asphalt by incorporating minor amounts of isotactic polypropylene [21]. The SBS macroradical, produced during processing or during service on the roof as a result of high heat, is terminated by the degrading polypropylene macroradical, thereby producing graft copolymers in situ. The SBS cross-linking and polypropylene degradation are thus retarded and such membranes have longer service life. Furthermore, incorporation of IPP slightly improves the room temperature stiffness. Such membranes are torch grade, which lends them to torching during installation. The melted compound serves as the adhesive. In contrast, the mop grade requires blown asphalt or a cold asphalt adhesive for installation on roof.

The most common asphalt roofing membrane has a polyethylene terephthalate (PET) or PET/Nylon 6 nonwoven mat of 170 to 200 g/yd^2. To the mat SBS- or polypropylene-modified compound is applied on a production line. Rollers remove excess compound and a water tray cools the membrane. The membranes are then cut to size and wound on spools. Placement of the mat, use of two thin mats, and use of dual compounds (on one side SBS- and on the other polypropylene-modified asphalt) are some variations for making other than basic products. White or colored ceramic-treated granite granules can be placed on the surface of one side of the membrane to face the elements of nature; these granules are for decoration and to improve durability.

VI. CHARACTERIZATION OF ASPHALT AND COMPOUNDS

A. Asphalt

A knowledge of the chemical makeup of straight cut asphalts or asphalt blends is essential for proper modification with either polypropylene or SBS thermoplastic elastomer. In Europe, membrane manufacturers custom blend their asphalt for modification in their plants. A single asphalt such as AC-5, AC-10, or a blend made by refineries is the production choice in the United States. Some manufacturers use separate stocks for SBS and polypropylene modifications. Although asphalts differ in composition from refinery to refinery depending upon the crude oil and refining, a proper modification design allows use of most of the asphalts in roofing applications.

The asphalt composition partially indicates the potential durability of the membrane in service, specifically in a harsh environment. Thus, asphalt composition, modification, and the application technique for installing membranes on the roof are important.

Iatroscan (thin-layer chromatography/flame ionization detector, TLC/FID) is a useful tool in pharmaceuticals and biotechnology. Figure 1 shows a typical

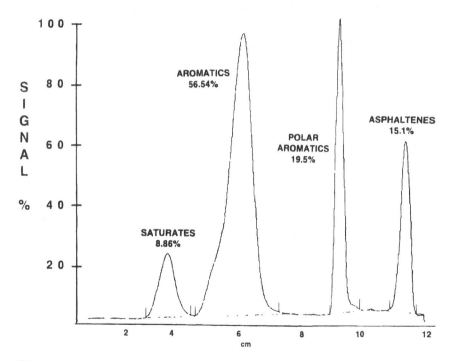

Figure 1 A typical Iatroscan (TLC/FID).

scan for an AC-7.5 asphalt. This asphalt contains 8.86% saturate, 56.54% aromatics, 19.5% polar aromatics, and 15.1% asphaltenes. Because of the variability in solvents used to dissolve asphalt and to develop the rods, results from different laboratories may vary. Results from the same laboratory are reliable for R&D and manufacturing provided the technique remains unchanged. Results are comparative and definitely not absolute.

Nuclear magnetic resonance (NMR) spectroscopy (^{1}H and ^{13}C) is an extremely useful tool for measuring aliphatic and aromatic contents in an asphalt. We used these methods to compare asphalts and select suitable asphalts for roofing membrane preparation [22]. The characterization of asphalt by high performance-gel permeation chromatography (HP-GPC), NMR, differential scanning calorimetry (DSC), dynamic mechanical analysis (DMA), modulated DSC, and other techniques was discussed at a recent American Chemical Society meeting [23].

Figure 2 shows DMA spectra of asphalt G. The glass transition temperature (T_g) of asphalts G and A is presented in Table 2. A T_g around 10°C is suitable for making roofing membrane.

B. Compounds

There is a need for reverse engineering aged membranes to determine changes in the composition as the sample ages during service. A method described by us for quantitatively analyzing polysulfide sealant is adaptable to polymer-modified asphalt compounds [24]. The filler is separated from polymer and asphalt by dissolving the compound and centrifuging. Soxhlet extraction can also be used to separate the filler and the binder. A third method is by thermogravimetric analysis (TGA). Thus. we know the filler content quantitatively. To determine type of filler. X-ray fluorescence (XRF) or X-ray dispersive energy analysis (EDAX) is suitable. Polymer and asphalt can be resolved by a solvent/nonsolvent technique or by utilizing the Flory θ temperature. Typical characterization methods, e.g., TLC/FID and NMR, can quantify the asphalt portion. Useful methods for characterizing the modifying polymer include GPC. DSC, DMA, and Fourier transform infrared (FTIR). The status of the polymer network can be seen by UV microscopy.

VII. ASPHALT, SECONDARY STRUCTURE: MECHANISMS

A. Polymer-Like Nature of Asphalt

The polar portion of asphalt contains negatively and positively charged groups along the polyaromatic rings. This molecular arrangement leads to a rodlike polymer structure similar to that of proteins, enzymes, and aromatic nylons. The rodlike structure imparts elasticity to asphalt. The neutral portion of asphalt,

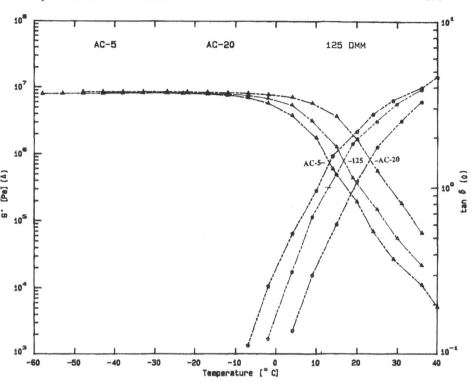

Figure 2   DMA spectra of asphalt G.

Table 2 T_g of Selected Asphalts

Asphalt	$T_g(°C)$
G (AC-5)	9.5
G (AC-125 Pen)	12.0
G (AC-20)	19.5
A (AC-5)	8.0
A (AC-10)	11.0
A (AC-20)	13.0

trapped inside rodlike structures, imparts viscosity. Heating destroys the secondary rodlike structure of asphalt. Of course, the destruction is reversible. The polymer nature and reversibility of the secondary structure of asphalt are shown Figure 3.

B. Mechanism of Phase Inversion

Asphalt per se is not a suitable material for single-ply type roofing. Asphalt is usually modified by polypropylene or SBS rubber to mitigate its shortcomings. In SBS modification, the styrene blocks dissolve in the trapped hydrocarbons and the butadiene chains snugly wrap around asphalt particles. Because styrene blocks become connected to an asphalt particle at both ends, the covering is very tight. Thus, we find that 7% of a medium molecular weight SBS containing 30% styrene is sufficient to wrap around the asphalt (Figure 4). If the molecular weight of SBS is higher, the available wrapping chain length increases, so less rubber polymer is required. For example, the molecular weight of SBS polymer A is 93,000, whereas the molecular weights of polymers B and C are 262,000 and 131,000. Naturally, more polymer A will be required than C to achieve nearly the same modification results. We will need the least amount of polymer B.

Polyolefins, e.g., polypropylene and polyethylene, are insoluble in aromatic hydrocarbons at low temperatures. They dissolve in aromatics, e.g., xylene around 125°C. If the modifying polymer dissolves in asphalt, improvements in properties are only marginal. In solid form, polypropylene chains are coiled significantly. Upon melting, the chains uncoil and extend. Although dipole-dipole forces between polypropylene and asphalt are weak, the large number of these forces can lead to covering the asphalt particles. The coverings with these chains are

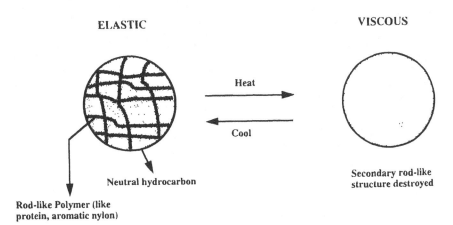

Figure 3   Asphalt: polymer, secondary structure.

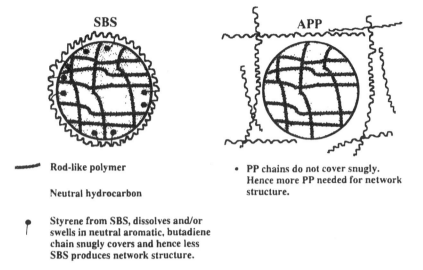

SBS

APP

━━ Rod-like polymer

Neutral hydrocarbon

• PP chains do not cover snugly. Hence more PP needed for network structure.

⌐ Styrene from SBS, dissolves and/or swells in neutral aromatic, butadiene chain snugly covers and hence less SBS produces network structure.

Figure 4   Asphalt modification by SBS and polypropylene.

less intimate than with SBS. Consequently, we need more polypropylene (around 20%) to make a membrane.

C.   Stability Considerations

Colloidal or polymer network stability is important in asphalt modification. For dispersions to be stable, it is imperative to provide a repulsion barrier between the particles so that the attractive forces do not overcome the energy of thermal motion [25]. Dispersions of neutral particles flocculate rapidly because of the long-range attractive forces. In particle dispersions, the total potential is the sum of the energies of attraction and and the energies of repulsion. To achieve a potential energy maximum, repulsive potential energy between particles is achieved by electrostatic or steric stabilization.

Interparticle repulsion due to coulombic forces between two particles is a function of the dielectric constant of the continuous medium (polymer is the continuous phase in the modified asphalt). The surface potential develops by methods such as adsorption and ionization of potential determining ions. We know asphalt contains both positively and negatively charged groups. These lead to the rodlike structure. We have not determined the isoelectric point of asphalt. The asphalt particles can carry either positive or negative charges depending on the pH. The surface charges can easily be manipulated, specifically in aqueous systems.

The term steric is not used above in an organic chemical sense of restriction of motion; rather it has thermodynamic implications. Interpenetration of particles through Brownian motion (in membranes modified as a result of high heat in a laboratory oven or the solar collector effect on a roof) results in compression of the modifying polymer or its chain segments. This compression produces a change in the free energy given by the Gibbs-Helmholtz equation.

$$\Delta F_R = \Delta H_R - T\Delta S_R$$

There are three possibilities regarding the change of free energy.

Case I: $\Delta F_R = 0$ as if no polymer is present
Case II: $\Delta F_R = $ positive indicating polymer stabilizes the particles and does not sensitize them
Case III: $\Delta F_R = $ negative indicating polymer sensitizes the particles and destabilizes them

In modified membranes, a negative ΔF_R is important for network formation. However, a positive or a less negative change in free energy will help during service on the roof (Figure 5).

D. Role of Carboxylation: Retardation of Phase Reversion

We found that inclusion of a carboxylated monomer, oligomer, or polymer improved the thermostability of calcium carbonate–filled polypropylated asphalt. These carboxylated molecules act as a dispersant. After dispersing calcium carbonate, they react in situ and become the calcium salt of the acid.

The carboxylated molecules and the calcium salt both act as polymer network stabilizers. Calcium carbonate–filled polypropylated asphalt membranes with and without carboxylated molecules were heat treated at 70°C for 3 months. After 2 weeks, oil-like liquid dripped in membranes not containing the dispersant. This became more pronounced after 4 weeks. Membranes with dispersant did not drip and maintained their form stability. Membranes without dispersant also showed compound flow, sag, and form instability.

Polypropylated asphalt is nearly nonpolar. This hot compound, containing small amounts of COOH group, will orient when chilled by water. The hydrocarbon tail will orient toward the compound whereas the polar carboxyl head will orient toward water. The net result is highly organized sealing of the membrane, which improves thermostability (Figure 5).

Very slight carboxylation of polypropylated asphalt will provide resistance to phase reversion during service. The process is shown on page 382.

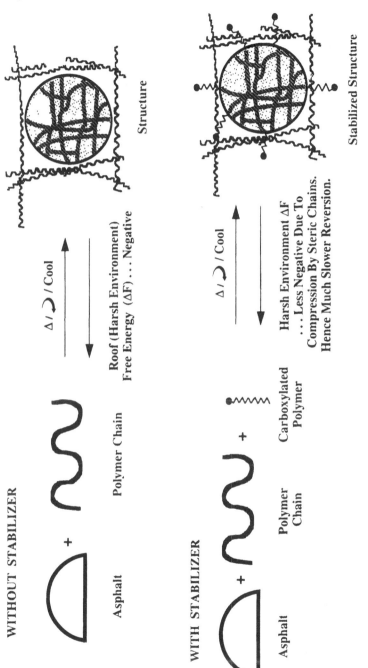

Figure 5 Structure stabilization.

Noncarboxylated   High/low temperature/RH during service
membrane   ──────────────────────────────────────→
                      Phase reversion initiation
                      Phase reversion propagation
                      Freeing of asphalt in membrane
                      Reduction of performance
                      Surface defects
                      Failure

Very lightly   High/low temperature/RH during service
carboxylated   ──────────────────────────────────────→
membrane   Phase reversion retardation/inhibition
                      Mitigation of asphalt freeing
                      Minimal performance reduction
                      Less surface defects
                      Life and performance prolonged

VIII. CONCLUSIONS

The techniques of NMR, GPC, DMA, and TLC/FID are particularly useful for characterization of asphalt used in roofing membrane. These tools are essential in the development of newer products and they are also useful in determining membrane durability.

Asphalt has a secondary structure that is destroyed by heat. Upon cooling, the structure reappears. The shortcomings of asphalt, namely rutting and high temperature dependence, are improved considerably by polymer modification. At a low polymer level, the asphalt-polymer polyblends produce slight improvement. At higher modification levels, however, polymer network formation occurs, producing dramatic improvements in properties. We have proposed a mechanism of polymer network formation. In addition, we presented a mechanism of reversion on roofs due to the harsh environment. To improve thermostability, minor amounts of carboxyl groups help. We also present reasons why slight carboxylation improves thermostability.

REFERENCES

1. A. M. Usmani, Polymer Modification of Asphalt, Research Institute, King Fahd University of Petroleum and Minerals, Dhahran, Saudi Arabia, July 1982 and July 1996.
2. D. N. Little, J. W. Button, R. M. White, E. K. Ensley, Y. Kim, and S. J. Ahmed, Investigation of Asphalt Additives, FHWA Report RD-871001, 1987.
3. S. A. Piazza, A. Arcozzi, and C. Verga, Rubber Chem. Technol., **53**, 994 (1980).
4. H. Abraham, *Asphalt and Allied Substances*, 6th Ed., 1962.
5. N. Akmal and A. M. Usmani, in *Recent Advances in Polyolefins*, (R. B. Seymour, Ed), Springer-Verlag, Berlin (1993).

6. G. Holden, E. T. Bishop, and N. R. Legge, *J. Polym. Sci. C*, **26**, 37 (1969).
7. J. M. Tancrede and G. M. Marchand, *Adhesive Age*, **27** (June 1994).
8. G. Ourisson, P. Albrecht, and M. Rohmer, *Sci. Am.*, **250**, 44 (August 1984).
9. P. J. Champagne et al., *Fuel*, **64**, 423 (1985).
10. R. J. Cierc and M. J. O'Neel, The Mass Spectrometric Analysis of Asphalt—A Preliminary Investigation, in *Symposium on Fundamental Nature of Asphalt*, American Chemical Society, Division of Petroleum Chemistry, Vol. 5, p. 4-A (1980).
11. E. J. Gallegos, High Resolution Mass Spectrometry of Asphalt Fractions, in *Proceedings of the Seventh World Petroleum Congress*, Vol. 4, p. 249 (1987).
12. R. H. Fish et al., *Anal. Chem.*, **58**, 2452 (1984).
13. K. H. Altgeit, *J. Appl. Polym. Sci.*, **9**, 3389 (1965).
14. P. W. Jennings, J. A. S. Pribanic, K. R. Dawson, and C. E. Bricca, *Prep. Div. Petrol. Chem. Am. Chem. Soc.*, **28**, 915 (1981).
15. P. W. Jennings et al., FHWA Report MT-DOH-7930, 1980.
16. Y. Yamamoto and T. Kewanobe, *J. Jpn. Petrol Inst.*, **27**, 373 (1984).
17. J. E. Ray, K. M. Oliver, and J. C. Wainwright, in *Advances in Analytical Chemistry in the Petroleum Industry*, (G. B. Crump, Ed.), Wiley, New York (1982).
18. G. van Gooswilligen, *Int. J. Roofing Technol.*, **1**, 45 (1989).
19. G. King, H. King, and B. Brule, presented at the American Chemical Society Rubber Division Meeting, Las Vegas, 1990.
20. A. M. Usmani et al., U.S. Patent 5,516,817, to Bridgestone/Firestone (May 14, 1996).
21. A. M. Usmani et al., U.S. Patent, to Bridgestone/Firestone (1996).
22. A. M. Usmani, UDC, unpublished work, 1995.
23. A. M. Usmani, *Polym. News*, **21**, 283 (1996).
24. A. M. Usmani, *Polym. Plast. Technol. Eng.*, **19**, 185 (1982).
25. A. M. Usmani and I. O. Salyer, *J. Appl. Polym. Sci.*, **23**, 381 (1979).

19

Production, Identification, and Application of Crumb Rubber Modifiers for Asphalt Pavements

Michael Wm. Rouse
Rouse Rubber Industries, Inc., Vicksburg, Mississippi

I. INTRODUCTION

A. Background

The use of crumb rubber modifiers (CRMs) in asphaltic paving applications is receiving increased attention from federal and state agencies, design engineers, road contractors, and asphalt suppliers. As a result, the changes being seen by the pavement industry are improved mix design processes, advancements in equipment designs, and greater communications between the contractor, equipment manufacturers, government agencies, asphalt suppliers, and modifier suppliers. There is no doubt that government regulatory issues, such as the Clean Air Act, Intermodal Surface Transportation Efficiency Act (ISTEA), and new design procedures from the Strategic Highway Research Program (SHRP) are the driving forces behind major changes that are occurring in the pavement industry. The pavement engineer is faced with the realization that performance-based mixtures are becoming the main focus in the road building industry. The SHRP program has and will bring about a greater need for more sophisticated instruments that will test the mix designs, qualifying them to meet the performance specifications required.

    The use of recycled tire rubber (CRMs) in asphaltic blends to enhance mix design performance is becoming a viable alternative in modifier selection. There

385

are new testing systems that have the ability to pinpoint the critical physical and chemical properties needed in today's asphalt market. If roads prepared from rubber-modified asphalt cement (RMAC) fail to meet the new SHRP performance specifications, then CRMs in road mix designs will have a very short usage cycle. Economic and performance data should be the criteria determining whether CRMs have a place in the construction, rehabilitation, and maintenance of today's highways.

B. Highway Traffic

Figure 1 illustrates the reason why America's road infrastructure is deteriorating at an exponential rate. Vehicle traffic, or road miles traveled per vehicle, is nearly doubling each year in relation to road construction and maintenance, according to Federal Highway Administration statistics.

This amount of vehicular traffic puts undue stress on our existing roads to the point that new road construction and rehabilitation cannot keep pace under present funding for new construction and maintenance. For today's road engineers and technologists to meet the new demands being placed on our highways, new hot mix asphalt designs and asphalt binder specifications need to be developed to meet this challenge. The conventional asphalt cement (AC) produced by a refinery needs to be enhanced to improve the binder qualities in the hot mix asphalt binder (HMA).

An asphalt binder is defined as the substance used to hold the aggregate particles together. Hot mix asphalt concrete is a designed aggregate and asphalt cement mixture produced in a hot mix plant (batch, drum, drum/batch, or double barrel) where the aggregates are dried, heated, mixed with (liquid) asphalt cement, and then transported, placed, and compacted while still at an elevated temperature

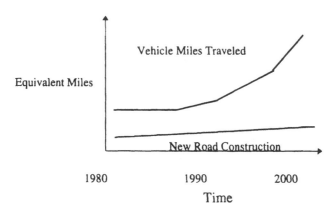

Figure 1 Vehicle traffic and road construction.

of 257°F (125°C) or 275°F (135°C) to give a durable and fatigue-resistant pavement course.

These conditions appear to present an ideal picture of how a pavement should perform. However, with the tremendous pounding to the surface and internal structure of the pavement by increased vehicular traffic, many of the designs incorporated in today's pavements cannot withstand the increased loads. Pavement deterioration, such as stress cracking, rutting, raveling, and shoveling, is the physical result of increased traffic loads. Environmental conditions, such as wide temperature variations and moisture, also play an important role in the service life of pavements.

Pavements need rehabilitation time to recover from severe overloading. This "self-recovery" phenomenon is called healing. Small stress cracks will bond or heal during off-peak traffic hours. However, around-the-clock traffic patterns do not allow the self-healing process to occur, and accelerated deterioration and failure in the pavement are being noted. To provide resistance to the stress and to eliminate failures, and thereby enhance highway performance, chemical modifiers should be added to conventional asphalt binders (Figure 2).

Today's tire is designed for esthetics, comfort, performance, and control. The tire itself is a geometrically a torus, structurally a high-performance composite, mechanically a flexible-membrane pneumatic container, and chemically consists of long chains of macromolecules. In terms of chemistry, the tire polymers are very random in nature versus block polymers, which tend to be linear and branched (see Figure 3). It is the random vulcanization of these styrene-butadiene rubber/natural rubber (SBR/NR) polymers that creates a tire that seems indestructible. Figure 4 illustrates the construction of a typical radial tire.

The tire used on today's automobile has taken many years to evolve into its present composition, durability, and service. The tire is designed for ridability and long-term performance in life and mileage as reflected in Figure 5. The same characteristics, inherent in a tire, can be translated into highway performance through gained pavement life cycle. After the tire has seen its useful life on a

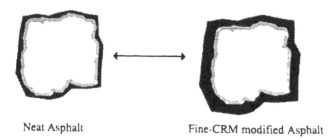

Neat Asphalt   Fine-CRM modified Asphalt

Figure 2   Comparison of film thickness with fine CRM over a neat asphalt.

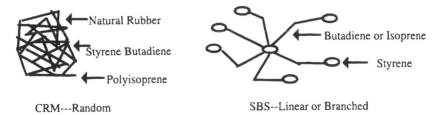

CRM---Random   SBS--Linear or Branched

Figure 3   Random CRM configuration versus a block polymer.

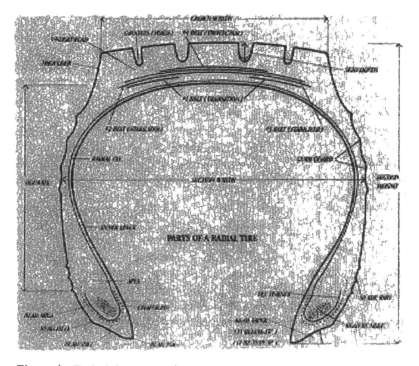

Figure 4   Typical tire construction.

vehicle, it can be processed into a high-quality CRM and be incorporated into the matrix of the neat asphalt.

II.   ASPHALT RUBBER HISTORY

The development of asphalt rubber (AR) has a long history. Table 1 is a brief synopsis of the major developments leading to today's use of high-quality CRMs

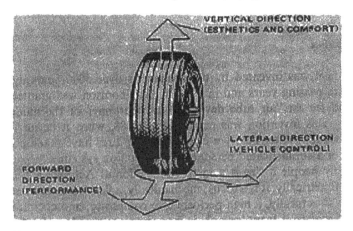

Figure 5 Tire design dynamics.

as modifiers in asphalt cements. U.S. Rubber Reclaiming (Vicksburg, Mississippi) has seen a very rough path to demonstrate the place of CRM as a major modifier to enhance pavement performance.

Mandating rubber use or specifying any other polymer modifiers for asphalt cements is inappropriate. Action by state and federal agencies to specify a particular polymer or even a recipe specification is poor practice, reflects favoritism, and lack of willingness to set performance guidelines, and does not reflect the open arena for achieving the optimum benefits to the taxpayer through appropriate research. The shortcut approach, such as specifying recipes and/or specifying a particular polymer, should be discontinued and focus should be placed on developing performance methods. Specifications should be based on life cycle costs and performance-based data. Each polymer additive like CRM must stand on its own merits. With the implementation of SHRP-PG grades and soon Superpave, the role of modifiers in pavements should be evaluated on a more even playing field based on life cycle costs.

III. TERMINOLOGY

The U.S. Federal Highway Administration (FHWA) has identified crumb rubber produced from scrap tires for use in highway pavements as crumb rubber modifiers. CRMs are incorporated into HMA binders by either the wet or the dry process. Figure 6 shows the FHWA's standard CRM terminology, process, and product terminology.

Interestingly, only the fine CRM meets the full range of process and product applications. In general, coarse CRM is considered to be CRM materials greater

Table 1   History of Asphalt Rubber

1843	British patent for modified bitumen with natural rubber.
1844	British patent bitumen rubber mixture in road surfacing.
1920	Research in rubberized roads is documented in England, South Africa, Japan, and Java.
1925	Goodyear developed first solid rubber block road with a 25-year life span.
1929	Southeast Asians, British, and Dutch installed solid block natural rubber roads to reduce noise and vibrations.
1933	Paper by the Russians Pospkhov and Fokin entitled "Mixture of Asphalt and Synthetic Rubber" is published.
1936	Holland paved a road with rubberized asphalt. The Germans utilized this road extensively as a highway for military traffic. Still is good repair in 1950. (Believed to be first wet process application.)
1937	*Rubber World* published rubber and asphalt mixtures.
1939	*Rubber World* published bitumen-rubber mixtures for road construction.
1940–1950	U.S. Rubber Reclaiming and Midwest Rubber Reclaiming began to market vulcanized crumb rubber for asphaltic hot pour joint seals.
1950	Massachusetts Department of Transportation utilizes discrete chunks of tire rubber as road surface aggregate. (Believed to be first dry process application.)
1951	Ramflex (reclaim or devulcanized) was developed by U.S. Rubber Reclaiming for asphaltic road applications. U.S. Rubber Reclaiming develops an aggregate coating technique using asphalt and special ground rubber.
Mid-1950s	Uniroyal and Midwest Rubber Reclaiming market powdered or flaked rubber for dispersion in molten asphalt. This heat treatment softened and converted the rubber to a dispersible gel. Hard rubber core may still remain. Thomas Johnson of U.S. Rubber Reclaiming patents his rubberized athletic surfaces using ground rubber.
1950–1960	New York, Pennsylvania, and Massachusetts led the nation in the use of Ramflex in hot pour joint seals and rubberized asphalt applications. Vulcanized ground rubber powders became a major component of prepackaged joint sealants.
1961	Massachusetts utilizes rubberized asphalt in bridge decks.
1964	Early work was conducted by Charles McDonald. He used information obtained from and known by U.S. Rubber Reclaiming since 1952.
1960–1970	Ground vulcanized rubber, reclaimed rubber, and latexes are proved to be valuable as asphalt modifiers. Usage is low due to economic factors, poor distribution systems, and the availability of low-priced asphalt.
1970–1985	Charles McDonald patents, B. J. Huff patents, B. D. LaGrone patents, Don Neilson and J. Bagley patents all demonstrate the value of ground rubber and reclaimed rubber as asphalt modifiers. Technology of these patents allows the road designer to utilize products in road applications such as chip seals, tack coats, open graded mixes, dense graded mixes, SAMI, and SAM.

Table 1 Continued

1975	G. Lind patents PlusRide, a process similar to the old chunk rubber theory used in Massachusetts in 1950.
1984–1986	U.S. Rubber Reclaiming develops the fine-CRM wet grinding process, which provides extremely fine vulcanized rubber powder for asphalt compounding applications.
1988	Florida Department of Transportation evaluates fine rubber powders for use in all asphalt applications, resulting in mandated use of recycled tire rubber. The process is also known as the continuous or Rouse process.
1989	Florida Department of Transportation applies fine CRM in rubber-modified asphalt cement (RMAC) in dense and open grade wearing courses. Generic dry process techniques applied.
1990	Asphalt rubber membrane innerlayer (ARMI) with fine CRM is terminal blended in Florida. Mixtures using fine CRM, called UltraFine, tested by the Asphalt Institute, demonstrate that fine CRM is an effective modifier, improving temperature susceptibility for a wide range of asphalt cements.
1993	Terminal blended performance-based asphalts (PBAs) with fine CRM (−80 mesh) are a demonstrated success by U.S. Oil & Refining.
1995	New surface-modified fine CRMs, called Builder series, are introduced by Rouse Rubber Industries to further enhance specific high- and low-temperature performance and adhesion of asphalt cements while meeting all the SHRP, Superpave, and AASHTO procedures.

than 40 mesh (420 μm) in diameter and fine CRM to be less than 40 mesh. More information can be found in the FHWA report to Congress entitled "A Study of the Use of Recycled Paving Materials," June 1993, and "Design Procedures and Construction Practices," FHWA-SA-92-022. In addition, for asphalt rubber membranes, the use of either a coarse or fine CRM will have a dramatic effect

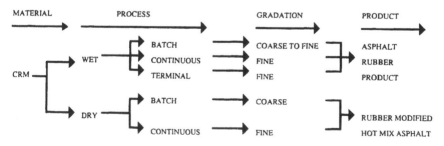

Figure 6 Crumb rubber modifier relationship terminology.

on cost and ease of use by the asphalt paving contractor. This is illustrated in Figure 7, which shows that with 2 fine CRM (minus 80 mesh), ambient ground tire rubber with a 200 mesh (74 μm) mean particle size, called UltraSAM and UltraSAMI, can be produced continuously or at the terminal for field applications with reductions in energy required.

Although a great deal of work has focused on the dry process wherein the coarse CRM is viewed as an aggregate replacement in a portion of the hot mix, this area is not discussed. Since 1985, the major use of CRM has been in open and gap graded mixes, even though asphalt rubber became recognized through the work initiated in the 1950s. AR mixes provided thicker films of binder, and initial reports relay improvement in durability and crack resistance. The mechanism of AR technology or modifiers in general was not fully understood or well documented in the early work by New York and Massachusetts transportation authorities, even though these states used significant quantities of reclaimed rubber in road pavements. Today Florida, California, and Arizona use significant quantities of CRM to improve pavement performance. Rubber-modified asphalt binders have been used in surface treatments, stress-absorbing membrane interlayers, seals, and as an important component in hot mix asphalt mixtures.

IV. CRM MANUFACTURING AND PROCESSES

The manufacture of CRM materials varies according to the quality of the end products desired by the pavement engineer. The better the quality and the greater the fineness of the CRM product, the greater the capital investment for processing such products. In addition, the support services required to ensure product quality need to be included in the final cost of the CRM to a highway project or asphalt blender. However, with improved fine-CRM production methods, costs to the user of fine CRM have been kept stable and competitive to those of other modifier manufacturers. The reasons why recycled products, such as tire rubber, has not

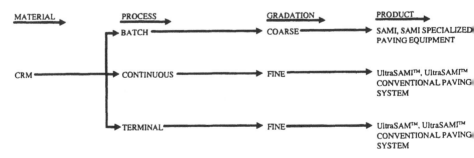

Figure 7   Asphalt rubber membrane terminology.

been met with greater enthusiasm by highway departments and industries has been the lack of dedicated recycling professionals and inadequate financial commitments, along with inadequate documentation to ensure product quality and performance. Service and cost-competitiveness in relation to other recycled materials and virgin polymers are other reasons for slow growth of the recycling industry in supplying a high-quality CRM. Table 2 summarizes the three parameters necessary for CRM products to be competitive and accepted by industry and highway officials. It takes a commitment by industry and government to define and experiment with recycled materials, such as CRMs, to determine their place in the road construction industry.

A. CRM Processes

There are several ways to produce CRM material economically and practically, depending on the desired end product mesh size and configuration. The most common ways are listed in Table 3. Each of these processes requires a basic feedstock of tires, tire parts, or other types of surplus rubber materials. This table focuses on scrap tires in the form of buffings, tread rubber pieces, tire chunks, or shredded tire chips from $\frac{1}{2}$ inch to several inches in size, which are derived from whole tires. During the process, all the types are reduced to what are referred to as mesh sizes.

Each of these processes must be evaluated in terms of economics, particle size distributions, and morphologies. As a rule, the surface area or morphology increases with decreasing particle size, especially with the ambient wet grinding technology. Also, the finer the particle size, the cleaner the CRM can become of foreign matter such as fiber, steel, and other inerts. The cleaner the CRM, the

Table 2 CRM Acceptance Criteria

Provide performance and quality
Ensure environmental/safety concerns
Provide cost-effectiveness

Table 3 Common CRM Process Capabilities

CRM process	Resultant CRM gradation
Basic screening	1/4 inch to 40 mesh
Ambient grinding	1/4 inch to 40 mesh
Cryogenic grinding	1/4 inch to 60 mesh
Wet grinding	30 mesh to 325 mesh

greater the capital investment. The fine CRM (−80 mesh) rubber is very clean, handles like a mineral filler, is compatible in all types of asphalt equipment, and reacts at normal HMA temperatures to produce a homogeneous asphalt-rubber mix.

1. Basic Screening Process

The most common and simplest feedstock for producing a CRM material is the screening of buffings. Buffings are produced when tires are prepared for retreading, normally truck and bus tires. The resulting particles are of various sizes from very small to long thin strips. The bulk of the particles are circular in shape and may be about 1/8 to 1 inch long. The exact configuration depends on the buffing method. This particle gradation mesh is accomplished through a basic screening process by which all the particles pass through or are retained on a certain size mesh screen. The mesh size refers to the number of openings in the screen for each square inch of the screen's area. For example, if it is a 40 mesh, there will be 40 openings in the wire or wire cloth every square inch, through which a particle of 40 mesh size or smaller will pass. If 80 mesh, there will be 80 openings, etc. The procedure for determining the specification for mesh is set by the American Society for Testing and Materials (ASTM) and will be discussed in more detail later. Essentially, no metal or cord is found in this material. However, tire buffings have a very limited supply based on the the number of truck tires recapped. Without additional grinding of the buffing dust material, only a small percent of it will be fine enough to meet the criteria for optimum modification of asphalt.

The CRM producers, with only the basic screening process, generally lack the quality control, equipment, and personnel to ensure product uniformity and are unable to assist the HMA or terminal operator, asphalt contractors, and Department of Transportation (DOT) engineers in the use and application of CRM materials.

In some operations, larger pieces of tread rubber result in long strips or in larger but elongated chunks. Often, this is called tread peel. There is little or no tire cord, although sometimes the peeling operation does access the cord material in the tread area. Care must be taken to remove this fibrous material because it can cause material handling problems, asphalt nozzle and filter plugging, and increased absorption of the AC binder by the fiber inherent in tires. Most CRM systems usually do not remove the inerts such as stones, metals, or other debris in the form of wood, paper, or metals, which generally go through the screens into the final product.

Whole tires are processed in many different ways. A common method is to run the whole tire through a shredder or chopper and then to a granulator or cracker to reduce the material to a suitable sized feedstock about $\frac{1}{2}$ to 3 inches in size. It is common to cut up the tire bead and all other parts of the tire together;

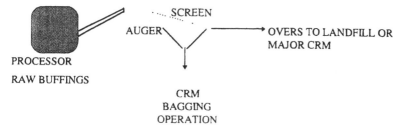

Figure 8 Basic screen process diagram.

although in some operations the bead is removed first and disposed of in various ways. During these preliminary stages, some wire and fabric are removed. Normally, the wire and fabric materials are sent to a landfill.

2. Ambient Grinding

The complete ambient grinding process is usually accomplished with open mills, granulators, extruders, or hammer mills. All grinding takes place at room temperature (also referred to as ambient conditions). The actual particle reduction is accomplished by a tearing, ripping, or smashing action. This creates a particle with a rough surface, perhaps described as spongy. A schematic of a typical open mill grinding system and granulator appears in Figure 9.

The ambient process consists of a series of open mills or granulators, screeners, conveyors, and various types of magnets to remove any steel wire or other steel particles. Many combinations of machinery configurations are found in today's modern tire-rubber grinding facilities. A fabric removal system is also required if the material starts out as whole tire chunks. In addition, there may be stone or inert object removal equipment (air tables) in the overall system.

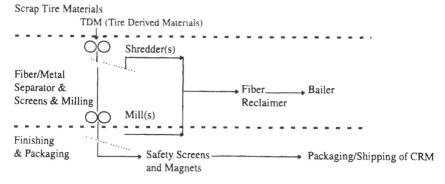

Figure 9 Basic ambient grinding system diagram.

The ambient process is fairly efficient, the oldest known method in the production of the coarser mesh products, and usually results in the lowest cost for CRM greater than 40 mesh in size. The most common is known as a cracker or two roll (open) mill with grooves on the surfaces and the rolls turning into each other at different ratio speeds to achieve a grinding action and maximize efficiency.

3. Cryogenic Grinding Process

In this process, the scrap tire rubber is cooled to its embrittlement point by the use of liquid nitrogen, often called LIN or LN_2. This causes the rubber to become very brittle and is easily fractured in a hammer mill. The resulting surface is very clean faceted or glasslike, resulting in less particle surface area than for ambiently ground materials. A schematic of a basic cryogenic grinding process appears in Figure 10. This process consists of a precooler to chill the material, a grinder, appropriate screens, conveyors, and a fiber separation system. The capital investment cost for the cryogenic process in relation to an ambient system is generally less, because fewer grinding mills and related handling equipment are required. However, the operating costs, due to the cost of liquid nitrogen, make cryogenic products more costly than the ambient produced CRM, especially for meshes finer than 40 mesh (425 µm). Research is being performed to determine whether the surface morphology of a cryogenically ground CRM performs differently from that of ambiently ground CRM for the same given gradation.

4. Wet Grinding System

In this process, the rubber granules, after their initial reduction to a −10 to −20 mesh, are made into a wet slurry and passed between milling stones, producing a very uniform, high-surface-area particle with broad, predetermined, and predictable particle size distribution ranges. A schematic of an overall basic wet grinding

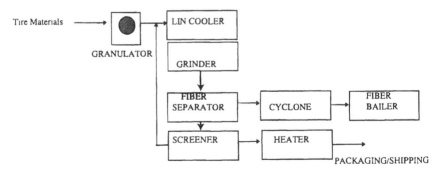

Figure 10   Basic cryogenic system diagram.

operation is shown in Figure 11. This system requires very effective drying equipment to control moisture content, which is not required for the other two processes. The wet grinding system is a patented and sophisticated technology that produces very high quality, uniform CRM products with defined particle gradations.

V. TECHNICAL APPLICATIONS

The recycling of waste rubber from pneumatic tires and from other sources is of increasing importance to society. The scrap tire problem should be evaluated as a possible opportunity to help improve road life. The form of recycled waste rubber discussed here is ground automotive and truck tire rubber, which has found a wide range of applications based on economic advantages alone. Ground rubber has also been used in applications based on its unique properties when added to other materials including plastics, friction materials for brake pads, mats, playground surfaces, athletic surfaces, rubber-modified asphalts, tires, and as a soil amendment. Plastics with added ground rubber compounds give desired properties, such as increased flexibility and impact resistance, which pure plastics may not possess. In many circumstances, rubber helps reduce overall costs in friction products, such as brake shoes.

Ground rubber has been used in the tire industry by utilizing a certain percentage of the rubber as a filler or extender in tires. Its use can reduce cost and embrace the highest reuse theory for tire rubber recycling.

VI. RAW MATERIAL TYPES

The production of CRM involves selecting, cleaning, and reducing scrap rubber to a desired particle size distribution and surface morphology. The CRM particle

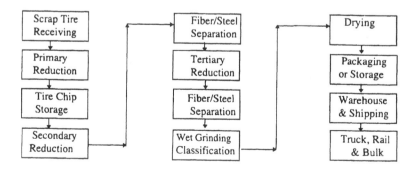

Figure 11   Wet grinding system diagram.

distribution curves are based on the process used for reduction. There are several different types of ground rubber products depending on the parent rubber from which it was derived. The type of ground rubber produced is limited only by the parent compounds of the raw materials used. The most common types used in rubber compounding and other applications may be simply classified into the following categories.

A. Automotive Tire Rubber

1. Whole tire ground rubber is derived from automotive and truck tires from which fiber, metal, and other inerts have been removed. Whole tire buffings generated from the retreading plants that grind or "buff" the sidewalls, as well as the tread, are also included in this category. The rubber is then ground to the desired particle size distribution. Some common examples of these are 10, 12, 40, and 80 mesh products and also selected mesh cuts such as 10/30 (a rubber particle passing a 10 mesh screen but retained on a 30 mesh screen).
2. Tread ground rubber is derived from truck, bus, and/or passenger treads in the form of peelings (peel) or made by utilizing buffing particles generated from automotive tire retreading. The material is then processed to desired particle sizes. Some common examples of these are 10, 20, 30, 80, and 120 mesh products.

B. Nonautomotive Tire Rubber

Airplane or off-the-road tire ground rubber is derived from airplane tires or tires for off-the-road (OTR) use, such as heavy equipment tires. Airplane tire rubber typically comes in the form of buffings. Both of these types have chemical contents significantly different from those of automotive tires. These differences show up in natural rubber and/or ash content.

VII. CHEMICAL COMPOSITION

The chemical composition of ground rubber is based on two major groupings: automotive tire rubber and nonautomotive tire rubber. These groupings are important because, in the past, asphaltic applications have been predominantly limited to automotive truck tire rubber. The users of ground rubber for asphalt applications generally specify truck tire or high natural rubber material.

Generally speaking, if truck and passenger tires are ground as a composite, that is, in the ratio in which they are produced and discarded, a uniform chemical composition can be obtained. However, it is well known that truck tires contain a greater percentage of natural rubber (NR). NR is easier to disperse and more compatible with most neat asphalts. This is why truck and bus tires are preferred over passenger tire rubber.

Nonautomotive tire rubber typically shows different results from those obtained for automotive tire rubber when analyzed for acetone extract, ash, carbon black, and rubber hydrocarbon. These differences can be used by suppliers and users to qualify materials when attempting to determine whether or not a correct source of raw material (i.e., automotive tire rubber) was used. They can also be used to determine whether a finished ground rubber product has possible contamination even when the correct raw material has been used. These nonautomotive types, when tested in the same manner as automotive tire rubber, yield results different from those for CRM produced from truck, bus, and automotive tires. In all cases, the synthetic-to-natural ratio of the total rubber hydrocarbon in the certificate of analysis (COA) for the CRM should be reported.

VIII. CHEMICAL IDENTIFICATION OF RECYCLED TIRE RUBBER

The chemical identification of components in recycled automotive tire rubber (hereafter referred to as tire rubber) and other industrial scrap rubber involves the application of testing methods described in ASTM standards D-297 [10], "Standard Test Methods for Rubber Products—Chemical Analysis" and D-3677 [11] "Standard Test Methods for Rubber—Identification by Infrared Spectrophotometer." These methods are designed to identify chemical components in a rubber compound and are widely used in recycled tire rubber analysis. It is a common practice that only some components, which are most important for the application purposes, are tested. It is common to test for acetone extract, ash, carbon black, rubber hydrocarbon, and natural rubber. New developments in analytical test instruments are setting new frontiers in more precise and reliable measurements, such as the Fourier transform infrared spectrophotometer (FTIR) and the thermogravimetric analyzer (TGA), which will be discussed. Results of all these tests are expressed as percent by weight of sample tested.

An additional important chemical test for asphalt applications is that of the moisture content, which is not a part of the tire compound itself. In the ground rubber, it will be listed as a specification with an upper limit of 0.5 to 1.0% by weight.

A. Testing Methods

Industry test methods have been developed from ASTM test methods with varying degrees of modifications. Modifications have been made to create the most practical test methods for laboratories while retaining a high degree of repeatability and reproducibility. These testing methods for tire rubber analysis are briefly outlined as follows.

1. Acetone Extraction Test

This test method covers the determination of the percentage of acetone extract, which is a measure of the amount of oils used in the tire rubber compound. Acetone extraction removes mineral oils or waxes, free sulfur, acetone-soluble plasticizers, processing aids, acetone-soluble antioxidants, organic accelerators or their decomposition products, and fatty acids from tire rubber compound. It also may remove part of bituminous substances, high-molecular-mass hydrocarbons, and soaps found in rubber compounds.

A brief description of the experimental procedure for the acetone extraction test is as follows. A weighed specimen of approximately 2 g is placed in a filter paper and the filter paper is folded to make a sample to fit the extraction cup (a thimble). The sample is placed in the thimble and the thimble into a condenser. The sample is extracted for 16 h to obtain the acetone extraction solution, in which the processing oils and other acetone-soluble materials are dissolved. The acetone is evaporated off over a hot plate to obtain the weight of acetone extract.

2. Ash Content Test

The ash content test covers the determination of the amount of inorganic fillers such as zinc oxide, clays, and other foreign contaminants such as dirt and metals in recycled tire rubber. A weighed specimen of approximately 2 g is placed in a crucible and ashed or burned in a muffle furnace at 1382°F (750°C) overnight. The mass of residue in the crucible represents the ash content.

3. Carbon Black Test (Nitric Acid Digestion Test Method)

This carbon black test method determines the amount of carbon black in tire rubber by a nitric acid digestion/oxidation method. Tire rubber is digested/oxidized in hot nitric acid and oxidized into fragments soluble in water. The carbon and the acid-insoluble fillers are filtered off, wash, dried, and weighed. The carbon is then burned off and the loss of mass represents carbon black.

4. Determination of Rubber Hydrocarbon Content

The rubber hydrocarbon content (RHC) refers to the percentage of high-molecular-weight polymers, such as vulcanized natural rubber (NR), synthetic natural rubber (polyisoprene, or IR), polystyrene-butadiene (SBR), polybutadiene (BR), and butyl rubber (IIR) insoluble in acetone based on the original sample. Symbolically, this can be represented as RHC = NR + SBR + BR + etc.

5. Natural Rubber Test

Two methods for the natural rubber test are described in ASTM D-297, "Standard Test Methods for Rubber Products—Chemical Analysis" and in D-3677 [11], "Standard Test Methods for Rubber—Identification by Infrared Spectrophotometer." In ASTM D-297, natural rubber and/or synthetic natural rubber (polyisoprene) is allowed to digest/oxidize the rubber compound in chromic acid. The double bonds in the rubber polymer are oxidized into acetic acid. The acetic acid formed is separated by distillation and determined by titration of distillate after carbon dioxide has been removed by aeration. The natural rubber content is calculated on the basis of the quantitative oxidation of polyisoprene. The chemical reactions are listed below:

1. Oxidation of natural rubber:

$$(-CH_2-CH_3=CH-CH_2-)_n \xrightarrow{CRO_3/H_2SO_3} nCH_3-COOH$$

2. Titrating using sodium hydroxide:

$$nCH_3-COOH + nNaOH \rightarrow nCH_3COONa + nH_2O$$

The limitation of this test method is that when natural rubber, polybutadiene, polystyrene-butadiene, and butyl rubber instead of only natural rubber coexist in a rubber compound at the same time, which is the usual case for recycled tire rubber, the results may not reflect the actual natural rubber content correctly. This is due to the probable oxidation of a small percentage of double bonds in the polymer system. This oxidation testing method cannot differentiate isoprene from other double-bond polymers.

The following are the basic chemical structures of polybutadiene, polystyrene-butadiene, and butyl rubber:

1. The chemical structure of polybutadiene is

$$(-CH_2-CH=CH-CH_2-)_n$$

2. The chemical structure of polystyrene-butadiene is

$$[(CH_2-CH=CH-CH_2-)_x(CH_2-CH-)_y]_n$$
$$|$$
$$CH_6H_5$$

3. The chemical structure of butyl rubber is

$$\begin{array}{cc} CH_3 & \\ | & \\ [-CH_2-C-)_{50}(CH_2C=CH-CH_2-)]_n \\ | & | \\ CH_3 & CH_3 \end{array}$$

Tire rubber usually contains natural rubber, polybutadiene, polystyrene-butadiene, and butyl or halogenated butyl rubber, and other polymers may be present. The suggested testing method for natural rubber contents described in ASTM standard D-3677 is more appropriate for recycled tire rubber compounds. "Standard Test Methods for Rubber Identification by Infrared Spectrophotometer" described in ASTM standard D-3677 involves the use of an infrared spectrometer and the examination of infrared absorption diagrams of rubber films or pyrolysis products (pyrolysates). Comparisons of the infrared absorption values of characteristic bonds and groups for standard samples and the unknown rubber sample lead to the natural rubber content determination. The methods can be used to identify other polymers such as polybutadiene, polystyrene-butadiene, and butyl rubber. However, butyl rubber is very hard to determine in a CRM. Therefore, the CRM producer should take special care not to have any additional butyl entering the CRM process system from such sources as inner tubes or curing bladders.

6. Thermogravimetric Analyzer

The TGA measures the change in weight of a sample as it is heated in a controlled environment. The atmosphere is controlled by selection of purge gas and a temperature-controlled furnace. As the temperature increases, volatile components are liberated and thermal degradation occurs, both resulting in weight loss. This weight loss can be reported as either direct weight in milligrams or percent of weight of the original sample. In addition, the first derivative or rate of weight loss (DTG) is recorded to amplify small but significant changes in the shape of the weight loss curve.

In TGA analysis of elastomers, a heating rate of 20°C/min is used with an isothermal region to separate the oils and plasticizers from the polymer. The sample is then heated at the same rate to fragment and then volatilize the polymer. The purge gas of the sample chamber is changed from nitrogen to oxygen to oxidize the carbon black and leave a residue of inorganic fillers. Inspection of the DT6 curve gives information on the polymer components of a blend. The decomposition of NR takes place at a lower temperature than the decomposition of SBR or ethylene-propylene-diene monomer (EPDM) and can be distinguished by looking at the DT6 curve. As indicated in Figure 12, the volatiles (acetone extract) volatilize first followed by the polymer (NR, then SBR/BR) and then carbon black, and the remaining weight is residue (ash).

TGA gives results similar to those obtained with ASTM D-297, which has been the standard test method in the rubber industry for years. Tests indicate that the TGA gives a volatile content an average of 5% higher than ASTM D-297 acetone extract results. This is due to difficult separation of the volatiles and the polymers in the sample as well as some plasticizers' insolubility in acetone.

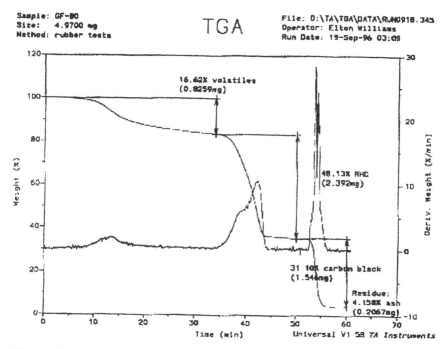

Figure 12   TGA printout of a typical fine CRM showing pyrolysis curve.

7.   Fourier Transfer Infrared Spectrophotometer

An FTIR instrument complete with a TGA/IR accessory is a highly advanced scientific instrument. Using an infrared laser, it scans a sample from 4000 to 0 cm^{-1} looking for changes in absorbancy. The changes in absorbancy are caused by different types of chemical bonds in the compound. It is then possible to reference the resulting spectrum against known spectra to determine the type of compound being studied.

Using the TGA accessory, it is possible to examine each type of rubber in a sample separately as it is being pyrolyzed. These instruments, used in tandem, increase the certainty that the product being shipped is the type of rubber requested by the customer. Figure 13 is an example of a fine-CRM FTIR graph of wavelength versus absorbancy.

B.   CRM Chemical Dynamics

Tire rubber is primarily a composite of a number of blends of natural rubber, synthetic rubbers, carbon black, and other additives. Tire rubber, fiber, and steel belting are the key elements of today's tire. Figure 4 shows that the various parts

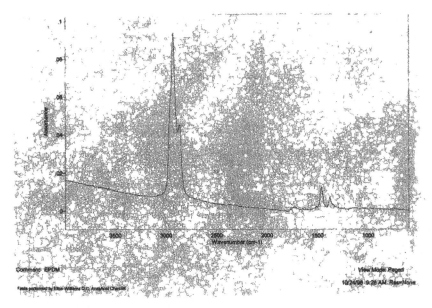

Figure 13   FTIR spectrum of a typical fine CRM.

of the tire construction require specific rubber properties, e.g., flexible sidewalls and abrasion-resistant tread. These various parts of the tire contain rubber with different amounts of natural and synthetic components. Natural rubber provides elastic properties and synthetic rubbers improve the compound's thermal stability. In addition, there are differences between types of tires, such as passenger, light truck, and heavy duty truck-tires. The composition of recycled tire rubber depends not only on the original recipe of the tire rubber but also on the source of the discarded tires. Recycled tire rubber contains foreign objects, such as stone, sand, clay, fiber, and even metals. These foreign objects may affect the composition and quality of recycled tire rubber products. Therefore, the asphalt technician should require a detailed certificate of analysis for each shipment and compare the data with the original specifications to ensure that these contaminants have been removed.

The average tire contains about 45% rubber hydrocarbon with the remainder being carbon black, oils, waxes, fillers, antioxidants, antiozants, sulfur, and curing systems. The RHC portion of a tire can be analyzed by major tire grades, as noted in Table 4. However, the elastomeric components in a CRM undergo dramatic changes in a hot asphaltic or bitumen solution, as noted in Table 5.

Table 4 Rubber Hydrocarbon Contents of Various Tire Types

Tire Category	Type	Polymer percentages
Passenger	Radial	35% natural rubber, 65% synthetic rubber
	Bias	15% natural rubber, 85% synthetic rubber
Truck	Radial	65% natural rubber, 35% synthetic rubber
	Bias	30% natural rubber, 70% synthetic rubber
OTR	Radial	100% natural rubber
	Bias	100% natural rubber

Table 5 Degradation of Various Common Elastomers Found in CRM

Polymer	Change
Natural rubber	Scission (softens)
Polyisoprene	Scission (softens)
Polychloroprene	Cross-linking and scission (hardens)
SBR	Cross-linking and scission (hardens)
NBR	Cross-linking (hardens)
BR	Cross-linking (hardens)
IIR	Scission (softens)
EPM	Cross-linking and scission (hardens)
EPDM	Cross-linking and scission (hardens)

This is the basic reason why CRM can provide both high- and low-temperature performance to a given neat asphalt. The sissioning elastomers provide low T_g or low-temperature performance and the cross-linking rubbers provide the stiffness or high-temperature benefit.

Producers have used chemical analysis of tire rubber for many years as an effective means of quality control. A large number of test results have been accumulated and a database of the chemical analyses data has been compiled. Table 6 lists the chemical compositions for passenger and light truck tread rubber. Table 7 lists the chemical compositions for heavy truck tread rubber, and Table 8 lists the chemical compositions for whole passenger tires.

Of the recycled whole tire rubber, about 80% comes from heavy truck tires and about 20% from passenger and light truck tires. The data in Tables 6 through 8 are based on a statistical analysis of our test results for the recycled truck stream for a recent year in our database of chemical analysis for passenger/light truck tire rubber, heavy truck tire rubber, and whole tire rubber. The mean value

Table 6 Chemical Compositions for Passenger/Light Truck Tread Rubber

Composition	Mean (%)	Standard deviation (%)	Min. (%)	Max. (%)
Acetone extract	17.2	1.12	15.5	19.1
Ash	4.18	0.51	3.9	5.4
Carbon black rubber	32.7	1.72	30.4	35.5
Rubber hydrocarbon	42.9	1.45	41.5	44.4
Natural rubber	17.8	—	15.0	22.2

Table 7 Chemical Compositions for Truck and Bus Tread Rubber

Composition	Mean (%)	Standard deviation (%)	Min. (%)	Max. (%)
Acetone extract	11.4	1.1	10.0	16.1
Ash	5.1	0.54	3.7	6.6
Carbon black rubber	33.2	1.45	30.	36.9
Rubber hydrocarbon	50.2	2.01	45.0	54.9
Natural rubber	42.3	—	40.	45.0

Table 8 Chemical Compositions for Whole Passenger Tire Rubber

Composition	Mean (%)	Standard deviation (%)	Min. (%)	Max. (%)
Acetone extract	15.1	1.33	13.1	17.5
Ash	5.0	0.60	4.0	6.1
Carbon black rubber	32.0	1.03	30.5	33.2
Rubber hydrocarbon	47.9	1.33	44.9	49.7
Natural rubber	18.1	—	14.	22.0

can be viewed as a typical value. The standard deviation provides a basis for estimation of variation in tire rubber composition and testing procedures. The maximum and minimum values obtained during testing determine the possible range of composition variation.

The data in the preceding tables reflect the chemical composition of tire rubber. Comparing the data, the acetone extract for passenger and light truck tread rubber is higher than that for heavy truck tread rubber and whole tire rubber. The rubber hydrocarbon content for passenger/light truck tread rubber is lower than that for truck and bus tread rubber and whole tire rubber. The natural rubber content for truck tires is higher than for passenger tires.

C. Sieve Analysis Procedure

The following is the new revised ASTM sieve analysis.

Revised ASTM D5644-94

Standard Test Method For Rubber Compounding Materials

Determination of Particle Size Distribution of Recycled Vulcanized Particulate Rubber[1]

This standard is issued under the fixed designation D 5644; the number immediately following the designation indicates the year of original adoption or, in the case of a revision, the year of last revision. A number in parentheses indicates the year of last reapproval. A superscript epsilon (E) indicates an editorial change since the last revision or reapproval.

1.0 *Scope*

1.1 This test method covers the determination of the particle size distribution of recycled vulcanized particulate rubber products.

1.2 The values stated in SI units are to be regarded as the standard. The values in parentheses are for information only.

1.3 This standard does not purport to address all of the safety concerns, if any, associated with its use. It is the responsibility of the user of this standard to establish appropriate safety and health practices and determine the applicability of regulatory limitations prior to use.

2.0 *Referenced Documents*

2.1 ASTM Standard

E11 Specification for Wire-Cloth Sieves for Testing Purposes[2]

D 5603-94Standard Classification for Rubber Compounding— Recycled Vulcanized Particulate Rubber

3.0 *Significance and Use*

3.1 The particulate size distribution of vulcanized particulate rubber is used for the purpose of assigning a product mesh size designation.

3.2 The product designation for mesh size is based on the size designation screen which allows a range for the upper limit retained from 0 to 5.0% up to 850 μm (20 mesh) particles and 0–10% for finer than 850 μm. No rubber particles are retained on the top (zero) screen. See 6.3, 6.4, and Note 1.

---

[1] This method is under the jurisdiction of ASTM Committee D-11 on Rubber and is the direct responsibility of Subcommittee D 11.26 on Recycled Rubber.

[2] *Annual Book of ASTM Standards*, Vol. 14.02.

4.0   *Apparatus*

4.1   Mechanical Sieve Shaker[3]—This is a mechanically operated sieve shaker which imparts a uniform rotary and tapping motion to a stack of 22-mm (8-in.) sieves in accordance with 4.2. The sieve shaker should be adjusted to accommodate a stack of sieves, receiver pan, and cover plate. The bottom stops should be adjusted to give a clearance of 1.55 mm (0.061 in.) between the bottom plate and the screens so that the screens will be free to rotate. The sieve shaker machine shall be powered with an electric motor producing 101 to 103 rads/s (1725 to 1750 r/min). This will produce 140 to 160 raps/min and 280 to 320 rotary motions/min. The cover plate shall be fitted with a cork stopper which shall extend from 3 to 9 mm (1/8 to 3/8 in.) above the metal recess. At no time shall a rubber, wood, or other material other than cork be permitted.

4.2   Standard Sieves, stainless steel or brass, 200 mm (8 in.) in diameter conforming to Specification E 11. The sieve set should include a lid and a bottom pan.

4.3   Scale, with a sensitivity of 0.1 g.

4.4   Soft Brush.

4.5   Jar, capacity of 55 cm^3 (1 pint) with large opening.

4.6   Rubber Balls,[4] with a diameter from 25 mm (1 in.) to 50 mm (2 in.). Enough balls are needed to have 2 balls per sieve.

5.0   *Procedure*

5.1   Select test screens appropriate to the particle size distribution of the individual sample being tested. A set of two to six sieves and a receiver pan are normally used. Actual number of sieves to be agreed upon by vendor and customer.

5.2   Clean with brush (see 4.4) making sure all particles are removed.

5.3   Stack test screens in order of increasing mesh size with smallest number on top (coarsest), and highest number on bottom (finest). For products of 425 μm (40 mesh) or finer, add two rubber balls per sieve. For products coarser than 425 μm (40 mesh), the use of rubber balls is optional.

5.4   Add bottom receiver pan to stack.

5.5   Obtain approximately 150 to 200 g of vulcanized particulate rubber from the lot.

5.6   Prepare a 100-g specimen as follows:

5.6.1   Weigh 100 g of specimen.

---

[3] The Ro-Tap Sieve Shaker meets the specified conditions and has been found satisfactory for this purpose. They are available through many scientific laboratory suppliers.

[4] Available from many sieve manufacturing suppliers.

5.6.2 Weigh talc according to product gradation designation. For products designated coarser than 250 μm (50 mesh), weigh 5.0 g of talc. For products designated 250 μm (50 mesh), or finer, weight 15.0 g talc.

5.6.3 Add talc to specimen.

5.6.4 Mix thoroughly by placing talc and sample in a 500-cm^3 (1-pint) jar and shaking for a minimum of 1 min, until agglomerates are broken and talc is uniformly mixed.

5.7 Place the specimen on the top sieve and place a cover on the stack.

5.8 Place the stack in the shaker.

5.9 Activate the shaker for 10 min for products designated coarser than 300 μm (50 mesh). For products designated 300 μm (50 mesh) or finer, activate the shaker for 45 min.

5.10 After the shaker completes the appropriate cycle, remove the stack.

5.11 Starting with the top sieve, remove the screened fraction by gently tapping its contents to one side and pouring the contents on the scale and recording its mass to the nearest 0.1 g. Any mass less than 0.1 g should be recorded as trace.

5.12 Brush any material adhering to the bottom of the screen onto the next finer screen.

5.13 Zero the scale in preparation for weighing the retained contents of the next screen.

5.14 Repeat 5.11 to 5.12 until all sieves in the stack and the bottom pan have been emptied, weighed, and recorded. This gives percent retained on each screen.

6.0 *Calculation*

6.1 The sum of the masses of each fraction from the sieves and the bottom pan shall not be less than the original mass of the sample plus weight of talc less 2 gms or greater than the original mass of the sample plus 100% of talc added. Repeat test if either of these conditions occur.

6.2 To adjust for the addition of talc to the specimen, the mass of the contents of the bottom pan is adjusted by the following calculation. Sum the total mass of the contents of each sieve mass including the bottom pan and subtract 100 g. This is the calculated value of the weight of the mass of the talc in the bottom pan. Subtract this value from the bottom pan contents value previously weighed and recorded. The final value obtained is the adjusted bottom pan contents mass for rubber.

6.3 The top screen (zero screen) selected shall be one in which no rubber particles are retained. This "zero percent retained screen" is designated in ASTM practice D 5603-94 Table 1.

6.4 The second screen in the sieve deck is for product designation and can contain 9 to 5% rubber particles for up to 850 μm (20 mesh) and 0–10% for finer than 850 μm (20 mesh).

NOTE: 1—An example of 6.3 and 6.4 is if an 850 μm (20 mesh) sieve contains zero rubber particles and a 600 μm (30 mesh) sieve contains 3% rubber particles, then the proper product designation for this material is 600 μm (30 mesh) particulate rubber product.

7.0 *Report*

7.1 Report the following information:

7.1.1 Proper identification of samples,

7.1.2 Identification of each sieve used,

7.1.3 The residue mass on each sieve,

7.1.4 The mass on the bottom pan and its adjusted mass, and

7.1.5 The product mesh size determined for the sample.

8.0 *Precision and Bias*

8.1 Round-robin testing will be conducted and precision and bias statements will be balloted for inclusion when testing is completed.

9.0 *Keywords*

9.1 Particle size distribution; recycled vulcanized particulate rubber

The American Society for Testing and Materials takes no position respecting the validity of any patent rights asserted in connection with any item mentioned in this standard. Users of this standard are expressly advised that determination of the validity of any such patent rights, and the risk of infringement of such rights, are entirely their own responsibility.

This standard is subject to revision at any time by the responsible technical committee and must be reviewed every five years and if not revised, either reapproved or withdrawn. Your comments are invited either for revision of this standard or for additional standards and should be addressed to ASTM Headquarters, your comments will receive careful consideration at a meeting of the responsible technical committee, which you may attend. If you feel that your comments have not received a fair hearing, you should make your views known to the ASTM Committee on Standards, 1916 Race St., Philadelphia, PA 19103.

1. Product Designation Based on Particle Distribution

The purpose of the sieve test is to assign a product mesh size designation. Relative to the mesh size, the product designation will be based on the first sieve that allows an upper limit retained of 5.0 to 10.0% on that screen. Each designated mesh size will also have a larger than nominal screen opening size on which no percent retention is allowed. Additional screens, up to a total of six screens (including the two previous), can be specified as agreed to between the vendor and customer to obtain a particular particle size distribution. For example, a material with no percentage retained allowed on the 20 mesh and 0.0 to 5.0%

retained on the 30 mesh screen would be classified as a 30 mesh product. Table 9 through Table 11 show the classification of particle size designations according to the preceding definition.

2. Laser Analysis

Today, modern particle analyzers reflect the state of the art in scientific instruments designed to measure the consistency of the particle size. They use laser diffraction to measure the particle size of materials. A report is generated giving a distribution curve of the different particle sizes in the product in micrometers, mesh sizes, or whatever units necessary.

This instrument can be installed in a production line. As an in-line tester it allows close monitoring of the particle size distribution of a fine-CRM product as it is being made in a wet CRM system. This gives the advantage of being able to catch problems early, thereby limiting off-specification material or reprocessing. The particle size analyzer used in conjunction with sieve analysis and other quality control tests can greatly improve confidence in the particle size classification of products.

Table 9   Designation −10 Mesh Whole Tire Rubber % Retained

Sieve size	8	10	16	20	Bottom pan
LSK[a] (%)	0	0	Open	28.0	Open
USL[a] (%)	0	5.0	Open	Open	Open

[a] LSL, lower specification limit; USL, upper specification limit.

Table 10   Designation −30 Mesh Tread Rubber % Retained

Sieve size	20	30	40	50	Bottom pan
LSK[a] (%)	0	0	0	25.0	Open
USL[a] (%)	0	10.0	Open	Open	Open

[a] LSL, lower specification limit; USL, upper specification limit.

Table 11   Designation −40 Mesh Tread Rubber % Retained

Sieve size	20	30	40	50	Bottom pan
LSK[a] (%)	0	0	0	25.0	Open
USL[a] (%)	0	10.0	Open	Open	Open

[a] LSL, lower specification limit; USL, upper specification limit.

Figure 14 is a typical printout of a fine CRM using a Malvern particle analyzer.

3. Surface Characteristics

Currently, the impact of surface characteristics of the ground tire rubber on combinations of the materials with asphalt is not well documented. For example, material that is "clean faceted" or produced by cryogenic methods reacts less quickly with asphalt than material with a rougher or greater surface area. In general, the rule "the finer the better" is desired for all types of asphaltic applications, especially the wet asphalt processes.

A BET surface area analyzer is a highly advanced scientific instrument that measures the surface area of a powder by measuring the change in conductivity of a carrier gas across an electrode. Liquid nitrogen is used to supercool a gas mixture of krypton and helium as it passes through a sample cell containing the powder. The low temperature causes the krypton gas to be absorbed by the surface of the rubber powder. After the liquid nitrogen has been removed, hot air is

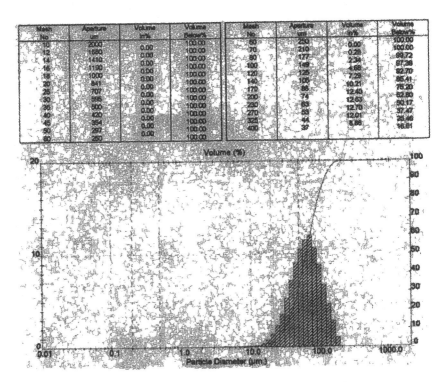

Figure 14   Particle distribution from a laser particle analyzer.

blown over the sample cell to cause rapid desorbtion of the krypton gas. The amount of krypton gas held in the sample is measured by the change in conductivity of the air as it passes over the sensor. The more gas held in the sample, the greater the surface area. This helps to guarantee that the product has the highest possible surface area to react with a customer's raw material.

Chemical analysis methods for CRM-modified asphalts, as for the other asphalt modifiers, can employ a number of standard and nonstandard techniques available to today's asphalt technician and researcher. Table 12 list some of the major methods for analyzing AR.

4. Quality Control/Assurance Procedures

Supplier Audits. A standard practice for assuring the quality of incoming raw materials is to perform a supplier audit. The purpose of the audit is to verify that the supplier has the necessary controls in place to produce the raw material. During the audit such things as quality control procedures for the process and testing areas, shipping, and release procedures and nonconforming material identification, segregation, and disposition need to be verified. Discrepancies, if any, would be noted and corrective action procedures implemented to prevent users from receiving out-of-specification materials.

Testing. Written procedures for the testing of chemical and physical properties need to be established and used. These procedures should be collected into a manual and made readily available to those performing the testing. These procedures should be available upon request to customers and auditors. The procedures should include appropriate information for the recording of the results.

Table 12 Chemical Analysis for CRM-Modified Asphalt Binders

Standard techniques
TLC-FID Iatroscan
Rostler and Sternberg
Extraction
Schweyer and Taxler
Corbett
Nonstandard techniques
NMR, nuclear magnetic resonance
GPC, photodiode array
GPC (UV, IR)
IEC, ion exchange chromotography
MDSC, modulated differential scanning calorimetry
FTIR, Fourier transform infrared spectroscopy
TGA, thermal gravimetric analysis

These results are distributed in the form of either certification of compliance (COC) or certification of analysis (COA), depending on the request of the customer. It is appropriate, as discussed previously, that the COC or COA be based on ASTM. These certifications are kept on file and also distributed to the customer in a manner and at intervals agreed upon between the customer and producer.

5. Statistical Analysis

Data are compiled regarding particle size distribution for performing statistical analysis to determine the consistency of the process in manufacturing the ground rubber. From these data, basic statistical calculations are performed, such as the mean, range, and standard deviation. These calculations provide information on within-lot, lot-to-lot, and between-production-site variation. Additional Cp, Cpk, and hypothesis testing can be performed as deemed necessary by the producer to provide such information as process capabilities or other product-related information.

6. Laboratory Quality Assurance

It is necessary to have procedures in place to ensure that the laboratory test results are reliable. This requires procedures that document test results, chemicals used and their sources, and regular periodic testing of a known standard or material to use as a reference. The reference material should be obtained by random selection of a known lot of production of tire rubber material. From this lot, a random sample is selected. By testing the same material again and again over time, it is possible to detect the ability of a particular test to give consistent results. If more than two testing facilities are involved and they tested material from the same sample, comparing the results between laboratories can determine whether there is significant variation between them in the results obtained. Finally, by performing these tests daily and weekly, the results can be used to help detect whether there is a problem with the testing that could be the cause of an unexpected result.

7. Potential Problems

Several areas for potential problems with the combining of asphalt and ground rubber exist. These potential problems are addressed as follows.

High Ash Content. It is possible to have material that yields a high-ash test result. This can happen for several reasons. The ash result can be due to contamination of the feedstock material with sand, dirt, or other high-inorganic material. This can have a negative effect on the reaction between the rubber and the asphalt and/or the user's finished product characteristics. High ash can also be obtained if the material is produced from or contaminated with wrong feedstock materials. Some rubber compounds are heavily loaded with heavy fillers, such as clays and whiting, and therefore yield a higher ash test result than is typical for a ground tire rubber.

Incorrect Rubber Feedstock. It is also possible that the product delivered to the user could have been manufactured from the wrong feed stockrubber material. In this case, typically, the reaction with the asphalt is different from that with tire rubber, leading to potential mixing problems. The use of a different material and, therefore, a different polymer than expected can yield different finished product characteristics. Depending on the polymer in the material, there could be unknown worker safety and/or environmental concerns. It is imperative the chemical analysis procedures be used by the CRM manufacturer to ensure that the CRM meets the stated specifications.

High Moisture Content. High moisture content is revealed by moisture testing and can be a significant problem. If the moisture content is high in the ground rubber, it can cause foaming and flashing on mixing with hot asphalt. This can lead to overflow of the mixing vessel, creating the possibility of worker injury or environmental problems. High CRM moisture contents can result from improper processing, storing, packaging, and/or shipping of the ground rubber. High moisture contents can inhibit handling and storage at the asphalt producer's job site or terminal. All CRM should have maximum moisture content of 1% or less by weight at the time of use.

Incorrect Particle Distribution Incorrect particle distribution can create problems for the user. Problems may vary depending on the particular user application. With the wet process, material with an incorrect particle distribution can either overreact or underreact with the asphalt compared with the expected reaction result. If the application is a spray application. such as a membrane interlayer, surface membrane, or slurry seal, large particles can result in plugging of the nozzles in the spraying apparatus or filters or reduce adhesion to aggregates. In the dry process. incorrect particle distribution in the ground rubber can cause incorrect final gradation when mixed with the aggregate. This can lead to unacceptable user-end product characteristics and pavement failures.

IX. PACKAGING AND DISTRIBUTION

A. Packaging

Packaging of CRM can be in the form of:

Plastic or paper bags, normally 50 lb (22.2 kg) per bag placed on 2200-lb skids and stretch wrapped for storing inside or outside.
1000- to 2000-lb Gaylord boxes, super sacks, or totes.
Bulk shipments, normally handled by bulk pneumatic transport trucks (38,000 to 44,000 lb per truck load) or rail to 60,000 lb per container car.

If CRM materials are going to be competitive with other asphalt modifiers, shipment of CRM will eventually have to be accomplished by bulk handling methods. The handling of plastic bags or paper bags, boxes, stretch wrap, and skids presents a waste disposal problem, not to mention the handling problem of unloading, unwrapping, opening, and adding CRM to an HMA or refinery system. Generally, only the fine CRMs, such as fine -80 mesh CRM, allow the asphalt plant or terminal operator to receive CRM bulk shipments without ever having to handle CRM powder. Successful bulk trials by the states of Florida, Kansas, Oklahoma, Iowa, and Arkansas have demonstrated that bulk pneumatic transport saves money, time, and frustration. Conventional mineral filler silos, such as built by Astec Industries, and standard pneumatic handling systems and transportation hauling equipment are required. In addition, the fine tire powders do not segregate with the fines settling to the bottom of blending and storage tanks, as is the case with coarse CRM, thus eliminating the need to remix CRM materials at the job site to ensure CRM homogeneity.

B. Asphalt Rubber Mixture Determination

Research and development evaluation has been conducted on determining the properties of asphalt rubber blend since the 1950s. Until the commercialism of the fine tire powders by Rouse Rubber Industries in 1988, achieving predictable viscosity blends at a given temperature, rubber concentration and AC crude type were not possible. Today, desired viscosities can be achieved by using a -80 mesh rubber without the use of extender oils exceeding heating cycles or high-torque blending. Laboratory work has shown that a -80 mesh CRM with a 200 mesh mean particle size reacts much more quickly than a -20 mesh CRM. It is important to check the compatibility of any rubber with the asphalt proposed for each project because chemical components of crudes differ. The Brookfield viscometer (shear flow between cyclinders) is an excellent instrument available today to determine the viscosity of a CRM mixture at a given temperature. Fine CRM works very well with the capillary viscometer (shear flow in a capillary tube). Other devices for shear flow measurement include the Scheweyer rheometer (shear flow between parallel plates) and the dynamic shear rheometer (shear flow between parallel plates).

C. Fine-CRM Asphalt Cement Applications

Many types of applications exist in the asphalt paving industry today. The challenge is to develop inexpensive and effective rubber modifiers to meet today's challenging demands. With the enactment of the Intermodal Surface Transportation Economy Act of 1991 (ISTEA) and the new emphasis on performance grade (PG) asphalts by the Federal Highway Department of Transportation through the Strategic Highway Research Program, the demand for cost-effective, performing tire rubber modifiers is critical to the success of using tire rubber in pavement CRM applications. This challenge has been addressed with the development of a fine tire powder, called UltraFine GF-80A. GF-80A is an ambiently produced tire rubber powder that passes the 80 mesh sieve (174 μm) screen with a resulting 200 mesh (74 μm) mean particle size and high surface area of ≈1 m^2/g. The combination of these properties allows the asphalt producer, distributor, or hot mix plant operator to handle, disperse, react, and produce a very uniform, stable and flowable rubber-modified asphalt cement (RMAC) economically in all types of asphaltic processes and for all applications. The physical properties of a fine 80 mesh CRM also allow the inherent properties of the tire rubber to be imparted to the asphalt binder itself to achieve asphalt modification.

D. Dispersion and Reactivity

Figure 15 illustrates the percent retained on each screen for a series of CRMs. This in turn gives a very good visual indication of the mean particle size. For example, the 80 CRM (fine-CRM) reflects a mean particle size of 200 mesh (74 μm), Figure 16 is indicative of how rubber gradations can be plotted and reported in S configuration.

Note that Figures 15 and 16 reflect typical ambient grinding methods. These curves may vary significantly if other CRM processes are employed. Also, surface-treated CRMs may not respond to the sample cure duplication or values retained because of the surface treatment involved.

Therefore, the degree of reactivity or dissolution is a direct function of surface area and particle size. The smaller the particle size, the greater the

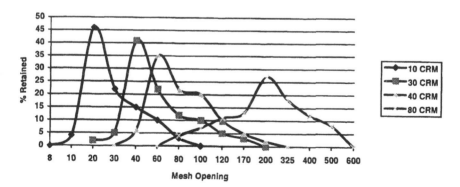

Figure 15   CRM gradation comparisons of a coarse and fine CRM.

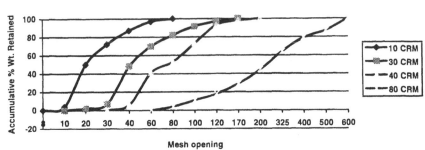

Mesh opening

Figure 16   *S* curves for CRM types.

reactivity, and thus less capital and operating costs are involved. Normally, the rate of interaction or dissolution of a CRM doubles for every 15 to 25°F increase in temperature.

Figure 17 gives a general indication of the time required to react various-sized rubber particles, adjusted to the mean particle size. By referring to Table 13, the average particle size can be used. The larger the particle, the greater the time required to digest or the limiting degree of reactivity. Reactivity or solvation can also be hampered by high heating temperatures, which can lead to environ-

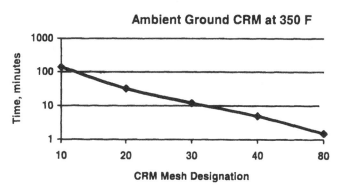

Figure 17   Stability of CRM particle size versus time in neat asphalts.

Table 13   Comparison of Mean Particle Size to CRM Gradation

	Mesh				
CRM designation size	10	20	30	40	80
Mean particle size, μm	710	300	240	200	74

mental and polymer degradation problems. Extender oils, which are generally aromatic in nature, can greatly affect the performance of the final rubber-modified asphalt cement and should be avoided or used with caution. However, using Table 13, one can calculate the estimated stability of tire for various CRM grades based on the mean particle diameter at 350°F. Figure 17 reflects the time required to reach stability (consistent viscosity) in an asphaltic solution by applying the formula $T_1 = T_2(D_1^2/D_2^2)$, where T is the temperature at time of addition to react the rubber particle and D is the average particle diameter; $T_1 = T_2$ for an isothermal condition at 350°F. For example, a -20 mesh rubber with a mean particle diameter of 50 mesh (297 μm) would require 20 min to react. Therefore, the reaction time for a -80 mesh rubber particle (fine CRM) with a 200 mesh (74 μm) mean particle size over a -50 mesh particle would be determined by substituting in the preceding formula, $T_{50} = T_{200}$ (297²/74²), resulting in a -80 mesh ambient ground rubber reacting 16.1 times faster than a -20 mesh rubber in 75 s versus 20 min. The reactivity is greatly enhanced as the surface area increases and, most important, the greater surface area results in less potential settling for the unreacted CRM particles.

To better understand the differences in ambient ground CRM and why the gradation curves play such an important role in dispersion, compatibility, and settling characteristics, refer to Figures 15 and 16. The coarser the gradation, the greater the digesting time and incompatibility with the neat asphalt medium. The use of aromatic additives and longer and higher cooking times are required to compensate for large CRM particles results in loss of physical properties, potential emission problems, and safety and health problems. Excessive heat for the coarse and medium CRM additives can also degrade the CRM polymers, and this overheating results in lose of their effectiveness over time.

X. APPLICATIONS

As noted in Figure 6, CRM materials can be designed for both the wet and dry processes. The dry process is more restrictive in application and more demanding on the part of the contractor to ensure proper blending in the hot mix asphalt unit. Historically, chunky or coarse rubber, that is, $\frac{1}{4}$ inch to 30 mesh gradation rubber, is used in an aggregate or gap gradation mix applications. Care has to be taken to ensure that the rubber gradation is strictly adhered to and that contaminants, such as tire fiber, are minimal by percentage allowable so as not to interfere with the absorption of asphalt by the CRM materials. As discussed earlier, a fine CRM can be added as a mineral filler continuously in a batch, drum, or double-barrel HMA plant very easily with precision. It can also be delivered and stored at the HMA plant site in bulk standard lime silos. This eliminates the need for handling and disposing of paper or plastic bags, pallets, and banding materials.

In the wet process applications, rubber less than 40 mesh in maximum size eliminates the need for adding unnecessary aromatic extender oils to the asphalt/ CRM batch or other materials such as natural rubber scrap or from other sources such as tennis ball scrap. Processing with coarse CRM requires long cooking times at elevated temperatures of 375 to 425°F (190 to 218°C) to achieve an asphalt-rubber mixture and is directly reflected in the particle size relationship. The coarse-CRM wet technologies also result in an asphalt rubber product with a very short shelf life. Specialized pumps and batch mixing equipment are also needed. Fine-CRM asphalt cement can be prepared at a refinery asphalt terminal or at the plant site with minimum complications. This asphalt rubber will have a long shelf life and a very high assurance of meeting DOT performance. Experience to date has shown that the continuous dry or wet method using fine rubber has achieved or exceeded pavement engineer's expectations.

The states of Oregon and Washington have tested and applied fine CRM as an alternative in the performance-based asphalt PBA-6, a refinery blended asphalt. PBAs are modified asphalt for severe climate changes, and they are evaluating RMACs to meet PBA-6GR asphalt for milder climate applications. The PBA specifications were developed by the West Coast User Producer Group for use in the Pacific Northwest. Table 14 is a comparison of PBA-6 and PBA-6GR products being produced by U.S. Oil and Refining Company, Tacoma, Washington. Note the SHRP grades regular maximum particle size or growth.

XII. APPLICATION EXAMPLES

A number of options for each of the wet or dry processes, using fine CRM, are available for today's highway engineers. Figures 18, 19, and 20 reflect the processing methods now commercially available along with the associated texture,

Table 14 PBA-6GR and PBA-6

Test	PBA-6GR	PBA-6
Abs. visc. 14°F, P	450 min.	450 min.
Kin. visc. 77°F, dmm	2000 min.	2000 min.
Penetration, 392°F, dmm	2000 max.	2000 max.
Penetration, 177°F, dmm	Report	Report
RFTO	PBA-6GR	PBA-6
Penetration, 392°F, dmm	30 min.	30 min.
Abs. visc., 140°F, P	5000 max.	5000 max.
Kin. visc., 275°F, cSt	275 min.	275 min.
Duct., 77°F, 5 cm/min, cm	Report	60 min.

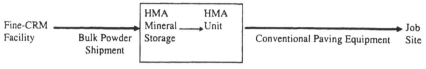

Features & Benefits

1. Use existing HMA equipment
2. Simplest of all CRM applications
3. Least costly of all CRM technologies
4. Maintain rubber gradation & fast reacting

5. No additional personnel or equipment required
6. Bulk handling (no disposal of pkg. waste materials)
7. Use conventional mix designs & standard test procedures
8. Indefinite shelf life, greatest flexibility for RMAC grades
9. Higher drop temperature of the HMA to ensure interaction

Figure 18 Dry method processing using fine-CRM (−80 mesh) powders for HMA RMACs.

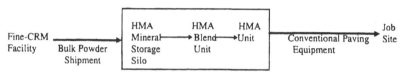

Features & Benefits

1. No additional personnel required.
2. Use existing HMA & paving equipment.
3. Maintain rubber gradation & very fast interactions.
4. Simplest of all wet modifier HMA plant technologies.

5. Bulk handling (no disposal of pkg. waste materials.
6. Use conventional mix designs & std. test procedures.
7. Least costly of all wet rubber HMA plant technologies.
8. Excellent storage life, minimal polymer degradation/swelling.

Figure 19 Wet method processing using fine-CRM (−80 mesh) powders for HMA plant RMACs.

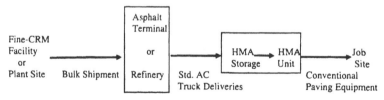

Features & Benefits

1. Liquid handling only.
2. No additional personnel required.
3. Use existing HMA & paving equipment.
4. Guaranteed DPT RMAC liquid specifications.

5. Most ideal of all wet rubber technologies.
6. Least costly of all wet rubber technologies.
7. Use conventional mix designs & std. test procedures.
8. Excellent shelf life, min. polymer degradation, & ability to receive performance based DOT asphalts.
9. Ability to add or blend other pavement polymers for preparing PBA or SHRP blend grades.

Figure 20 Asphalt refinery or terminal processing terminal blend using fine-CRM (−80 mesh) powders for asphalt applications.

benefits, and trials completed or in the process of being completed by the state. In general, fine CRM is the same price as or less costly than other polymer additives to achieve SHRP and Superpave performance. Each method noted has its own distinct advantages, flexibility, and economics.

The fine-CRM dry method is probably the simplest and most economical but is limited to producing asphalt concrete cements. However, the new Superpave mixture tests can be used. The fine CRM can produce all types of RMACs, including membrane interlayers and surface treatments, but requires additional equipment. The fine-CRM method provides the hot mix plant operator and the state DOTs with the quality assurance of a modified asphalt binder product that will meet the new FHWA-SHRP-PG specifications.

The ground rubber modifiers should be selected for pavement applications where premium asphalts are required. Such applications include high-traffic areas and those with extreme weather changes. The features and benefits of CRM can be realized only with test sections 1 mile or greater in length where visual as well as long-term evaluations by standard techniques are utilized.

A great deal of confusion has centered around the various rubber technologies for hot mix asphalt (HMA) that are being presented to the asphalt industry. The "dry method" technology generally consists of replacing a portion of the aggregates with a similar gradation as illustrated in Figure 21. The "wet method" technologies call for the rubber being dispersed and interacted with the asphalt cement (AC) binder prior to addition to the aggregates.

Figure 22 reflects the potential failure effect during expansion and contraction in final HMA pavements. If the AR is not fully dispersed in the AC, the resulting dry mix CRM will absorb oil and other solvents emitted to the surface of the pavements as a result of vehicle traffic, swell the CRM, and cause loss of CRM and deterioration of the pavement itself.

In selecting a rubber for any type of asphaltic application, particle size, gradation, source and types of rubber, technology, testing procedures (existing or modified), and the asphalt itself need to be taken into consideration. However, the bottom line will be the actual cost-effectiveness and performance of scrap tire rubber in pavement designs. Asphalt pavements have to compete with other road building materials such as Portland concrete cement (PCC).

Figure 23 summarizes the current understanding of interaction between the asphalt and rubber. The implementation of the SHRP binder tests will provide further understanding of how rubber interacts with specific asphalt grades and crude oil sources.

The foregoing was intended to give a general overview of CRMs role in hot mix asphalt mixtures. Today, as paving technologist search for additives that enhance road performance, such as CRMs, an understanding of how this additive rubber interacts with the binder and the aggregates is essential to achieving optimum pavement design and construction.

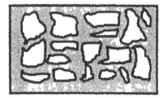

Conventional HMA Design

Plus-Ride™ Design Aggregate Substitute   Generic Design Aggregate Gradation

Arizona Design - Gelled State   Florida (Continuous, Rouse) Design
US Rubber Reclaiming, McDonald   Free Flowing - Dispersed State
Patents and Discoveries

Figure 21   HMA asphalt rubber comparisons.

Normal Applied HMA   Dry Gap or Aggregate Application   Dry Gap or Aggregate
                        After Initial Application   Application after Time and
                                                             Effect on Unreacted Particles

Figure 22   Illustration of why coarse dry CRM mixes fail.

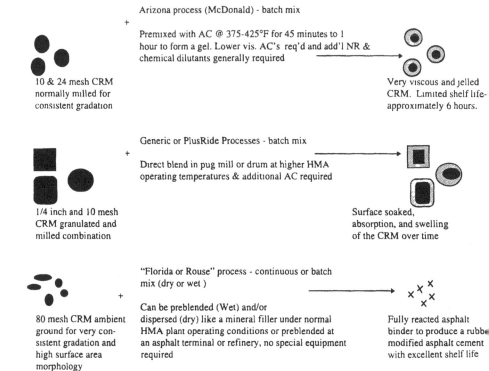

Figure 23   Comparison of various rubber technologies.

XII.   BINDER AND MIXTURE TESTING

Any modified asphalt will have to meet the new criteria set up by the American Association of State Highway and Transportation Officials (AASHTO)–Federal Highway program. This program, called the Strategic Highway Research Program, spent $50 million to develop new ways to test and specify asphalt binders and asphaltic mixtures.

In 1987 the SHRP began developing new tests for measuring the physical properties of asphalt. One of the results of this research is the new provisional performance grade binder specifications that were written with a set of tests to match these criteria. The resulting document is called the binder specification because it is intended to function equally well for all modified as well as unmodified asphalts.

The new SHRP tests measure the physical properties that can be related directly to field performance by engineering principles. Pavement sections needed to be laid and evaluated over very long periods of time, usually 15 years, in order to determine whether a new additive or modifier was effective in improving the life cycles. The new SHRP test methods are designed to produce long-term performance data in the laboratory within days. These tests are identified in Table 15.

In addition to the new SHRP binder test, a new generation of equipment has been designed and developed around the new binder-mixture specifications. That is, once the asphalt binder has been modified and added to an aggregate mix, the binder as well as the new mixture will be tested. Fortunately, this will allow the addition of ground rubber in a dry form to the hot mixes to be analyzed to determine exactly what performance would be forthcoming. For the wet method technologies, such as the Arizona, bitumen, and Rouse UltraFine processes, most of the SHRP binder specifications can be used. Only the Rouse and bitumen processes can apply all the SHRP binder specifications with no modifications. The main difficulties in handling the coarser, digested rubber in the wet process are in the rolling thin-film oven test and gaping or spacing on the dynamic shear rheometer (DSR). The challenge for the RMAC/CRM is to show that the products can meet the new SHRP-PG specifications. To date, the only products that have been shown to be able to use all the new SHRP test equipment previously described are the continuous or fine rubber powders developed by Rouse Rubber Industries.

The new SHRP asphalt binder specifications are intended to control permanent deformation, low-temperature cracking, and fatigue cracking in asphalt pavements. The SHRP parameters accomplish this by controlling various physical properties measured with the instruments described previously. The major important difference is that the physical properties remain constant for all grades, but the temperatures at which these properties must be achieved vary depending on the climate in which the binder is expected to serve.

Two forms of testing are associated with the new SHRP specifications.

1. Performance testing
2. Classification testing

Table 15   FHWA-SHRP Binder Tests

Procedure	Purpose
Dynamic shear Rheometer (DSR)	High/intermediate temperatures
Rotational viscometer (RV)	Measure properties at high temperatures
Bending beam rheometer (BBR)	Measure properties at low temperatures
Direct tension tester (DTT)	Measure properties at low temperatures
Rolling thin-film oven test (RTFO)	Stimulate hardening "durability characteristics"
Pressure aging vessel (PAV)	Stimulate hardening "durability characteristics"

In both cases, a series of tests are performed on the binder sample and a decision is made on the basis of the results of these tests. Performance testing answers the question, "Does the material meet all the requirements of a PG specification?" In other words, specification performance testing is an accept/reject form of testing in which the properties of the binder sample are compared with the required properties of a single grade.

On the other hand, classification testing answers the question, "What specification grade or grades does the sample meet?" In this case, a coordinated series of tests must be formed to classify the unknown material according to the SHRP binder specification.

In general, R&D engineers in state departments of transportation are usually concerned with performance testing. During a paving project, an agency laboratory will often collect and test an asphalt sample to make sure it conforms to the appropriate specification. If the tests do not achieve the specified values, the sample is reported to be out of spec. No further testing is necessary. However, research and development laboratories usually perform classification testing. That is, if an asphalt being tested does not meet the requirements of a specific grade, the laboratory continues testing to see whether the asphalt might meet another grade. Although classification and performance testing share all the same tests, they vary in the decisions based on each test result. Classification testing is a trial-and-error process. Thus, when performing SHRP binder tests on an unknown performance grade asphalt sample, the technician observes the results and decides the direction in which the tests will proceed and usually also performs the same test at another temperature. The main important difference between the SHRP binder specifications and those currently used is the binder sample will often meet several performance grades. Therefore, the development of SHRP binder tests to complement the new Superpave mixture test procedures will be the true test for CRM in pavement materials.

As noted, the SHRP test procedures can be applied to fine CRM. Table 16 reflects the increase in both the high- and low-temperature properties of various neat asphalts. Because of the uniqueness of properties of tire rubber, both high- and low-temperature performance can be realized at one time. For example, an AAMCO AC2.5 neat asphalt has an initial value of 46–28. With the addition of 7.5% fine CRM a 58–34 SHRP grade is produced. At 15% a 70–34 can be produced. Figure 24 graphically reflects the values listed in Table 16. Particularly noteworthy is Coastal's AC-30 neat AC or 64–22 becoming 82–34 with the addition of 15% fine CRM.

Portable blending units, such as the unit illustrated in Figure 25, can be used at job sites or, more important, at terminal sites. This unit constructed by Rouse Rubber Industries is used as a demonstration unit at various sites to demonstrate the producibility and ease of CRM modification. Since the fine CRMs are flowable and can be handled in bulk, the terminal operator or HMA

Table 16 SHRP-PG Data for Modified Produced Fine CRM

Asphalt type[a]	PG property (°C)	
	High temp.	Low temp.
AC-2.5-C	46	−28
AC-2.5-C w/10%	52	−34
AC-2.5-C w/20%	58	−34
AC-2.5-C w/20%-B	58	−40
AC-2.5A	46	−28
AC-2.5-A w/7.5%	58	−34
AC-2.5-A w/15%	70	−34
AC-2.5-A w/15%-B	70	−40
AC-2.5-A w/20%	70	−34
AC-5-C	58	−28
AC-5-C w/7.5%	58	−28
AC-5-C w/15%	70	−34
AR-1000-S	46	−28
AR-1000-S w/15%	64	−34
AC-7.5-ANS	42	−28
AC-7.5-AND w/12%	64	−34
AC-10-C	58	−22
AC-10-C w/7.5%	64	−22
AC-10-C w/10%-B	64	−28
AC-10-C w/15%	70	−28
AC-20-CI	64	−22
AC-20-C	64	−22
AC-20-C w/7.5%	70	−22
AC-20-C w/7.5%B	70	−28
AC-20-CI w/10%	70	−28
AC-20-C w/15%	70	−28
AC-20-CI w/15%	82	−34
AC-30-C	64	−22
AC-30-C w/7.5%	76	−22
AC-30-C w/15%	82	−28
AC-30-C w/15%-B	82	−34

[a] B, Borderline value; C, coastal; A, AAMCO; CI, Citgo; S, Shell; ANS, Alaska North Slope.

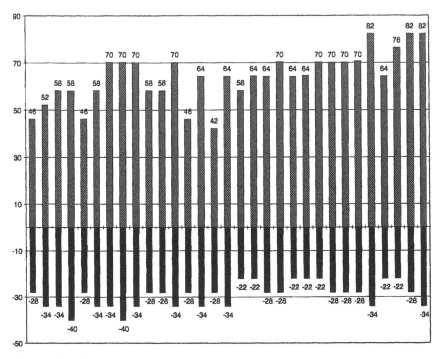

Figure 24 Illustration of Table 16.

operator can use standard asphalt cement equipment without need for any special type of pumps or high temperatures for handing the modified AR. For permanent installations, a simple blending tank with low-shear mixing and jacketing is all that is required, along with a screw auger or pneumatic handling, similar to systems for lime.

A. Terminal Blending of Fine CRM as an Asphalt Modifier

In addition to portable blending units that are available in today's market from such industries as Astec, Inc. (Chattanooga, TN) (see Figure 26) and Reichel and Drews (Chicago, IL) and a unit similar to a demonstration unit by Rouse Rubber Industries (Vicksburg, MS) (see Figure 25), the system for terminal blending of fine CRM into a high-quality SHRP or PG grade is feasible and economical. A typical example of a refinery setup is illustrated in Figure 27, which shows a low-shear fine-CRM terminal blending system.

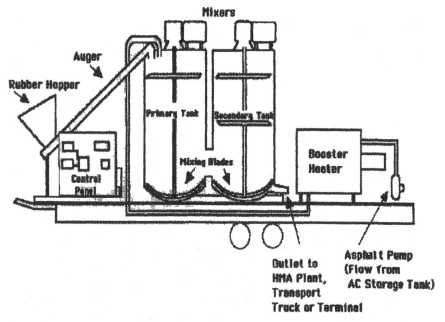

Figure 25   Portable continuous AR blending unit.

XIII. ACHIEVING HIGH- AND LOW-TEMPERATURE PROPERTIES WITH FINE CRM

A. Background

To meet the new high- and low-temperature requirements of SHRP grades in a given geographic area, the development of specific CRM will be required. In addition, for the given traffic or road area, it will be necessary to look at the type of traffic movement, slow or fast, and whether it is an interstate, downtown core area, rural, etc. The CRM-modified asphalt must also be adaptable to projects using open-graded friction courses (OGFCs) and stone mastic asphalt (SMA) with and without fiber. It can also be used for maintenance and rehabilitation projects. A CRM can be used as an overlay on PCCs or a PCC that has been rubberized or crack seated.

Work has been completed on numerous projects across North America. These projects show that with coarse CRMs, a lighter viscosity oil or an AC 2.5 to 10 is generally required. This is also true for latexes, the block polymers (linear and branched), and the polyolefins. With fine CRMs, a heavier viscosity oil such as AC-20 or 30 can be utilized with direct savings to a paving contractor.

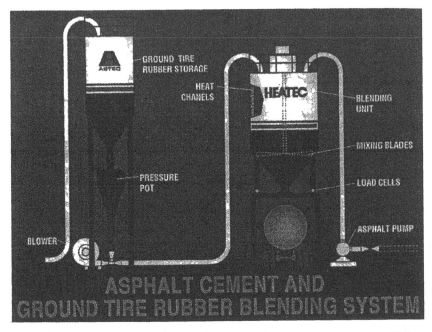

Figure 26 Terminal blending system. (Courtesy of Astec, Inc., Chattanooga, TN.)

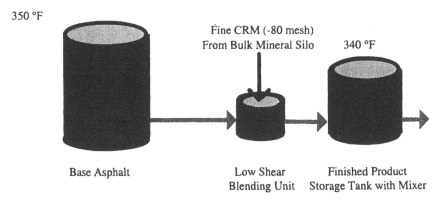

Figure 27 Terminal blending of find CRM to produce a high-quality SHRP binder.

For the coarse CRMs, high temperatures are needed with constant agitation over long periods of time. The fine CRMs result in reaction times of less than 1 h and generally no antistrip agents are required, reducing the specter of health and environmental issues.

Long-term storage capabilities of fine CRM have been realized as noted in Figure 28 by stable viscosities after blending. Although this figure reflects 60 min, holding the storage temperature at 300 to 330°F does not interfere in product degradation; holding times of several weeks are not unusual.

B. Research

The general logic is a tire rubber with a mean particle size of 200 mesh (74 μm) and passing the 80 mesh sieve (174 μm) in combination with a very high surface area morphology, allowing this tire rubber particle to react or solvate within the asphalt binder very quickly if properly dispersed. Normally, blending temperatures of 340 to 350°F are recommended.

Slightly higher mix asphalt or terminal operating temperatures are recommended for modifying the AC binder. Lower viscosity asphalt binders and the addition of extender oils are not recommended. Higher temperatures can result in fume emissions (health considerations) and CRM and asphalt degradation. As noted, in colder climates, the standard lower viscosity AC is used. Work by the Florida Department of Transportation illustrates the dispersion of CRMs of various sizes in a standard AC-30 at various temperatures as noted in Figure 29. The result is uniform, stable, and flowable rubber-modified asphalt cement that imparts the properties of tire rubber to the asphalt binder matrix.

Fine rubber disperses and to a limited extent reacts or solvates very readily to form a stable asphalt-rubber modified mixture. As a general rule, 15 to 25% less rubber is required with a −80 mesh rubber than with a −40 or −24 mesh to achieve equivalent viscosity values while achieving a stable viscosity mixture at lower temperatures. The advantage of the fine CRM is the ease of dispersion and imparting the properties of the tire rubber readily to the asphalt matrix. This

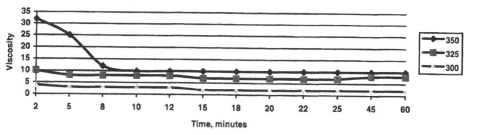

Figure 28 Time versus viscosity for a 10% fine CRM in an AC-30.

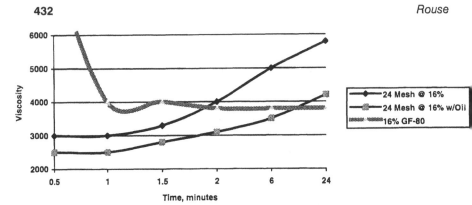

Figure 29   Comparisons of coarse CRM (24 mesh) and fine CRM (GF-80) in an AC-30.

also simplifies the equipment required to assist the terminal or hot mix plant operator in adding tire rubber to the AC binder reliably, economically, and systematically.

Both the coarse and fine CRM gradations can result in excellent rubber asphalt mixtures as long as the preparation procedures, tire rubber concentration by particle size, holding times, and temperature limits are strictly adhered to and maintained. Figure 30, based on work by Auburn University and the Florida DOT, illustrates this phenomenon.

Slightly different viscosity curves will be generated for various neat asphalts as CRM is added. However, working with a known consistent neat asphalt will result in repeatable viscosity curves. Please note that rubber is crude specific, and each crude source must be evaluated for its particular properties and chemical composition to achieve RMAC by some procedure at a temperature within a time frame using certain ACs. As noted, these findings support the use of the fine rubber. However, similar results can be obtained by adding extender oils to the AC-30.

As discussed, coarse CRM swells and forms a gel-like material. In order to control the viscosity, an extender oil and/or extended cooking times are required when it is reacted with asphalt at approximately 400°F. The coarse swelled rubber-asphalt blend is made more easily using an extender oil. The extender oil and extended heating periods change the properties of the asphalt-rubber mix as illustrated. Extended high heat brings about degradation of the rubber tire poly-mers. This is not the case with fine rubber when the temperature is held relatively constant at normal AC binder temperatures.

To minimize potential CRM polymer degradation, contractors, AC distribu-tors, refiners, or others should achieve interaction at approximately 350°F and maintain the storage temperature at 320°F or less. Holding time for the Arizona coarse wet process (−24 mesh) is usually limited to a maximum of 6 h, and the

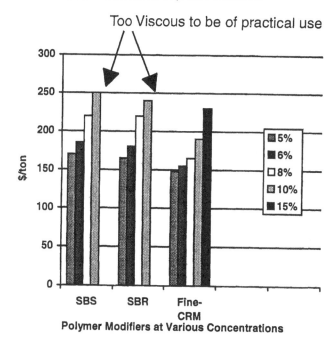

Figure 30 Cost analysis.

fine wet or continuous process (−80 mesh) has a minimum of 3 weeks holding time with negligible settling of rubber particles. The combination of high surface area and lightweight particles follows Stokes' law for settlement of micrometer-size particles. The lighter the particle and greater its surface area, the greater the resistance to settling. It is generally recommended that with the fine CRM, the binder be prepared and turned or agitated occasionally with a standard AC pump once per day when the HMA plant is in operation. A compromise of 1–2 days holding time for RMAC prepared with a medium CRM (40 mesh) would be in order for this gradation. However, compromising leads to a wider variation in the final performance.

Table 17 summarizes the benefits of using a fine CRM as a modifier for nearly all types of neat asphalts when produced at a central site that has the storage capabilities for multiple neat asphalt grades and related SHRP testing equipment and qualified personnel.

C. Polymer Comparison and Field Handling

The characteristics of a fine CRM (−80 mesh rubber) available today and of other polymers are summarized in Table 18. A realistic evaluation must be made

Table 17 Benefits of Terminal and Refinery
Prepared PBA and SHRP-PG Asphalts

Long-term storage capability
Viscosity asphalts stable and pumpable
No separation problems
No polymer breakdown
No field blending
Manufactured at terminal or refinery
Increased high production capability
AC is certified prior to shipment or blending
Many agencies now requiring terminal blending

Table 18 Polymer Type Comparisons in Field Applications

	Polymer type		
Characteristic	SBS	SBR	Fine CRM (ultrafine)
Property	Modified	Modified	Modified
Appearance	Glossy	Glossy	Dull
Overnight storage	Poor	Poor	Good
Appearance	High	Very high	Normal
Paving machine buildup			
Flights	High	Very high	Normal
Hopper	High	Very high	Normal
Shovels, lutes, etc,	High	Very high	Normal
"Spiderwebs"	High	Very high	Little to none

of field practices for blending and storing block polymers, latex, or fine CRM. Table 19 shows results typical of those experienced in the field by hot mix plant operators or blenders.

In order to meet the new demands for improving the performance of neat asphalts of PG grades, high concentrations of modifiers will be required. Based on laboratory observations and experience in the field, high concentrations of latex, SBR, neoprene, and polyolefins are required. These are difficult to handle at elevated concentrations and must receive special storage and handling. These factors will need to be considered when making economic justifications.

Since CRM is very crude specific, care must be taken to evaluate the CRM in terms of its physical and chemical attributes. Compatibility determinations can be made using the DSR and the softening point for samples produced from a cigar tube test sample. The finer the particle, the better the compatibility. Also, treating the CRM at the point of manufacture can greatly improve compatibility.

Table 19 Blending and Storage Characteristics

	Polymer type		
Property	SBS	SBR	Fine CRM (ultrafine)
Mix temperature, °F	350–380	350–380	340–360
Mix method	High shear	Low shear	Low shear
Tire required for incorporation	4+ h	4+ h	Immediate
		Water boiloff required	10–30 min
Manufacturing method	Batch	Batch	Batch or continuous
Storage temperature, °F	350+	350+	320–330
Storage stability	Asphalt type dependent	Asphalt type dependent	2–3 weeks No settlement

However, care must be taken to test the treated or untreated CRM in the specific unmodified AC. In some cases compatibly will be reduced. Table 20 illustrates several cases using the DSR and softening point to predict compatibility.

For a Venezuelan crude asphalt (ABL) there is poor compatibility, versus excellent compatibility for a Wyoming Sour (AAB) using 15% for a fine CRM. However, good compatibility could be achieved with the ABL using a treated fine CRM. The opposite can also hold true, as discovered by a refiner in the Midwest who tested fine treated and untreated CRM. Note that the DSR values can vary significantly from the softening point (SP) measurements. Therefore, selection of the appropriate compatibility test procedure is important.

D. Cost Comparisons

A reasonable correlation can be made with the three most common type of modifiers, SBS, latex, and fine CRM. Coarse CRMs are not considered. Because

Table 20 DSR and Softening Point Compatibilities

	DSR (kPa)			SP (°F)		
Asphalt	Top	Bottom	ΔkPa	Top	Bottom	Δ°F
AAB	0.80	0.73	0.07			
ABL	2.28	0.67	1.61			
Treated ABL	1.46	1.34	0.12			
Sample 1, untreated	1.62	1.86	0.24	143	154	11
Sample 2, treated	3.08	1.13	1.95	140	148	8

coarse CRMs require a change in the SHRP test procedures, they do not meet the ASHTO settling tests and in general are recipe driven and not performance based. Using the cost comparisons of modified asphalts in Table 21, we can arrive at a cost analysis on a per-ton basis, reflected in Figure 30. Normally, the higher loading of SBS can result in a reduction in low-temperature performance of the binder, and high concentration of latex for low-temperature performance produces a product too viscous to handle.

Tables 18, 19, and 21 refers to mixing and handling properties in relation to final cost. The lower cost modifiers do not reflect practical use in the field as illustrated in Figure 30.

E. Potential Benefits and Suggested Evaluation Plan

One of the most important benefits of RMAC-HMA is the increase in the film thickness on the surfaces of aggregates. This increased viscoelastic behavior is demonstrated by reduced runoff but does not interfere with agglomeration of the aggregates or HMA. Figure 2 represents a typical improvement in film thickness when fine-CRM is used as a modifier.

Increased film thickness can also be an aid in stone mastic asphalts (SMAs), where expensive modifiers and fibers may not be warranted. Certainly, the combination of fibers and blending of other polymers may achieve performance not yet realized for paving projects.

The use of fine CRMs significantly improve both high- and low-end performance. Data shown in Figure 31 demonstrate that neat grades are improved to higher grades by use of a fine CRM on both high- and low-temperature sides. Asphalt modified with SBR, the block polymers, and the polyolefins show a drop or reversion in low-temperature performance. A suggested plan for using a fine-CRM is outlined in Table 22. The key benefits that can be expected from using a fine CRM are summarized in Table 23.

Table 21   Cost Comparisons

	AC-10 neat	6% SBS	6% SBR (70% by wt. Solids)	15% fine CRM (ultrafine GF-80A)
PG grade	52–22	64–22	64–34	70–34
Asphalt, $/ton	$125	$118	$118	$107
Polymer cost, $/ton	0	85	157	105
Total cost, $/ton	$125	$203	$275	$212

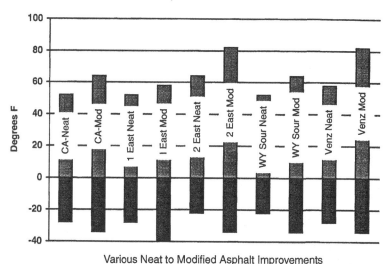

Figure 31 Comparison of PG graded asphalts from different crude sources with fine CRM.

Table 22 Suggested Plan of Action Using a Fine CRM

Confirm the PG grade through laboratory testing by
  Obtaining a local available asphalt
  Determine the % fine CRM required
  Are any antistrips required?
  Make pilot plant run
  Tank blending
  HMA mix and laydown
Scale up to a full production run

XV. CRITICAL FACTORS FOR SPECIFYING OR IDENTIFYING CRM

A. Identification

The use and identification of CRM for pavement applications, research, and future identification should include the information noted in Table 24.

    In nearly all cases the CRM rubber chemistry can be different depending on the type of tires used as well as the gradation. The neat asphalt source can vary in chemical composition as well as rheological behavior, and clearly identifying the CRM source and type of CRM will assist the asphalt technician

Table 23   Benefits of CRM-Modified Asphalts

Meet PG grade requirements with a fine CRM by
   A favorable cost factor
   Use existing asphalt supply
   One polymer only required
   Minimal blending requirements
   Minimal reaction time
   No modification to test procedures
   No undue hardship in AC storage
   No undue hardship in laydown procedures
Adds long-term viscoelastic properties to asphalt
   Since CRM contains antioxidants, antiozants
   CRM highly resistant to degradation
   Adds long-term viscoelastic properties to HMA by
      Improving resistance to pavement deformation (rutting)
      Improving resistance to pavement low-temperature cracking

Table 24   CRM Source Information

1. CRM type: passenger, truck, OTR, or combination of type tires. Designate if produced from buffings.
2. CRM processor: company and location.
3. CRM process: wet ambient, cryogenic, extruder, two-roll mills, combination or other process descriptions.
4. CRM gradations: percent passing each designated sieve or using laser techniques.
5. CRM chemical analysis: percent RHC percent with synthetic to natural rubber ratio, moisture, carbon black, extractable and ash analysis with surface area and specific gravity. Information should be part of the certificate of analysis and be included as part of the asphalt binder test data.

and researcher in reproducibility of test work and comparisons for future work. In all cases the ideal CRM asphalt modifier must possess the attributes noted in Table 25, based on the viscoelastic parameter measured in the Superpave binder analysis system as shown in Figure 32.

B.   Major CRM Advantages Overall

The use of coarse CRM is well documented. Agencies that use course CRM do so because they can observe a positive outcome with the applied AR mixes. For any modifier to achieve complete success, the disadvantages it must be minimized and inconstancies corrected. This could include equipment, HMA production,

Table 25   Ideal CRM Asphalt Binder

High elasticity (G') at high temperatures	$\rightarrow$	resistance to flow at high temperature
High viscosity (G'') at low temperatures	$\rightarrow$	Resistance to low-temperature cracking
Temperature-dependent modules	$\rightarrow$	Low-temperature susceptibility

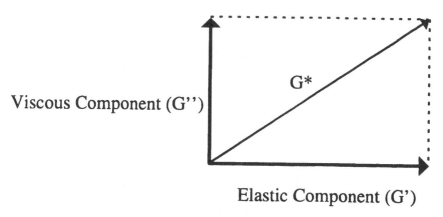

Viscous Component (G'')

G*

Elastic Component (G')

Figure 32   Viscoelastic parameter of a CRM.

and application in the field. The advantages of CRMs within themselves need to be considered prior to comparisons with competing polymers. In general, coarse and medium-sized CRMs can be compared with fine CRM. Fine CRMs can compete on a level playing field with other polymers such as SBS, latex, EVA, and others where the coarse or medium CRM may not. Table 26 lists some of the major advantages and disadvantages of coarse CRM compared with other modifiers and fine CRM.

C. Observations

The SHRP Superpave binder specification can be used for measuring the perfor-mance-based rheological properties of CRM in the asphalt binders. The use of the solubility test would exclude CRM from the specifications. Fine CRM may be the exception, because no adjustment needs to be made in any of the SHRP binder tests in the gap openings. It should be noted that the maximum rubber particle size (180 μm) is less than the maximum particle size allowed (250 μm) by AASHTO TP5 for filled systems. The solubility test is not a good parameter to include as a standard, because many other polymers, especially the polyolefins, would be excluded although they also impart key properties to an asphalt mix.

Table 26 Advantages and Disadvantages of CRM

Advantages
  Use of a scrap material
  Highly engineered material with antioxidants, antiozants, carbon-black, and rubber
    polymers
  Improvement in fatigue life
  Durability is maintained on thin lifts or wearing surfaces
  Reduced reflective cracking
Disadvantages of coarse CRM
  Generally more expensive than modified or polymer-modified mixtures
  High capital cost for equipment modification
  Short shelf life of the CRM as with most modified polymers
  Lack of experience with CRM mixtures and misguided perceptions
  Variability of the product in terms of consistency of chemistry and gradation includ-
    ing quality
Advantages of fine CRM (−80 mesh rubber)
  Very competitive with all types of modified asphalts
  Minimum capital investment required and utilizes existing asphalt handling equipment
  Excellent shelf life with minimum settling—only light agitation required for some crudes
  Education and experience relatively easy to understand versus the other modifiers
  Minimum variability and uniformity in chemistry and gradation of fine CRM

At higher temperatures the properties of the CRM are a function of asphalt itself and rubber type. More work is required on how CRM affects resistance to permanent deformation. The increase in $G^*/\sin \delta$ is very noticeable when coarse CRMs are used. This is reflected in pumping conditions at standard operating temperatures. Normally, temperatures of 375°F or greater are needed, for both the coarse and medium-sized CRMs. Fine CRM at a concentration of 10 to 15% is very pumpable in the 320 to 345°F range. Lowering and raising the tank storage temperature of the fine CRM have been successfully accomplished at HMA plants and refineries.

D. Performance

The addition of fine CRM can increase failure strain above that of the softer asphalt through the fundamental mechanism of dissipating energy at crack tips. The addition of CRM increases low-temperature toughness and fracture energy, mainly through enhanced shear yielding, which increases plastic deformation at crack tips. Finally, the increase in fracture toughness is governed by the CRM content, particle size, and interfacial strength.

XVI. SUMMARY AND CONCLUSIONS

Scrap tire rubber, as a marketable material, must be subjected to the same scrutiny as any other asphalt additives. As experience is gained and information is shared by supplier, pavement engineers, and end users, there will be greater confidence and increased usage of CRMs. Development of new and better test methods and procedures in accordance with the Strategic Highway Research Program, Superpave, and improved process and application techniques will also increase the use of CRM materials and its cost-effectiveness. Crumb rubber–modified material should be used in applications where binders are required in high-traffic conditions or extreme climate changes. The tire was designed and manufactured to ride on America's pavement. It only seems logical that when the tire has completed its useful life, it could be transformed into a CRM that would take advantage of its inherent properties for enhancing the pavement. If properly processed and applied, crumb rubber modifiers will improve paving performance and safety by being excellent and cost-effective modifiers for today's highway pavement industry. All these benefits can be gained while solving an environmental problem and utilizing a strategic resource, the scrap tire itself.

REFERENCES

1. Comparison of Various Road Surface Applications, Rouse Rubber Industries, Bulletin No. 602.
2. F. L. Roberts et al., *Hot Mix Asphalt Materials, Mixture Design, and Construction*, 1st ed., NAPA Education Foundation, Lanham, MD (1991).
3. ASTM D-5644 test procedure.
4. Laboratory Procedure for Preparing Rubberized Asphalt from UltraFine™ GF-80, Rouse Rubber Industries, Bulletin No. 636.
5. Standard Methods for Rubber Products-Chemical Analysis, ASTM Designation D-297-81, *Annual Book of ASTM Standards*, Vol. 09.01, *Rubber, Natural, and Synthetic—General Test Methods; Carbon Black* (1990).
6. Joint Cooperative Paper between Rouse Rubber Industries and Baker Rubber for FHWA CRM Workshop.
7. *Manual on Test Sieving Methods*, ASTM Comm. 17756 E-29 on Particle Site Measurement, Philadelphia, (1985).
8. ASTM C136, Standard Method for Sieve Analysis of Fine and Coarse Aggregates, *Annual Book of ASTM Standards*, Vol. 04.02, (1991).
9. Commercial Equipment for UltraFine™ Rubber Modified Asphalts, Courtesy of ASTEC Industries, Chattanooga, TN.
10. Design Procedures and Contractor Practices, U.S. Department of Transportation, Federal Highway Administration, FHWA-SA-92-022.
11. Better Pavements are the Mother of Invention, *Asphalt Contractor*, January, 11–14 (1994).

12. Discussion with M. Hitzman, FHWA, Washington, DC, January 1993; Evaluation of Rubber Modified Asphalt Demonstration Projects, Ministry of Environment and Energy, February 1994 (PIBS 2857).

13. Report to Congress: A Study of the Use of Recycled Paving Material, June 1993, FHWA-RD-93-147 or EPA/600/R93/095.

14. H. F. Waller, *Use of Waste Materials in Hot-Mix Asphalt*, ASTM, Philadelphia (1993).

15. Wardlaw/Schuller, *Polymer Modified Asphalt Binders*, STP 1108, Philadelphia (1992).

16. UltraFine™ Asphalt Cements Applications, Rouse Rubber Industries Bulletin No. 629.

16. Report to Congress: A Study of the Use of Recycled Paving Material, June 1993, FHWA-RD-93-147 or EPA/600/R93/095; M. W. Rouse, Production, Identification, and Application of Crumb Rubber Modifiers (CRM) for Asphalt Pavements, SASHTO Meeting, August 1993.

17. Report to Congress: A Study of the Use of Recycled Paving Material, June 1993, FHWA-RD-93-147 or EPA/600/R93/095.

18. M. Morton, *Rubber Technology*, 2nd ed., Van Nostrand Reinhold, New York (1973).

19. M. W. Rouse, Development and Application of Superfine Tire Powders for Rubber Compounding, *Rubber World*, June (1992).

20. Discussion with M. Heitzman, FHWA, Washington, DC, January 1993.

21. UltraFine™ Asphalt Cements, Rouse Rubber Industries Bulletin No. 628.

22. Florida Likes Fine Ground Rubber (Contractors Can Buy It at the Terminal), *Asphalt Contractor*, March (1992).

23. HMA Design Consideration, Rouse Rubber Industries Bulletin No. 630.

24. HMA Rubber Particle Dynamics, Rouse Rubber Industries Bulletin No. 631.

25. UltraFine™ Rubber Modified Asphalt Cement Projects, Rouse Rubber Industries Bulletin No. 634.

26. *Background of SHRP Asphalt Binder Test Methods*, Asphalt Institute Research Center, Lexington, KY (1993).

27. Polymer Modifiers for Improved Performance of Asphalt Mixtures, Sheraton Colony Square Hotel, Atlanta, November 2, 1993.

28. Performance Based Asphalt for Information, Courtesy of U.S. Oil and Refining, Tacoma, WA.

29. SHRP Binder Grading System Based on Pavement Design Temperature, *Asphalt Technol. News*, **5**(2), 1–5, (1993).

30. N. Bloomquest et al., Engineering and Environmental Aspects of Recycled Materials for Highway Contractor, FHWD-RD-93-088, June 1993.

31. M. W. Rouse, Impact of the 1991 Intermodal Surface Transportation Act on Scrap Tire Disposal, *Scrap Tire News*, April (1992).

32. Facts/Directory, 28th ed., *Modern Tire Dealer*, **75**(1), 19–47 (1994).

33. Background of SHRP Asphalt Binder Test Methods, Asphalt Institute Research Center, Lexington, KY, October 1996.

34. F. R. Erich, *Science and Technology of Rubber*, Academic Press, New York (1978).

35. *Hot Mix Asphalt Materials, Mixture Design and Construction* National Center for Asphalt Technology, NAPA Education Foundation, Lanham, MO (1991).

36. Crumb Rubber Modifier (CRM) in Asphalt Pavement, U.S. Department of Transportation, Federal Highway Administration, Report No. FHWA-SA-95-056, September 1995.

20

Asphalt Coatings and Mastics for Roofing and Waterproofing

Eileen M. Dutton
Karnak Corporation, Clark, New Jersey

I. INTRODUCTION

Cold-applied coatings are not a new technology by any means; they have been in use for as long as history has been recorded. In the year 3000 BC Egyptians used cold-applied coatings (varnishes and enamels) made of beeswax, gelatin, and clay to waterproof their wooden boats. Early natives in the Americas and Europe used cold-applied mixtures of clay soil and water to waterproof their roofs and walls. In the year 1122 BC, the Chinese invented lacquer and used it to protect virtually every structure known to them. It is no surprise, then, that a technology that has been around for such a long time has evolved to yield such a sophisticated variety of coatings that serve virtually every purpose imaginable.

The use of asphalt in cold-applied coatings was brought to the attention of the coatings world by the issuance of the first patent for a cold-applied asphalt roof coating to Dr. G. I. Oliensis in 1921. In the 1930s cold-applied coatings were introduced commercially into the U.S. construction market. Since that time, asphalt coatings and mastics for roofing and waterproofing have been used successfully for many years in a wide variety of applications, including mainte-nance and cold-applied built-up roofing. Traditionally, cold-applied built-up roof-ing has been used as a reroof application rather than for new roof construction. However, as improvements have been made in the cold-applied coatings and adhesives, as well as the reinforcements, the use of cold-applied coatings and adhesives in new roofing is increasing. Some other reasons for the increase in cold-applied coatings and mastics can be found in Table 1.

Table 1   Advantages of Cold-Applied Coatings

Ease of application: may be sprayed, brushed, or troweled on
Provide excellent adhesion to most asphalt surfaces
Provide excellent resistance to vapor passage
Safe and easy storage: no hot kettles
Minimal equipment required on job site
Easy coordination of work
Decreased insurance liability costs

Several terms are used in describing asphalt coatings and mastics. The first set of terms is related to the viscosity of the product. The term coating refers to a material that has low to medium viscosity, and is capable of being applied at about 16–30 mils without much flow. A coating is typically fibered and may contain some fillers. An asphalt coating is typically applied by spray, roller, or brush A paint is a low-viscosity material and does not contain fiber. Paints can be applied in the same manner as a coating, but usually at a thickness of 8–12 mils. Mastics are very thick materials, usually heavily fibered, and are applied with a trowel in typical thicknesses ranging from 80 to 120 mils.Some other terms used to describe asphalt coatings are related to the use of the material. For example, adhesives can be a coating or trowel grade material and are used for bonding a variety of substrates, felts, or membranes. Waterproofing materials generally have very low moisture vapor permeability and, when applied to a substrate such as wood or concrete, can prevent the passage of water under hydrostatic pressure. When surfaces have a dampproofing coating or mastic applied to them, the coating prevents the passage of water in the absence of hydrostatic pressure.

II. SOLVENT-BORNE COATINGS

The most common type of asphalt coating is a solvent-borne material—that is, asphalt is mixed with a solvent to lower the viscosity. Fillers and fibers are added to obtain the correct consistency. Asbestos has been the fiber of choice in the past; however, because of concerns about asbestos-related health issues, the use of asbestos fiber in asphalt coatings and mastics has been decreasing since about 1985 and will probably continue to decrease. The fiber most often used in the coatings is cellulose, although a variety of inorganic and synthetic fibers are also used. These other fibers do not give the solvent-borne coatings the same consistency as asbestos, so other functional fillers, such as clays and talcs, have been used to obtain better properties.

The asphalt used in solvent-borne coatings can be a wide variety of materials. Air-oxidized asphalts are often used as they generally have improved sag resistance and weatherability compared with nonoxidized asphalts. These air-oxidized asphalts will usually meet the specifications of either ASTM D-312 [1], type I through IV (asphalt used in roofing), or ASTM D449 [2], type I through III (asphalt used in dampproofing and waterproofing). Other products commonly used are graded on their viscosity and generally meet specifications for paving grade asphalts ranging from AC-20 to AC-40 (ASTM D-3381) [3]. The air-oxidized asphalts are usually to used in areas where high ambient temperatures are expected or in coatings and mastics that are used on steep slopes. This is because the air-oxidized asphalts are less susceptible to flow than the paving grade asphalts. The paving grade asphalts are often used in below-grade coatings, where they will not be susceptible to wide changes in temperatures, or in roof coatings that are made for flat or low-slope applications, which require a more ductile asphalt. Polymer modification of asphalts for coatings and mastics began in the late 1970s to early 1980s in the United States. The introduction of polymer membranes in the 1970s [4] spurred the development of coatings with improved properties. The polymer chosen to modify the asphalt depends on the final properties desired. The most common modifiers for asphalt coatings and mastics are styrene-butadiene-styrene (SBS) block copolymers, styrene-butadiene rubber (SBR), and styrene-ethylene-butylene-styrene (SEBS) copolymers. Asphalt adhesives often use styrene-isoprene-styrene copolymer or even some acrylics. Butyl, neoprene, and urethanes are also fairly common additives. However, as with any asphalt modification, the compatibility of the polymer must be checked to be sure the polymer improves not only the initial characteristics but also the long-term, or weathered, characteristics [5,6].

Modification of the asphalt for coatings improves on weaknesses of standard asphalts. Tensile strength and elongation of a modified asphalt show a large percentage of improvement over standard asphalts. Typically, asphalts have 50–150% elongation at room temperature, whereas a polymer-modified asphalt can have elongations in the range 300–2000%. The tensile strength at break of the polymer-modified materials is generally two to five times greater than that of a nonmodified asphalt. The recovery of elastomeric asphalts is also far superior to that of typical asphalts. Traditional asphalts show very little, if any, recovery. Elastomeric asphalts can have up to 90–100% recovery.

The polymer-modified asphalt typically has a much wider service range than a nonmodified asphalt. The low-temperature properties and adhesive strengths of modified asphalts are often much improved. Two traditional tests that are run on asphalt are tests of ductility and cold-temperature flexibility. Ductility at low temperatures is increased 300–500% in the modified asphalt versus the typical nonmodified asphalt. Cold-temperature bend properties are often increased 60°F in the polymer-modified asphalt over the nonmodified asphalt. High-temperature properties of polymer-modified asphalts show marked improvements over those of typical asphalts. The polymer-modified asphalt will not flow at upper temperatures in the way a typical nonmodified asphalt will flow. The wider temperature range allows the polymer-modified asphalt to be used in high-slope applications as well as areas where the coating is required to be more ductile. Polymer modification of asphalts generally involves mixing a low-viscosity, nonoxidizied asphalt (roofing flux to AC-20) with 3 to 20% polymer. An oxidized asphalt generally has lower compatibility with polymers, so it is not usually used. Mixing the low-viscosity asphalt with between 6 and 20% asphalt can result in a high softening point and relatively soft penetration. In general, modification of an asphalt can impart better properties on either end of the service temperature range, improving on the upper-temperature sag resistance of the oxidized asphalt and the ductility at low temperatures of the roofing or paving flux.

Polymer-modified coatings and mastics are finding many uses, replacing nonmodified coatings (see Table 2) in some areas but also replacing hot-applied asphalts. One of the disadvantages of traditional cold-applied roof coatings is that they take a long time to set up in a roof. In a 1975 symposium of the American Society for Testing and Materials (ASTM), Davis and Krenick [7] presented a paper in which they showed that a three-ply cold-applied system reaches the National Bureau of Standards minimum tensile strength when the cold-applied adhesive retains only 6% of the volatiles. In research work done by the author, the polymer-modified asphalts were able to retain the same strength with 12–15% volatiles remaining in the cold-applied adhesive. This indicates that the polymer-modified adhesives will have less risk of wind blow-off or displacement by heavy rains.

A study in 1975 by the testing office for bitumen building materials and plastics of the University of Munich showed that the adhesive power of a modified bitumen adhesive applied in strips was far in excess of the German standard. These tests were confirmed by an independent study by the TNO in Delft, The Netherlands [8]. In fact, modified bitumen adhesives can be applied in strips, using less material than for hot-applied asphalt, and yield good strength results.

Table 2 Advantages of Polymer-Modified Asphalt Coatings and Mastics

---

Excellent resistance to thermal and/or mechanically induced stresses
Higher tensile strength and elongation characteristics
Outstanding resistance to acids and alkalis
Improved abrasion resistance
Improved fatigue resistance and recovery
Improved low-temperature flexibility
Improved ductility
Improved service temperature range

Although asbestos-free formulations are on the increase, the strength of the asbestos coating or mastic has been hard to surpass with the asbestos-free formulas. The use of polymers to modify the coatings and mastics has resulted in coatings that are far superior to asbestos coatings and mastics. The use of thickeners other than fibers in the asbestos-free formulations also results in better elongation than for the old asbestos formulations, in which the asbestos fibers contributed to the tensile strength, but caused the coating to fail at 100–150% elongation. The cost of these coatings is more than that of asbestos coatings, but with the general public's attitude toward asbestos use and the increase in polymer-modified membranes, the market for these coatings continues to grow.

III. WATER BORNE COATINGS AND MASTICS—BITUMINOUS EMULSIONS

As an alternative to the use of solvents, asphalt may be mixed with water by means of a process called emulsification. This process enables the particles of asphalt to be "suspended" in water in a semigel-like coating. These particles remain suspended until the coating is applied and the water evaporates, allowing the particles to coalesce or stick together to form a protective coating.

The first bituminous emulsions were made over 90 years ago using anionic emulsifiers. The first emulsions were used for dust laying. Records show that the first emulsified bitumen was used in New York in 1905, followed by usage in two other states, Illinois and Massachusetts, in 1906 [9]. Chemical emulsion technology has continued to expand, using both cationic and anionic emulsifiers. The largest use of chemical emulsions is still in paving; however, they do find some uses in industrial applications such as coating pipe and as protective water vapor barriers for highway bridges and culverts.

The use of clay as an emulsifier began in about 1910–1915. The early clay emulsions used various types of clay, but it was not until the early 1920s that bentonite was used as the emufulsifier. The use of bentonite imparted many favorable qualities to the emulsion. The earlier clay emulsions had ratios of asphalt to clay that were very low (3–5:1). When bentonite was used, the quantity of asphalt in relation to clay increased tremendously, with 15 to 30 times more asphalt than clay [10].

Colloid mills are often used to manufacture the base emulsion for cold-applied coatings. By far the most common type of emulsion for roofing and waterproofing involves the use of bentonite clay. The bentonite imparts several attributes to the asphalt emulsion that chemical emulsions do not. The clay coats the surface of asphalt particles and slows down the rate of oxidation and loss of volatiles [11]. Thus clay-stabilized emulsions weather by chalking or erosion of the clay before the asphalt. Also, the bentonite forms a matrix with the asphalt and improves the rheology of the low-viscosity base asphalt used in manufacturing the emulsion. The sag and flow resistance resemble those of an air-oxidized asphalt, yet much of the ductility of the low-viscosity asphalt is retained. In fire testing (ASTM E-108), the clay emulsion remains resistant to flow up to high temperatures and delays the fueling of the flame source, allowing passage of the flame test. Asphalt emulsion coatings are typically formulated with straight-run asphalts, fluxed propane-precipatated asphalt, or slightly oxidized asphalts. These asphalts emulsify very easily; highly oxidized asphalts are difficult to emulsify and usually not very stable. The clay and water are mixed together first, and the pH is controlled to enhance emulsification. A slightly acidic pH has been found to work best; the extent is dependent on the asphalt being used for emulsification. Typical pH treatments may involve acids, such as sulfuric, phosphoric, chromic, or citric acid. Acid salts such as aluminum sulfate, sodium acid phosphate, or potassium dichromate may also be used to adjust the pH. The acid salts are thought to change the characteristics of the dried emulsion film, more so than the acids, improving the dried film's resistance to moisture.

The amount of shear and temperatures are very dependent on the type of emulsion produced. A narrow range of temperatures is used to produce a consistent emulsion. This temperature is dependent on many parameters, including the type of asphalt used, the shear rate, and the percentage of emulsifier used. The emulsion must remain stable in the packaging but should break fairly quickly when applied to the substrate. Typically, the smaller the particle size, the more stable the emulsion. Good asphalt emulsions will break and not reemulsify upon immersion in water. Adequate shear can be important in meeting this parameter.

Polymer modification enhances the asphalt emulsion coatings. Asphalt emulsions are known for their slow break or cure, as well as their poor resistance to standing water. Polymer modification can cure both these problems. In addition to improvements in cure time and water resistance, properly formulated polymer-modified emulsions show improvements in water vapor permeability, with permeability ratings close to those of solvent-borne asphalt coatings and mastics. Typical polymers used in asphalt emulsions are acrylics, styrene-butadiene-styrene block copolymers, styrene-butadiene rubber, or butyl. Polymer-modified asphalt emulsions can be applied in the same manner as traditional asphalt emulsions, usually by spray, brush, or trowel. Improvements in the physical properties of polymer-modified asphalt emulsions over traditional asphalt emulsions can be seen in Table 3.

Polymer modification of asphalt emulsions takes place in one of three ways. The first is to postadd a polymer, typically a latex polymer, to the emulsion. The other methods are to comill the polymer with the asphalt or to predisperse the polymer in the asphalt. Postadding polymer to an asphalt emulsion typically results in only slightly improved physical properties. To get better dried film results, the other two methods of modification are used. Predispersion of the

Table 3 Physical Property Improvements of SBS-Modified Emulsions

Property	SBS-modified asphalt emulsion	Traditional asphalt emulsion
Tensile strength, psi	80	40
Elongation, %	300	75
Water vapor permeability. English perms	5	14
Drying time (ASTM D-2939), h	4	20

polymer in the asphalt gives better physical properties than the comill method, but either of these methods results in superior film properties.

In addition to clay and polymers, asphalt emulsions often contain fibers to meet certain specifications. ASTM D-1227, Standard Specification for Emulsified Asphalt Used as a Protective Coating for Roofing [12], requires the use of fibers in some of the classes of emulsion. Other fillers, such as calcium carbonate, mica, or talc, are often used, and cellulosic thickeners are also used to build viscosity or improve the weathering of the asphalt emulsions.

Asphalt emulsions are often used for dampproofing building exteriors, particularly below grade. This work involves confined spaces, where a contractor has little room and poor ventilation for working conditions. The asphalt emulsion cold-applied systems are often chosen over the solvent-borne asphalt coatings (see Table 4), as they have very low odor. Contractors are also pleased with the ease of troweling the emulsions over the solvent-based materials. The emulsions are very thixotropic—they will spread with little exertion, yet they can build a thick film without flowing.

Other uses for asphalt-clay emulsions are in cold-applied built-up roofing, metal and masonry coatings and dampproofing compounds, primer coats, flooring mastics, and insulation adhesives.

Table 4 Advantages of Asphalt Emulsions over Solvent-Borne Coatings

Environmentally safe
May be used on slightly damp surfaces
Comply with environmental regulations such as VOC and VOS
Tools may be cleaned with water while product is still wet, reduces use of solvents
Use in confined areas without the need for ventilation
Will not attack certain compounds that solvents will, such as polystyrene
Excellent weathering characteristics
Less cure time required before top coat can be applied
Growing market due to VOC regulations
Safe to use on projects where torches or other sources of ignition are present

IV. ALUMINUM-PIGMENTED ASPHALT COATINGS

Aluminum-pigmented asphalt roof coatings are different from most architectural paints and coatings in their formulation. They are also the most difficult cold-applied asphalt coating to formulate. The trick is to make a bright silver reflective coating out of a black base. The main reason for an aluminum-pigmented asphalt roof coating is to protect the roofing membrane underneath. Using an asphalt-based product to bind to asphalt roofing helps avert many compatibility problems found in roofing products. A light-colored or reflective coating is desirable as it reduces the heat load on the roof, preserving the membrane and reducing the cooling requirements for a building. The reduction in heat load is more important in some areas of the country than others and is very dependent on the amount of insulation in a roof. My experience has shown that a noninsulated roof coated with an asphalt aluminum will show a decrease in temperature between 10 and 35°F.

Aluminum-pigmented asphalt roof coatings also improve the fire rating of many commercial systems, gaining the system a Factory Mutual (FM) or Underwriters Laboratory (UL) class A, B, or C rating. An additional reason for using an aluminum-pigmented asphalt coating is the added moisture protection for the roofing membrane. Asphalt is used as the base for many dampproofing or waterproofing coatings, as it provides a good barrier, yielding low moisture vapor transmission rates. The aluminum-pigmented asphalt has dampproofing properties; however, as this product is applied in thin films, the extent of damp-proofing is limited.

When formulating an aluminum-pigmented asphalt roof coating, the first thing to decide is whether the material is the traditional solvent-based material or the newer emulsion-based coating. Also, is a specification grade material desired? ASTM has specifications for solvent-based aluminum-pigmented asphalt roof coatings and is working on a specification for the water-based products. ASTM specification D-2824 [13] covers all types of solvent-based asphalt alumi-nums, including nonfibered, asbestos-fibered, and fibered coatings containing no asbestos. The specifications are listed in Table 5.

The National Standard of Canada, CGSB/ONGC (the Canadian General Standards Board/Office does normes générales du Canada), has similar specifica-tions for aluminum-pigmented asphalt coatings, CAN/CGSB-37.42-M89, which can be found in Table 6.

The aluminum pigment is a very important part of the formulation. Alumi-num pigments are made from very pure aluminum metal, typically in the range of 99.0–99.99% aluminum. A major distinction between the aluminum pigment and nonmetallic pigments is in the geometry of the particle. Rather than being spherical or granular, as most pigments are, the aluminum pigment is actually a

Table 5 ASTM D-2824 Specifications

Composition limits	Type I. nonfibered	Type II, III, fibered
Water, % maximum	0.3	0.3
Nonvolatile matter	40	40
Insoluble in CS_2, %	40	50
NVM metallic	11	9
Aluminum, min. %		
Reflectance, % minimum	50	50
Consistency, Stormer, s/100 rev	20–30 (+100 g load)	15–90 (+300 g load)

Table 6 CGSB/ONGC Specifications

Compositional requirements	Type 1 nonfibered	Type 2, fibered
Aluminum metal, by mass, min. %	11	9
Solvent, by mass, max. %	60	60
Mineral and/or other stabilizers, by mass, %	0–5	5–12
Asphalt, by mass (by difference), min. %	24	20
Asphalt properties	54–93	54–93
softening point, °C(°F)	129–199	129–199
Penetration, dmm	10–60	10–60

flake. The flakes appear flat. with jagged edges, and the diameter is many times longer than the thickness.

The process for manufacturing aluminum paste is a wet ball mill process, referred to as the Hall process, named after the inventor. E. J. Hall of Columbia University (Figure 1). Atomized or chopped foil is placed in a ball mill, and particle size reduction is carried out in the presence of a suitable lubricant and an aliphatic hydrocarbon solvent. The lubricant keeps the flakes from melding together again. Milling times typically vary between 2 and 15 h, depending on the process. The milled slurry is screened for size, with the removal of fines (less than 325 mesh) being an important part in the leafing aluminum pastes that are used in the manufacture of asphalt aluminums. Particles of too small a size will not leaf properly and can result in a darker coating.

There are three main types of pigments: metal flake or powder for commercial aerospace uses, nonleafing pigments for automotive-type paints, and leafing pigments for roof coatings and other industrial uses (Figure 2). For a leafing grade aluminum, stearic acid is used as the lubricant throughout the milling process and then more aluminum is added at the end of the process. Nonleafing

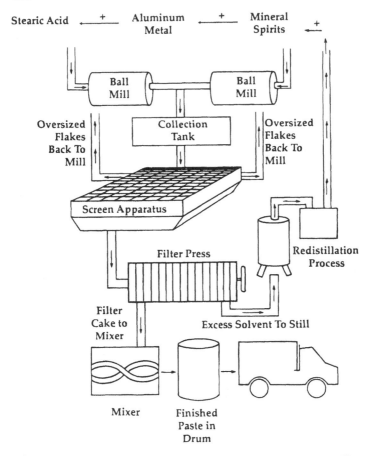

Stearic Acid ⟵ + ⟶ Aluminum ⟵ + ⟶ Mineral ⟵ + ⟶
                           Metal   Spirits

Figure 1 Flow diagram for the production of aluminum pigments. (Courtesy of Silber-line Manufacturing Company, Inc.)

grades generally have oleic acid as a lubricant, although they could have a small amount of stearic acid.

The slurry for asphalt aluminum pigment is pumped into a filter press for removal of most of the solvent. The press cake is then put into a mixer, where the metal content is adjusted using solvent (typically mineral spirits), and extra stearic acid is added to make the flakes leaf. Two to four percent total stearic acid is typically added for leaf stability. This is the process for manufacturing a solvent-based aluminum paste.

For pastes for aluminum emulsions the same general process is used, with one exception. After the filter press is put into the mixer, stearic acid is added,

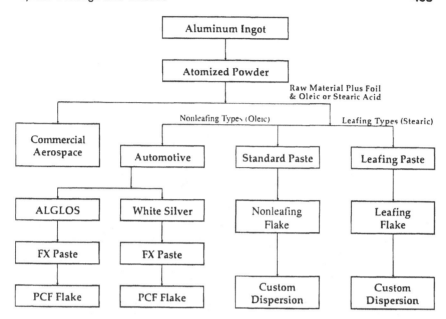

Figure 2   Product flow chart.

but rather than add plain mineral spirits, a material is added that will tie up the water so that it does not attack the aluminum flake. In some instances this is a solvent, such as nitropropane. However, aluminum emulsions are still in their infancy, and improvements are expected in the future.

Aluminum pastes are categorized into two types, type I, nonleafing, and type II, leafing, and three general classes by ASTM: class A, a fine grade; class B, a medium grade; and class C, a coarse grade. When considering the formulation of an asphalt aluminum roof coating, these grades are important. The finer the material, the better the hiding power and the gloss; however, the reflectivity decreases as the size becomes smaller. Therefore, you must weigh the desired properties and use the grade that meets your use. Most roofing grade materials use a class C flake, although the incorporation of the other classes may improve the performance on the roof.

The leafing aluminum (Figure 3) has better reflectivity than a nonleafing aluminum. This is because the flakes actually rise to the surface in a leafing material, with a large accumulation of flakes at the surface. This helps resist ultraviolet (UV), and the physical structure of the flake also acts as a barrier for air and moisture. A nonleafing grade (Figure 4) disperses throughout the coating, imparting highlights and gloss to a clear or light-colored base. When a dark base

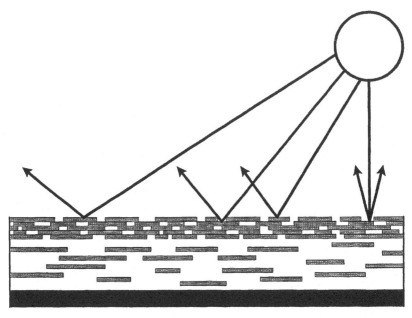

Figure 3   Leafing reflectivity depiction.

such as asphalt is used, the leafing grade aluminum is needed to see the silver color and gloss. A nonleafing aluminum would have a dingy gray color if used with asphalts.

The type of asphalt used is also important in an aluminum-pigmented asphalt roof coating. The Canadian specification (Table 6) gives some indication of the properties of the asphalt, but typically the range for softening point and penetration that works well is much more narrow. If asphalt used in the coating is not air oxidized enough, the asphalt will bleed through the aluminum, sometimes within days, otherwise in a few months. A highly oxidized asphalt can also cause problems, sometimes with darkening, but typically the coating will be too brittle and crack or delaminate on the roof. The range of softening point and penetration for an asphalt that is too brittle and one that makes an excellent coating is very narrow. Experience has shown that asphalt with a softening point between 155 and 190°F degree and penetration from about 10 to 30 dmm makes bright aluminum coatings. There are reports that show that 140 to 155°F softening points make bright aluminums, but in my experience these tend to be shorter lived and to give premature browning with age. When making a polymer-modified aluminum-pigmented asphalt coating and using high ratios of polymer to asphalt (up to 1:1 polymer to asphalt), the range of asphalt used

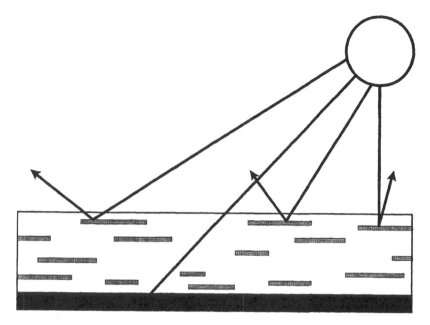

Figure 4   Nonleafing reflectivity depiction.

can widen significantly. Soft fluxes to highly oxidized material can be used, all with good aging and leafing results.

The source of asphalt is also important in an aluminum-pigmented asphalt coating, much as certain types of crude make good paving or roofing asphalt. Some Canadian asphalts and Venezuelan fluxes produce very good aging characteristics. However, since aluminum-pigmented asphalts typically use an oxidized material, the crude source is less critical, as the method of oxidation can often outweigh the crude source [14]. The new rheological tests, with modified aging tests (similar to SHRP methods [15]), can be used to identify the better sources of crudes, as well as methods of oxidation.

The choice of solvent in an aluminum-pigmented asphalt has changed in recent years. It was thought that a high aromatic solvent in the aluminum-pigmented asphalt would give a brighter product. Although this may be true to an extent, by the time materials are fully cured, the reflectance values for any solvent are about the same. Some work has shown that although lower aromatic solvents give slightly less bright initial films, they actually age better than blends made with high-aromatic solvents.

The fillers used in these coatings are very important. The amount of surface tension in the coating has a great deal to do with the reflectance of the blend. The choice of filler can increase the surface tension and actually increase reflectivity. Certain fillers whiten the coating, decreasing highlights caused by the aluminum pigment, and give a more uniform look to the coating. Fillers that are used in aluminum-pigmented asphalt roof coatings include diatomaceous earth, silica, wollastonite, calcium carbonate, mica, and talcs. Some people have recommended the use of clays; however, clays tend to inhibit the leafing of aluminums. There are several theories, one being that the clays can strip the stearic acid coating off the flake and cause the pigment to act as a nonleafing grade rather than a leafing grade.

Fillers are also important, as they can have tremendous effects on the settling of the coating. As the aluminum flakes are flat, certain types of fillers can cause the flake to settle harder than it normally would. Too much mica or calcium carbonate tends to cause hard settles. However, some types of filler can make the aluminum easier to stir back into solution.

Both the choice of fillers and that of fibers can improve chances of passing UL or FM fire testing. Logically, a rock wool or fiberglass fiber would not burn as much as a cellulose fiber. However, there are class A aluminum coatings in which cellulose fibers are used. Other types of fibers generally used are polypropylene, ceramic, asbestos, and glass wool. The fibers have many uses in a formula, the main uses being for reinforcement, viscosity, and appearance.

Moisture scavengers also play an important role in aluminum-pigmented asphalt coatings. Moisture can attack the aluminum flake, causing problems with optical properties. The reflectance can be lowered as the water attacks the aluminum flake. This reaction is exothermic, generating heat, which accelerates the reaction as time goes on. The first thing that happens is a buildup of hydrogen gas, which, if allowed to continue, can actually blow the lids off cans. It does not take much moisture for this to happen: even a trace amount of water can cause the reaction to occur at high temperatures. This small quantity can be found in the asphalt and the other raw materials used in the coating. The moisture scavengers tie up the excess water, keeping the aluminum bright and the evolution of hydrogen from developing. Typical moisture scavengers range from plaster of paris to silicon dioxides and even the use of molecular sieves in some formulations.

Modifiers are becoming more prevalent in the roofing market, and aluminum-pigmented asphalts are following the pattern. The modifiers improve the weatherability and reflectivity of the coating. The modifiers increase the surface tension of the coatings, improving the leafing characteristics and often increasing the gloss. Alkyds and alkyd resins have long been used in aluminum-pigmented asphalt coatings for gloss improvement. Hydrocarbon resins have also been found to improve the gloss.

Modifiers often improve the moisture resistance of the coating, although the moisture vapor permeability ratings are low in the asphalt aluminums. Rubber modifiers tend to lower the moisture vapor ratings the best. The typical rubber modifiers in aluminum-pigmented asphalts include SBS (styrene-butadiene-styrene) or SEBS (styrene-ethylene-butylene-styrene) block copolymers, and even some urethanes improve the moisture vapor permeability. The choice of asphalt specifications can also expand as modifiers are added to the formulation.

Rubber-modified aluminum-pigmented asphalts increase in importance as the use of modified sheets and modified built up roofing (BUR) continues to grow. A traditional aluminum-pigmented asphalt does not have the elongation of a modified membrane. Therefore, as the modified membrane expands and contracts with heat and freeze-thaw cycling, stress cracks can develop in the nonmodified asphalt aluminums, and adhesion loss can occur. The modified aluminum-pigmented asphalts are able to stretch with the modified membranes. The modified aluminum-pigmented asphalts also work well over traditional BUR and metal roofing.

A typical formulation for an aluminum-pigmented asphalt roof coating meeting the ASTM specification is shown in Table 7. If the formulation is modified, typically 10–50% of the asphalt used is actually a polymer.

Aluminum-pigmented asphalt emulsions are newer to the marketplace than the solvent-based coatings. The aluminum hydropastes have been manufactured for about 25 years, but the aluminum-pigmented asphalt emulsions have been on the market in areas for about 10–12 years. There is a stability problem with the hydropastes, not only in aluminum-pigmented asphalt emulsions but also in other uses, such as automotive paints. Because the industrial market is small compared with the automotive market, the improvements in hydropastes for aluminum-pigmented asphalt emulsions have been slow. However, depending on temperature, 1 to 2 years stability can be achieved.

Table 7 Typical ASTM Specification Grade Formulation

Raw material	Weight percent
Asphalt cutback (64% NV)	48
Rule 66 mineral spirits	20.3
Aluminum paste	17
Fiber-cellulose	6
Silicon dioxide (moisture scavenger)	0.7
Filler (diatomaceous earth/wollastonite)	8

The aluminum-pigmented asphalt emulsion usually contains the following materials: asphalt emulsion, clay or chemically stabilized, additional chemical or clay stabilizers, leafing hydropastes, fibers, and modifiers. The stability of the emulsion also affects the optical properties of the coating. Both the initial reflectivity and long-term reflectivity will diminish as the emulsion becomes less stabile. Aluminum-pigmented asphalt emulsions are good products to use where lower limits of volatile organic compounds (VOCs) are desired; however, they do have temperature limitations, as they cannot be subjected to freezing temperatures.

Mixing solvent- or water-based aluminum-pigmented asphalts in the manufacturing plant is a critical process. Care must be taken not to shear the aluminum flakes or cause them to bend, as this will decrease the leafing ability of the aluminum flakes. Therefore, low-shear mixers and pumps are needed in the plant. Premixed dispersions of the aluminum paste are recommended by most aluminum paste manufacturers. This helps alleviate shear but, more important, gives better dispersion of the aluminum paste in the coating, giving optimum brightness.

To obtain consistent quality and optical characteristics of the aluminum-pigmented asphalt, it is desirable to have accurate meters, such as mass-flow meters, or load cells to ensure that the same amount of raw materials is added to each batch. In this way batch-to-batch color variation is avoided. The filling operation is another area in which variation could happen, but it is easily avoided. Aluminum flake, mainly because of its high density, tends to settle out. Therefore, during the filling operation, the coating should be agitated lightly to ensure consistency while filling.

Application of the aluminum-pigmented asphalt is another important step in the brilliance and durability of the coating. Many things can affect the quality of the coating on the roof. As with any coating, the success or failure of the aluminum-pigmented asphalt coating depends largely on surface conditions, drainage, and application. These coatings are designed primarily for heat reduction, membrane protection, and decoration. Although the aluminum adds moisture protection, it is not designed to stop leaks or repair seams or blisters. Repairs should be made on any seams or other problem areas. The roof must be made sound before the application of any aluminum coating. Badly oxidized surfaces, where cracking and splitting is obvious, should be coated with an emulsion or solvent-based roof coating to repair the surface.

V. APPLICATION CONSIDERATIONS FOR COLD-APPLIED PAINTS, COATINGS, AND MASTICS

Surface contaminant that can affect adhesion include exudate, air- or water-borne contaminants, surface oxidation, and release or antiblock agents. Exudate is a soft brownish-yellow, glossy film that may form on the membrane surface. Exudate often results in a softening of the membrane or coating, bleed through of the surface coating, and loss of adhesion of the coating. Poorly mixed or incompatible modified sheets often exhibit exudation problems and adhesion loss.

Surface oxidation can also impair the adhesion of a coating to the substrate. Surface oxidation is often exhibited in the form of a brown or rust-colored film. "Tobacco juicing," the formation of water-soluble compounds from the asphalt (typically acids and ketones), is also caused by oxidation and will inhibit the adhesion of the cold-applied coatings.

To ensure adhesion, the surface must be free of loose debris, dirt, oils, and other materials that could cause loss of adhesion. Depending on the inspection of the surface, this may involve brooming or vacuuming to remove loose dirt or other dry material, power brooming to remove light contaminants (e.g., pollen, small amounts of light oil), power scrubbing for heavier contaminants, or, in the worst cases, cleaning by water blasting. Water blasting is used to take off heavy exudate, loosely bound particles, or old coating.

Roofs must be allowed to dry thoroughly before a solvent-borne coating is applied. However, if an asphalt emulsion is used, either a plain emulsion or an aluminum-pigmented asphalt emulsion, a slightly damp roof may be desirable for even spread of the emulsion. If the substrate is too dry and the humidity is low, the emulsion may break as it is applied to the substrate, resulting in poor adhesive properties.

To apply a top coat of an aluminum-pigmented asphalt, BURs and solvent-based coatings should be allowed to weather for 90 days of warm temperatures before being coated. Current asphalt emulsions can be coated right after curing (typically 5–14 days) without concern about staining. This is why some manufacturers recommend the use of an emulsion, either as a primer or repair coat, before an aluminum coating is applied.

Recent studies have shown that it is better to coat many modified membranes before they age, rather than after they age [16]. When coating a modified bitumen membrane with aluminum coating, the membrane manufacturer's recommendations should be followed. ARMA (Asphalt Roofing Manufacturers Association) and RCMA (Roof Coating Manufacturers Association) have published a brochure entitled "Evaluating and Preparing Modified Bitumen Membrane Roofing for Surface Coating Applications" [17] that goes into detail about the evaluation and preparation of the modified membranes before they are coated.

Areas where water ponds are detrimental to all cold-applied coatings, as the water causes premature coating failures. If a coating is under water, as occurs in a ponded area, the water may work its way under the coating, causing it to lift up and peel. Dirt can accumulate in depressions and inhibit the reflectance of the aluminum-pigmented asphalt coatings. Positive drainage is necessary to flush away any accumulations of surface dirt from the roof and keep the reflective properties of the aluminum-pigmented asphalt coating intact.

Once the surface is prepared, the cold-applied coatings must normally be mixed well before application. This is particularly important for paints and low-viscosity coatings, such as aluminum-pigmented asphalt coatings. The aluminum flake does settle, mainly because of its density, and if it is not mixed back in well, the coating will not be consistent on the roof. Coating manufacturers work hard to ensure a soft settlement, but a mechanical mixer using a paddle-type blade, such as a Jiffy Mixer blade designed for fibered products, not paint products, should be used to ensure that all the aluminum pigment is evenly dispersed. If dispersion of the coating is not complete, there may be streaking during application. Pigment left in the bottom of the pail will result in the coating not being as reflective as desired. Also, if some areas of the coating are low in binder (the asphalt), premature failure of the coating may occur with age.

When rolling or brushing an aluminum-pigmented asphalt coating, it is important to finish the strokes in relatively the same direction to achieve the best aesthetics. If the aluminum flakes are oriented in the same direction, they reflect light more uniformly. Some aluminum-pigmented asphalts are more forgiving than others, but this is still a good idea for any aluminum-pigmented asphalt coating. Also, the aluminum flake is fragile, and if brushed excessively, the flakes can bend and therefore not leaf properly.

Proper coverage is also important for the performance of the cold-applied coating. Low coverage will result in the coating failing prematurely, which could be attributed to pinholing and loss of adhesion due to moisture seeping under the coating and water or vapor intrusion through the substrate. Low coverage in the aluminum-pigmented asphalt coatings will result in low reflectivity, allowing the roofing membranes to reach higher temperatures and age prematurely.

Polymer-modified asphalt coatings, whether solvent or water borne, appear to be the future of cold-applied asphalt coatings. Although there are many polymer-modified asphalts on the market, the amount and type of polymerization are important. Currently, there are no consensus standards for polymer-modified cold-applied coatings, so most manufacturers cite current ASTM specifications. However, this does not give the user or specifier any idea about the extent of modification, percent elongation, or recovery. Standards need to be developed that are meaningful for the products' end use, whether they be used on modified membranes as an adhesive or coating or as a cold-applied built-up roof. Existing ASTM standards can be found in Table 8 to enable the user to find the specifications needed to specify a cold-applied asphalt coating.

Table 8   ASTM Standards for Cold Applied Asphalt Coatings

Standard	Description
D-41	Specification for asphalt primer used in roofing, dampproofing, and waterproofing
D-1187	Specification for asphalt-base emulsions for use as protective coatings for metal
D-1227	Specification for emulsified asphalt used as a protective coating for roofing
D-2882	Specification for asphalt roof cement
D-2823	Specification for asphalt roof coatings
D-2824	Specification for aluminum-pigmented asphalt roof coatings
D-3019	Specification for lap cement used with asphalt roll roofing
D-3747	Specification for emulsified asphalt adhesive for adhering roof insulation
D-4479	Specification for asphalt roof coatings—asbestos free
D-4586	Specification for asphalt roof cement—asbestos free

REFERENCES

1. *ASTM Annual Book of Standards*, Vol. 4.04, *Roofing, Waterproofing, and Bituminous Materials*, American Society for Testing and Materials, Philadelphia (1995).
2. *ASTM Annual Book of Standards*, Vol. 4.04, *Roofing, Waterproofing, and Bituminous Materials*, American Society for Testing and Materials, Philadelphia (1995).
3. ASTM Annual Book of Standards, Vol. 4.03, *Road and Paving Materials; Vehicle-Pavement Systems*, American Society for Testing and Materials, Philadelphia (1995).
4. W. C. Cullen, Transitions in Roofing Technology, in *9th Conference on Roofing Technology*, May 4–5, 1989, National Institute of Standards and Testing/National Roofing Contractors Association, pp. 1–5.
5. E. W. Mertens and S. H. Greenfeld, Some Qualitative Effects of Composition and Processing on Weatherability of Coating Grade Asphalt, in *Syposium on Bituminous, Waterproofing and Roofing Materials*, pp. 20–28, ASTM STP 280, American Society for Testing and Materials, Philadelphia (1959).
6. J. P. Porcher, Advantages, Limitations and Selection of Modified Bitumen, in *8th Conference on Roofing Technology*, April 16–17, 1987, National Bureau of Standards/National Roofing Contractors Association, pp. 75–80.
7. D. A. Davis and M. P. Krenick, A New Cold Process Roofing System: Description and Performance, in *Roofing Systems*, pp. 3–12, ASTM STP 603, American Society of Testing and Materials, Philadelphia (1976).
8. W. Schumacher and E. Scherp, Cold-Laid, Exposed, Two Ply Modified Bitumen Roofing Systems, in *Roofing Research and Standards Development*, pp. 112–126, ASTM STP 959 (R. A. Critchell, ed.), American Society for Testing and Materials, Philadelphia (1987).

9. A. J. Day and E. C. Herbert, Anionic Asphalt Emulsions, Bituminous Materials, in *Asphalts, Tars and Pitches*, Vol. 2, Part 1 (A. J. Hoiberg, ed.), Wiley, New York (1965).

10. J. J. Drukker, Clay Emulsions, Bituminous Materials, in *Asphalts, Tars and Pitches*, Vol. 2, Part 1 (A. J. Hoiberg, ed.), Wiley, New York (1965).

11. K. Brzozowski, New Coatings as Components of Roofing Systems, in *9th Conference on Roofing Technology*, May 4–5, 1989, National Institute of Standards and Testing/ National Roofing Contractors Association, pp. 20–25.

12. *ASTM Annual Book of Standards*, Vol. 4.04, *Roofing, Waterproofing, and Bituminous Materials*, American Society for Testing and Materials, Philadelphia (1995).

13. *ASTM Annual Book of Standards*, Vol. 4.04, *Roofing, Waterproofing, and Bituminous Materials*, American Society for Testing and Materials, Philadelphia (1995).

14. E. W. Mertens and S. H. Greenfeld, Some Qualitative Effects of Composition and Processing on Weatherability of Coating Grade Asphalt, in *Symposium on Bituminous, Waterproofing and Roofing Materials*, pp. 20–28, ASTM STP 280, American Society for Testing and Materials, Philadelphia (1959).

15. To learn more about SHRP, contact the Federal Highway Administration (FHWA) Research and Technology Report Center, HRD-11, 6300 Georgetown Pike, McLean, VA 22101-2296, and ask for information about SHRP testing or the Strategic Highway Research Program Asphalt Research Output and Implementation Program.

16. J. D. Carlson, T. L. Smith, and J. E. Christian, Field Performance Research of APP Modified Bitumen Roof Membranes and Coatings, presented at the 10th Annual NRCA/NIST Conference, April 22–23, 1993, Gaithersburg, Maryland.

17. Roof Coating Manufacturers Association and Asphalt Roofing Manufacturers Association, *Evaluating and Preparing Modified Bitumen Roofing for Surface Coating Applications*, October 1994.

21

Polymer-Modified Asphalt Cement Used in the Road Construction Industry: Basic Principles

Bernard Brûlé

Entreprise Jean Lefebvre, Neuilly sur Seine, France

I. INTRODUCTION

Modified binders, obtained by adding thermoplastic macromolecular systems of plastomers or elastomers to traditional pure asphalt cements for road construction, have existed for more than 20 years. These products are still poorly understood scientifically. Significant research efforts must be undertaken to better understand the relationship between composition, structure, and properties of the binders. For the hot mixtures, we need to understand the relationships between their performances (rutting, fatigue, thermal cracking) and the characteristics of the modified binders as measured in the laboratory.

This chapter presents the current state of knowledge regarding these issues. We also provide a discussion of the basic principles of bitumen-polymer compatibility, the mechanism for modification, and the effect of the components on the properties of the modified binders. Lastly, we present several specific problems related to the use of these products.

Trial strips were made as early as 1963 in France [1]. Their purpose was to determine the behavior of asphalt modified by introduction, at the time of batching and within the batching unit of the coating plant, of different "rubbers," either natural or synthetic. The results were not good enough for the process to be developed further. Research then centered on the addition of thermoplastic polymers to bitumen to make either road binders suitable for high-density traffic or road binders providing increased durability or allowing reduced coarse thickness.

It should be noted that in France the public authorities provided active incentive as early as 1972. Oil companies and contractors were informed about modified binders and their intention to run trials on test strips. They even participated in research by studying possibilities for addition of thermoplastic polymers to bitumen (in and around 1975) [2], which was of help in selecting the most valuable classes of polymer.

II. GENERAL DATA ON POLYMER-MODIFIED BITUMENS

A. Objectives for the "Ideal" Modified Binder

An "ideal" binder should have enhanced cohesion and very low temperature susceptibility throughout the range of temperatures. Its susceptibility to loading time should be low, and its permanent deformation resistance, break strength, and fatigue characteristics should be high. At the same time, it should have at least the same adhesion qualities (active and passive) as traditional binders. Lastly, its aging characteristics should be good, for both application and service.

Figure 1 illustrates these points concerning temperature susceptibility. A linear relationship between consistency (penetration and/or viscosity) and temper-

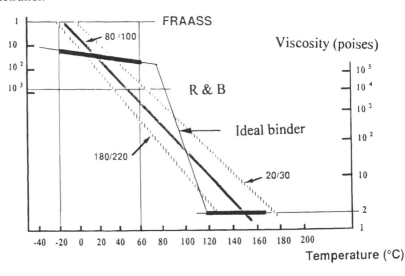

Figure 1   Temperature susceptibility of an "ideal" thermoplastic binder.

ature in the Heukelom data chart (from very low, $-20°C$ and lower, to very high temperature, $160°C$ and higher) is observed for neat bitumens. An ideal binder should, on the contrary, present a low temperature susceptibility in the range of temperature of use on the road. Still, a reasonable viscosity at application temperatures is required for pumping, mixing, and/or spraying. This means a high temperature susceptibility at intermediate temperatures, say between 80 and 120°C.

B. Compatibility

There is a complex relationship between the chemical composition of road bitumens, their colloidal structure, and their physical and rheological properties [3]. Anything that modifies the chemical composition of a bitumen unfailingly modifies its structure and, consequently, its properties.

If, without taking any special precautions, a road bitumen and a given thermoplastic polymer are mixed hot, one of the following three results is obtained:

1. The mix is heterogeneous. This is the most likely result, that is, where the polymer and the bitumen prove to be incompatible. In this case, the constituents in the mix separate and the mix has none of the characteristics of a road binder.
2. The mix is totally homogeneous, including the molecular level. This is the (infrequent) case of perfect compatibility. In this case the oils in the bitumen solvate the polymer perfectly and destroy any intermacromolecular interactions. The binder is extremely stable, but the modification of service qualities concerning those of the initial bitumen is very slight. Only the viscosity increases. This is therefore not the desired result.
3. The mix is microheterogeneous and is made up of two distinct finely interlocked phases. This is the compatibility sought, giving the bitumen genuinely modified properties. In such a system the compatible polymer "swells" by absorbing some of the oily fractions of the bitumen to form a polymer phase quite distinct from the residual bitumen phase. This comprises the heavy fractions of the binder (the rest of the oils, plus the resins and asphaltenes).

C. Compatible Polymers

Experience shows that few thermoplastic polymers are very compatible with road bitumens in the manner defined above. Industrial polymers used for bitumen modification are essentially of two types:

Plastomers. These are generally ethylene vinyl acetate (EVA) copolymers whose grades differ in accordance with their vinyl acetate content and their molecular weight (identified by the melt index, MI).
Thermoplastic elastomers. The best known are block styrene-butadiene-styrene (SBS) copolymers. They are differentiated by their styrene content, their molecular weight, and their configuration (linear or star). Styrene-isoprene-styrene copolymers are also used.

D. The Mechanism of Modification

Because a modified binder consists of two distinct phases, three cases must be considered:

Case I: The polymer content is low (less than 4%). In this case the bitumen is the continuous phase of the system, and the polymer phase is dispersed through it. With its lowered oil content, the bitumen phase has a correlatively higher asphaltene content; its cohesion and elasticity are both enhanced as a result. At high service temperatures (around 60°C), the stiffness modulus of the polymer phase is higher than that of the matrix. These reinforcing properties of the polymer phase enhance the mechanical performance of the binder. At low temperatures, the stiffness modulus of the dispersed phase is lower than that of the matrix, which reduces its brittleness. It can therefore be seen that the dispersed polymer phase enhances the properties of the binder both at high service temperatures and at low temperatures. In this case the choice of bitumen is the determining factor.

Case II: The polymer content is sufficiently high (more than 7%). In general, if the bitumen and polymer are correctly chosen the polymer phase will be the matrix of the system. This is, in fact, not a bitumen but a polymer plasticized by the oils in the bitumen in which the heavier fractions of the initial asphalt cement are dispersed. The properties of such a system are fundamentally different from those of a bitumen and depend essentially on those of the polymer. One should speak not of a polymer-modified bitumen but of a thermoplastic adhesive.

Case III: The polymer content is around 5%. This mixture may form microstructures in which the two phases are continuous and interlocked; such systems are generally difficult to control and pose stability problems (their micromorphology and properties often depend on temperature history).

Fluorescence optical microscopy [4] provides an effective vision of these microstructures at magnification of around 250 times. With this type of observation

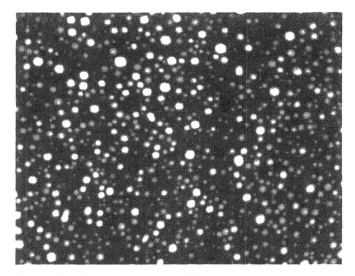

Figure 2   Microstructure of a modified binder with bitumen matrix.

the polymer phase is light and the bitumen phase is black. Figures 2, 3, and 4 are examples of three types of microstructures.

The microstructure of polymer-modified bitumens is extremely important insofar as there is obviously a close relationship between the microtexture of a modified bitumen and its physical properties [5]. It has been shown that polymer-modified bitumens with a constant polymer content made from bitumens of identical grade but of different origins could have substantially different microstructures and properties (especially low-temperature deformation) [6].

III. EFFECT OF POLYMER AND BITUMEN PARAMETERS ON THE PROPERTIES OF POLYMER-MODIFIED BITUMENS

A. Polymer Parameter

The effect of the molecular weight and chemical composition of SBS on the properties of modified bitumens has been studied by Diani and Gargani of Enichem [7]. The authors demonstrate that in the case of bitumens modified with 5% SBS, the determining factors are the molecular weight of the polystyrene (PS) and polybutadiene (PB) blocks. Figures 5 and 6 show the predominant effect of the molecular weight of the PS sequence on penetration at 25°C (Figure 5) and on the ring and ball softening point (Figure 6).

As far as viscosity at 180°C is concerned, the molecular weight of the polybutadiene sequence is seen to have a greater effect than the PS sequence, as shown by Figure 7.

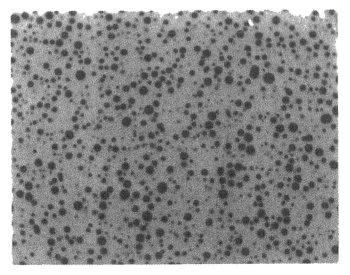

Figure 3   Microstructure of a modified binder with polymer matrix.

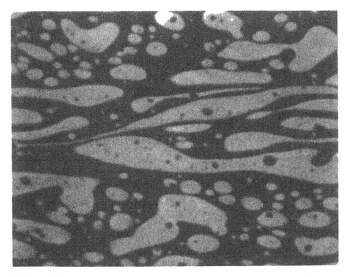

Figure 4   Microstructure of a modified binder with two continuous phases.

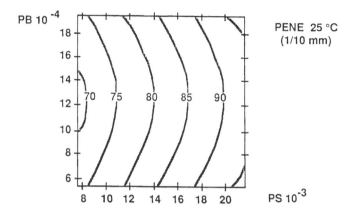

Figure 5   Effect of the molecular weight of the PS sequence on penetration of 5% SBS bitumen. (From Ref. 7.)

In the case of EVA-modified bitumens, it is seen that (Figures 8 and 9) penetration diminishes and the ring and ball softening point increases as the EVA content increases, and the effect is more pronounced with lower vinyl acetate contents in the copolymer.

B.   Bitumen Parameter

It has been shown, for SBS-modified bitumens [8], that the aromaticity of maltenes plays an important role in compatibility and properties. Figure 10 illustrates this.

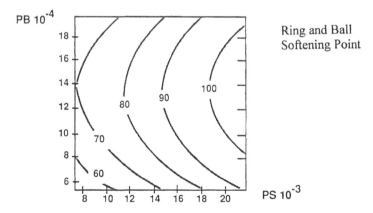

Figure 6   Effect of the molecular weight of the PS sequence on the ring and ball softening point of 5% SBS bitumen. (From Ref. 7.)

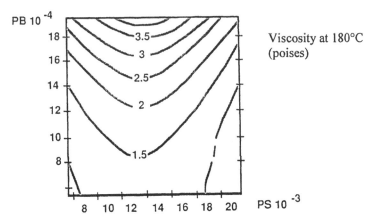

Figure 7 Effect of the molecular weight of the polybutadiene sequence on viscosity of SBS-modified bitumen at 180°C. (From Ref. 7.)

The work of Diani and Gargani [7] provides details on the effect of the generic composition of bitumens on the properties of SBS-modified bitumen mixes with high SBS contents used in the waterproofing industry. Among other things, it demonstrates the determining effect of the saturated oils content on the physical properties of modified binders (penetration and ring and ball softening point).

An equivalent study dealing with EVA-modified bitumens [9] demonstrates that the chemical composition of bitumens of the same grade plays a determining role in the physical properties of only slightly modified binders (binders with a

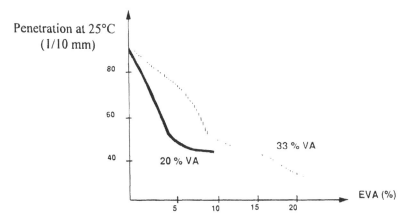

Figure 8 Effect of EVA content on penetration of EVA-modified bitumens.

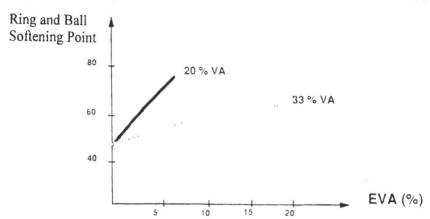

Figure 9   Effect of EVA content on the ring and ball softening point of EVA-modified bitumens.

bitumen matrix). This is illustrated by Figures 11 (penetration) and 12 (ring and ball softening point).

On the other hand, in the case of EVA-modified bitumens with a high polymer content (i.e., with a polymer matrix), the effect of the chemical composition of the bitumen is much less marked, as illustrated by Figures 13 and 14.

C. Special Cases

To the purely "physical" blends of bitumens and elastomeric or thermoplastic copolymers can be added:

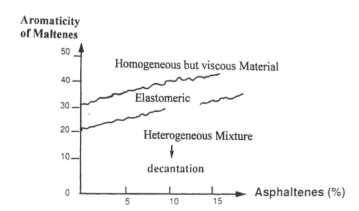

Figure 10   Effect of aromaticity of maltenes on the characteristics of SBS-modified bitumens. (From Ref. 8.)

Penetration at
25°C (1/10 mm)

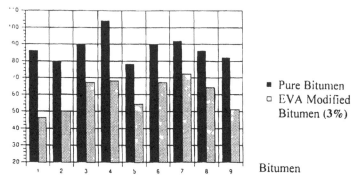

■ Pure Bitumen
□ EVA Modified
Bitumen (3%)

Bitumen

Figure 11   Effect of bitumen on the penetration of EVA-modified bitumens with a bitumen matrix.

Tire-crumb bitumens. The modifier is rubber crumbs obtained by scraping the tread of used tires. Its chemical composition and particle size distribution must be strictly controlled, and it must be made according to a very precise process if the expected results are to be obtained.

Binders obtained by in situ chemical modification of styrene-butadiene copolymers. This patented process optimizes the efficiency of added polymer and makes the product very stable on storage.

Ring and ball
Softening Point

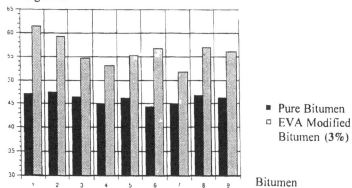

■ Pure Bitumen
□ EVA Modified
Bitumen (3%)

Bitumen

Figure 12   Effect of bitumen on the ring and ball softening point of EVA-modified bitumens with a bitumen matrix.

Penetration at
25°C (1/10 mm)

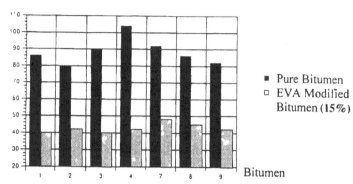

Figure 13   Effect of bitumen on the penetration of EVA-modified bitumens with a polymer matrix.

Ring and ball
Softening Point

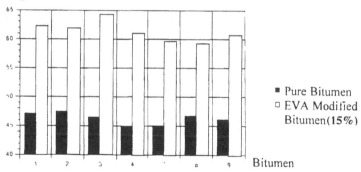

Figure 14   Effect of bitumen on the ring and ball softening point of EVA-modified bitumens with a polymer matrix.

IV. SPECIAL PROBLEMS

Industrial development of polymer-modified asphalt cements does not imply that all the associated problems have been overcome. We should say a few words on special problems associated with the issues of storage, the effect of past temperatures, and relationships between the characteristics of binders and the properties of special bituminous concrete.

A. Storage Stability

The storage stability of modified binders is an extremely important characteristic in actual use. Because these binders consist of two distinct phases, they are subject to the same physical principles as those governing the separation or sedimentation of bitumen emulsions (Stokes' law). In other words, the velocity of displacement of dispersed particles (polymer phase in the case of a bitumen matrix and bitumen phase in the case of a polymer matrix) increases as particle size increases, the difference in density between the two phases increases, and the viscosity of the continuous phase decreases.

Making polymer-modified bitumens with good storage stability therefore implicitly implies micronizing the dispersed phase (effect of the process and manufacturing process) or both phases of equal density at the storage temperature (effect of the chemical composition of the bitumen).

As an example, one can see (Figure 15) that, for EVA-modified asphalt cements, the storage stability of the binder depends strongly on the asphaltene content of the base asphalt cement (0% on storage stability test means that there is no separation of the polymer phase).

B. Influence of Past Temperatures

As stated above, some modified binders have a reasonably observable composite microstructure consisting of interlocking of a continuous bitumen phase and a continuous polymer phase (see Figure 4). Such structures can be obtained with polymer contents in the neighborhood of 4 to 7%, in general (if the "partners"

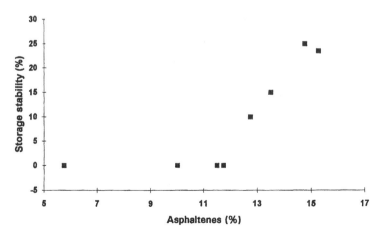

Figure 15   Effect of asphaltene content on the storage stability of EVA-modified asphalt cements.

in the mix are correctly chosen), and research at LCPC [10] shows that their microstructure and physical properties can depend on their past temperatures.

With such systems it is often seen that the existence of a continuous polymer phase giving the binder quite appreciably modified physical properties such as elongation ability at low temperature is associated with relatively rapid cooling (the term "quenching" is not inappropriate, by analogy with what happens in the field of metal or polymer alloys). On the other hand, programming of a relatively slow cooling speed (close to that observed for bituminous concrete on the road, i.e., approximately 30°C/h) results in a different microstructure, generally with a continuous bitumen phase and dispersed polymer phase. This modification of the microstructure of the same product, associated with slow cooling, is accompanied by a modification of properties: lower ring and ball softening point (which can drop as low as around 30°C), lower elastic characteristics, and possibility of deformation at low temperature.

C. Relationships Between the Characteristics of Modified Binders and the Properties of Special Bituminous Concrete

Many years of experience in the field of bituminous concrete using pure bitumens have led to the definition of empirical but reliable relationships between the characteristics of bitumens and the performance of bituminous concrete. This led, for example, to the association of penetration of bitumen and the stiffness modulus of bituminous concrete and of the ring and ball softening point and rutting. These relationships led to the development of pure-bitumen specifications and then to their evolution, providing more accurate prediction of the performance of bituminous concrete. Regarding this, the recent evolution of French specifications toward a reduced temperature range for the ring and ball softening point is particularly meaningful.

The natural tendency when developing modified binders has been to extrapolate the empirical notions acquired with the same indicators (like penetration and ring and ball softening point tests) and to interpret them in terms of performance (for a given consistency determined by penetration at 25°C, a high R&B softening point corresponds to low temperature susceptibility and high resistance to rutting). Although this extrapolation can be roughly applied in the case of slightly modified binders with bitumen matrices, it can be false in the case of binders with composite matrices or polymer matrices. One of the reasons for this lies in the effect of past temperatures on the properties of binders with composite matrices and can be explained by the fact that polymer-matrix binders are not actually bitumens but thermoplastic adhesives. It is now clear to all experts that the ring and ball softening point of a polymer-modified bitumen can in no way be used to predict the rutting potential of the bituminous concrete made from it [11,12].

There therefore remains the problem of the development of new tools for characterizing these binders (complex stiffness modulus, rheological behavior depending on temperature and loading time, etc.). While taking account of the effect of past temperatures, such tools should effectively predict binder performance. Until such methods appear, it would seem wise and reasonable to consider that modified binders are not pure bitumens and that as a result what we know about the latter cannot be applied to the former.

V. CONCLUSIONS

The polymer-modified binders obtained by mixing thermoplastic plastomeric or elastomeric macromolecular materials with traditional pure road bitumens have now been available for more than 20 years. To start with, they provided a satisfactory technical solution to certain special problems such as the application of waterproofing and wearing course complex on steel bridges with orthotropic decks. The application of the surface-dressing technique to roads with high traffic densities was a powerful driving force in launching and developing the widely used and reasonably priced (about twice the price of pure bitumen) modified binders with relatively low polymer contents that are now found on millions of square meters of road surfacings as BBTM (*béton bitumineux très mince*, very thin wearing course), BBUM (*béton bitumineux ultra mince*, ultrathin wearing course), and porous asphalt.

The potential market penetration of these modified binders in a country such as France appears to be around 10% of the total market for bituminous binders and would approach saturation if road-making techniques themselves evolved only slightly in the years to come.

Despite this, these products are relatively little known, scientifically speaking. Much research work remains to be done in order to better appreciate mix-structure-properties relationships for the binders themselves and, for the hot mixes produced with them, the relationships between performance (rutting, fatigue, thermal cracking, etc.) and the characteristics of the modified binders as measured in the laboratory.

It seems to be clear that research work developed in the SHRP program and devoted to the rheological characterization of the bituminous binders (e.g., dynamic shear rheometer, direct tension test, bending beam rheometer) should be very useful for establishing such relationships.

REFERENCES

1. Laboratoire Regional de Blois et Laboratoire Central, Section des Revêtements Hydrocarbonés, Essais de bitume-caoutchouc, *Bull. Liaison Lab. Ponts Chaussées*, **4**, 14–17 (1963).

2. G. Hameau and M. Druon, Contribution à l'étude physico-mécanique de mélanges de bitumes routiers et de polymères élastothermoplastiques, *Bull. Liaison Lab. Ponts Chaussées* , **81**, 121 (1976).

3. B. Brule, G. Ramond, and C. Such, Relationships Among Composition, Structure, and Properties of Road Asphalts, in *Asphaltenes and Asphalts*, 1 (T. F. Yen and G. V. Chiligarian, eds.), Elsevier Science, Amsterdam (1994).

4. B. Brule and M. Druon, La microscopie de fluorescence appliquée à l'observation de bitumes thermoplastiques, *Bull. Liaison Lab. Ponts Chaussées*, **79**, 11–14 (1975).

5. B. Brule, Y. Brion, and A. Tanguy, Paving Asphalt Polymer Blends: Relationship Between Composition, Structure and Properties, in *Proceedings Association of Asphalt Paving Technologists*, Vol. 57, Williamsburg, VA, February 29, March 1 and 2, 1988.

6. B. Brule, M. Druon, and Y. Brion, Micromorphologie et propriétés mécaniques d'un liant bitume-S-B-S: Influence de la préparation et du vieillissement, presented at RILEM, Essais sur liants et matériaux hydrocarbonés, Belgrade, September 1983.

7. E. Diani and L. Gargani, Bitumen Modification with Rubber, presented at Rubberex 87, Harrogate, June 1–5, 1987.

8. E. J. Van Beem and P. Brasser, Bituminous Binders of Improved Quality Containing "CARIFLEX" Thermoplastic Rubbers, *J. Inst. Petrol.*, **59** (566) (1973).

9. A. Madella, F. Le Bourlot, and B. Brule, Selection of Asphalt Cements for Asphalt Cements/EVA Blends, presented at International Symposium "Chemistry of Bitumens", Rome, June 5–8, 1991.

10. A. Donny, Liants bitumes-polymères. De la fabrication à la mise en oeuvre des enrobés: Influence de la nature du bitume sur leurs propriétés mécaniques, leur micromorphologie et leur stabilité thermique, *Etudes et recherches des Laboratoires des Ponts et Chaussées*, Série Chaussées, CR 15, LCPC (1991).

11. J. E. Poirier, M. Chappat, and J. Beauverd, Pour le client seul le résultat compte, *Rev. Gen. Routes Aerodromes*, **688** (September 1991).

12. G. N. King, H. W. King, O. Harders, P. Chaverot, and J. P. Planche, Influence of Asphalt Grade and Polymer Concentration on the High Temperature Performance of Polymer Modified Asphalt, Asphalt Paving Technology, *J. Assoc. Asphalt Paving Technol.*, **61** (1992).

22

Performance Characteristics of Asphalt Concrete Mixes Containing Conventional and Modified Asphalt Binders

Akhtarhusein A. Tayebali
North Carolina State University, Raleigh, North Carolina

John Harvey and Carl L. Monismith
University of California, Berkeley, Richmond, California

Alejandro J. Tanco
Cordoba, Argentina

Robert N. Doty
California Department of Transportation, Sacramento, California

I. SYNOPSIS

In recent years, modifiers have been incorporated in asphalt binders to improve their performance characteristics over a range of temperature conditions with the intent of mitigating distress in both cold and hot environments. To determine how to design effectively mixes with modified binders, a study of mixes containing asphalts produced in Montana and containing a number of different modifiers was conducted at the University of California, Berkeley. Mix evaluation included conventional test methods such as the Hveem stabilometer test and creep tests, as well as state-of-the-art performance-based test methods such as those developed for fatigue and permanent deformation characterization during the Strategic Highway Research Program.

Results of this study suggest that the conventional test methodologies may not be adequate for mix design for mixes containing modified binders. Performance-based binder-aggregate mix evaluation indicates that mix characteristics may or may not be improved by modification and are dependent on asphalt source and modifier type.

II. INTRODUCTION

Rutting of asphalt concrete pavement is a persistent problem facing the highway industry. Conventional asphalt cements used as binders in asphalt-aggregate concrete mixes, in some cases, have significant limitations with regard to performance over a wide range of temperatures. Although a harder virgin asphalt cement may mitigate rutting in hot climatic conditions, the same asphalt cement may be unsuitable for a colder climate due to the potential for thermal cracking. To encompass both temperature extremes, in recent years, modifiers have found increasing use in asphalt paving mixes with the intention of influencing the temperature susceptibility of asphalt binder to mitigate pavement distresses in both hot and cold climate. A vast variety of modifiers are available, and several questions that arise with regard to their use in asphalt mixes are the following:

1. Which modifiers are best suited for the pavement distress under consideration?
2. Is the suggested modifier(s) compatible with the base asphalt?
3. What modification level should be used?
4. What properties would best reflect the improvements in modified asphalt-aggregate mixes?
5. What factors significantly affect the determination of these properties?
6. What test equipment and procedures should be used to evaluate these properties?

Although the state of the technology with regard to evaluation of modified asphalt binders is well advanced, especially with the use of dynamic mechanical analysis to study the rheologic behavior of the binder, questions still remain as to what methodology should be used for the design and evaluation of the performance characteristics of modified asphalt-aggregate mixes.

Conventionally, the Hveem and Marshall methods of mix design are widely used to evaluate permanent deformation (rutting) characteristics of conventional asphalt-aggregate mixes. However, two major drawbacks of these existing methodologies are that (a) they are empirical methods that do not measure mix properties in fundamental engineering units and therefore do not lend themselves well to quantitative evaluation of permanent deformation, which may occur for a given set of conditions, and (b) in case of modified asphalt binders, both test methods are insensitive to the presence of modifiers in mixes and have been reported to give results inconsistent with observed pavement performance in some instances.

To determine how to design effectively mixes with modified binders, a study of mixes containing asphalts produced in Montana and containing a number of different modifiers was conducted at the University of California, Berkeley. Several conventional and modified asphalt-aggregate mixes were evaluated using conventional and unconventional test methodologies to evaluate the permanent deformation performance of these mixes. Test methodologies included Hveem and Marshall test methods, unconfined repeated axial and shear tests, shear creep test, and partially confined constant-height shear repeated load test.

III. OBJECTIVES OF THIS STUDY

The purpose of this study is to evaluate the performance characteristics of asphalt concrete mixes containing modified binders using conventional and unconventional asphalt aggregate mix design methods. Specifically, the objectives of this study are the following:

1. To evaluate the suitability of conventional mix design methods to distinguish the rutting propensity of mixes containing modified and unmodified asphalt binders
2. To evaluate unconventional as well as recently developed state-of-the-art test methods to distinguish the rutting propensity of mixes containing modified asphalt binders

IV. MATERIALS USED

A. Aggregate

Yellowstone River gravel from Billings, Montana, was used in this study. The aggregate gradation confirms to the Montana Department of Highways specification for 19 mm ($\frac{3}{4}$ in.) maximum aggregate size. The gradation as used in this study is shown in Table 1.

B. Asphalts and Modifiers

Two conventional asphalts were used: a 120/150 penetration asphalt from the Cenex Refinery in Laurel, Montana, and a 120/150 penetration asphalt from the Conoco Refinery in Billings, Montana. These conventional asphalts were modified using two modifiers, Kraton D4141G and Polybilt, selected on the basis of an earlier study by Armijo et al. [1]. Kraton-modified binders were prepared by

Table 1 Aggregate Gradation and Specifications Limits

	Percent passing		
		Specifications limits	
Sieve designation	Gradation used	High	Low
19 mm (3/4 in.)	100.0	100.0	100.0
12.7 mm (1/2 in.)	86.0	93.0	79.0
9.5 mm (3/8 in.)	75.0	82.0	68.0
4.75 mm (#4)	53.0	60.0	46.0
2.36 mm (#8)	40.0	46.0	34.0
1.18 mm (#16)	30.5	36.0	25.0
600 mm (#30)	22.5	27.5	17.5
300 mm (#50)	15.5	20.0	11.0
150 mm (#100)	10.5	14.0	7.0
75 mm (#200)	6.0	7.5	4.5

Shell Development Company using 6% Kraton by weight of asphalt. Polybilt-modified binders were prepared by Exxon Chemical Company using a 4% concentration by weight of asphalt for Cenex and 3.5% for the Conoco asphalt, respectively. Table 2 summarizes the physical properties of the conventional and modified binders discussed herein.

V. EFFECT OF MODIFIERS ON ASPHALT CEMENT PROPERTIES

The effects of modifiers on the asphalt cement properties were evaluated using three commonly used test methods standardized by the American Association of State Highway and Transportation Officials (AASHTO): (a) ring and ball softening point test (AASHTO T53), (b) penetration test (AASHTO T49), and (c) kinematic viscosity test (AASHTO T201). Test results for asphalt binders tested in both unaged and thin-film oven test (TFOT) aged condition are given in Table 2 and shown graphically in Figures 1 and 2. Test results indicate the following:

1. Both modifier types—Kraton and Polybilt—affect results from all three asphalt binder tests under consideration, with ring and ball softening point temperature and kinematic viscosity increasing and penetration values decreasing with addition of modifiers.
2. For both asphalt types, Cenex and Conoco, Kraton enhances the properties of both asphalt cements more than Polybilt; i.e., the ring and ball softening point temperatures and kinematic viscosity are higher and

Table 2 Properties of Conventional and Modified Asphalt Binders Used

Test type	Asphalt type					
	Unmodified Cenex	Kraton-modified Cenex	Polybilt-modified Cenex	Unmodified Conoco	Kraton-modified Conoco	Polybilt-modified Conoco
R&B softening point, °C (AASHTO T53)	46	72	56	45	76	53
Penetration at 25°C, dmm (AASHTO T49)	141	66	80	134	72	87
Kinematic viscosity at 135°C, cSt (AASHTO T201)	252	1139	479	235	952	436
R&B softening point after TFOT, °C (AASHTO T53)	50	70	57	49	79	59
Penetration at 25°C, dmm after TFOT (AASHTO T49)	79	64	66	86	67	59
Kinematic viscosity at 135°C, cSt after TFOT (AASHTO T201)	338	1291	653	251	1137	588

Source: Ref. 2.

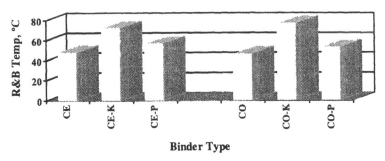

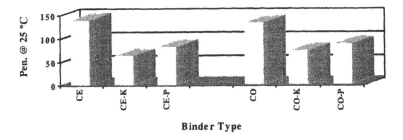

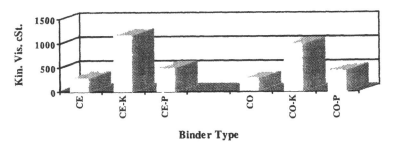

Figure 1   Effect of modifiers on binder properties, unaged condition.

  penetration values lower for Kraton-modified binders, followed by
  Polybilt-modified binders and the unmodified binders, respectively.
3. Thin-film oven aging has a greater effect on the measured properties
  of the unmodified binders compared with the modified binders.

In summary, the ring and ball softening point and kinematic viscosity test results
suggest that Kraton-modified binders are expected to better resist permanent
deformation (rutting) potential than Polybilt-modified binders or unmodified

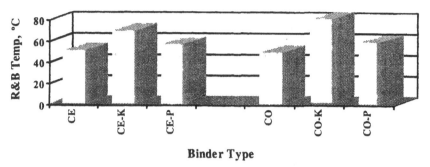

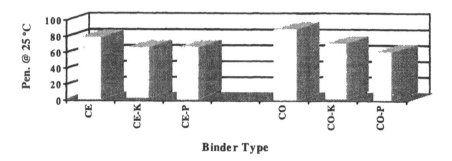

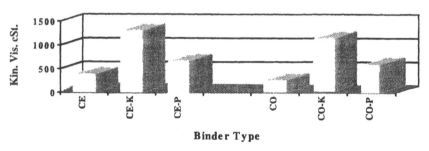

Figure 2   Effect of modifiers on binder properties, after TFOT aging.

binders. In turn, Polybilt-modified binders are expected to show better performance than the unmodified binders:

$$CE\text{-}K > CE\text{-}P > CE$$
$$CO\text{-}K > CO\text{-}P > CO$$

The question that arises, however, is whether the improvements indicated in the

modified binder properties also imply improvements in the permanent deformation characteristics of the asphalt mixes containing these modified binders. In
the following sections the performance of the modified binder-aggregate mixes is
evaluated using several conventional and unconventional test types. Performance
results for these mixes are then compared with the binder performance ranked
using the binder test results.

VI. USE OF CONVENTIONAL ASPHALT-AGGREGATE MIX
DESIGN METHODS FOR MODIFIED MIXES

With regard to asphalt-aggregate mix design, two methods—the Hveem (State
of California) and Marshall (Corp of Engineers) mix design methods—are the
most widely used. Both of these methodologies were developed during the 1940s
using empirical correlations between observed pavement performance under specific traffic conditions and certain stability and volumetric parameters. Performance characteristics of modified mixes using both methodologies and the
suitability of these methods for evaluating modified versus unmodified mixes
are discussed in the following sections.

A. Performance of Modified Mixes in Hveem Stabilometer Test

The procedures for the Hveem method of mix design have been standardized by
the California Department of Transportation (CALTRAN) as well as the American
Society for Testing and Materials (ASTM). Test procedures are found in ASTM
D-1560, *Resistance to Deformation and Cohesion of Bituminous Mixtures by
Means of Hveem Apparatus,* and ASTM D-1561, *Preparation of Bituminous
Mixture Test Specimen by Means of California Kneading Compactor* [3]. A
detailed test procedure has also been outlined in Asphalt Institute Manual MS-
2 [4].

The stability value in the Hveem procedure is determined using the Hveem
stabilometer, a closed-system triaxial compression test in which specimens 101.6
mm (4 in.) in diameter and 63.5 mm (2.5 in.) high compacted using California
kneading compactor are tested at a slow rate of deformation and at a temperature
of 60°C.

Table 3 shows the average Hveem stability S values (average of three
specimens) determined for asphalt-aggregate mixes containing 5.0, 5.5, and 6.0%
asphalt binders by the weight of aggregates. Results indicate that for a given
asphalt content there is only a slight difference between the stability values of
modified and unmodified mixes, with no consistent trend.

Figure 3 shows the stability values for mixes containing 5% binder. For
the Cenex asphalt, addition of modifiers results in very little change in the stability
values. For the Conoco asphalt, Kraton modification seems to give slightly higher

Table 3 Hveem Stabilometer Test Results for Asphalt-Aggregate Specimens
Containing Modified and Unmodified Binders

Asphalt type	Asphalt content (% by wt aggregate)	Average air voids (%)	Average Hveem stability (S)
Unmodified Cenex (CE)	5.0	3.5	33
	5.5	2.1	25
	6.0	1.7	19
Kraton-modified Cenex (CE-K)	5.0	4.1	32
	5.5	3.3	22
	6.0	2.2	13
Polybilt-modified Cenex (CE-P)	5.0	3.1	32
	5.5	1.7	25
	6.0	0.6	10
Unmodified Conoco (CO)	5.0	3.9	31
	5.5	2.6	22
	6.0	1.7	12
Kraton-modified Conoco (CO-K)	5.0	4.1	34
	5.5	2.5	27
	6.0	1.1	16
Polybilt-modified Conoco (CO-P)	5.0	3.3	29
	5.5	2.3	28
	6.0	0.8	16

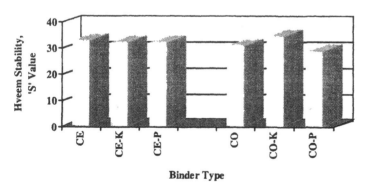

Figure 3 Hveem stability value versus binder type for modified and unmodified binders—5% binder content.

stability value and Polybilt modification gives a lower stability value compared with that for unmodified mixes. However, it should be noted that the differences in the stability values are due not only to the presence of modifiers but also to the differences in the air void contents of the specimens.

Results of an analysis of variance (ANOVA) conducted on the original data set [5,6] are shown in Table 4. ANOVA results indicate that the interaction between air void content and asphalt content has a significant effect on the stability values at the 99% significance level but the modifier type does not significantly affect the stability values at the 90% confidence level.

Even though the modifiers do not significantly affect the stability values, Pradhan [2] has reported that evaluation of the experimental field sections in Montana containing modified mixes showed significantly reduced rutting compared with the section containing unmodified mixes. It can therefore be concluded that Hveem stabilometer test is not sensitive to the modifiers used in this study. Similar results have also been reported by Tayebali et al. [7] for mixes containing binders with other polymer modifiers.

B. Performance of Modified Mixes in Marshall Stability Test

The procedures for the Marshall method of mix design have been standardized by the American Society for Testing and Materials. Test procedures are found in ASTM D-1559, *Resistance to Plastic Flow of Bituminous Mixtures Using Marshall Apparatus* [3]. A detailed test procedure has also been outlined in Asphalt Institute Manual MS-2 [4].

The stability values in the Marshall procedure are determined using the Marshall testing head. A compressive load is applied to a partially confined specimen 101.6 mm (4 in.) in diameter by 63.5 mm (2.5 in.) high, at a temperature of 60°C and at a constant rate of movement of the testing machine head of 50.8 mm (2 in.) per minute, until a maximum load is reached.

The Marshall stability testing for the materials used in this study was conducted at Montana State University. Detailed results have been reported in several publications [1,2,8,9] for normal dense graded mixes as well as large

Table 4   ANOVA Results for Hveem Stability Values

	F ratio		Significance level		Significant	
	Cenex	Conoco	Cenex	Conoco	Cenex	Conoco
Main factor level: mod- ifier type	2.34	2.94	0.12	0.08	No	No
Covariates: air voids– asphalt content	58.87	31.79	0.00	0.00	Yes	Yes

stone mixes. Figure 4 shows the average Marshall stability for the modified and unmodified mixes with an optimum asphalt content. For the Cenex asphalt, Kraton and Polybilt modification results in a slight reduction in stability values. For Conoco asphalt, modification results in a slight increase in stability values. It should be noted that all mixes show stability values more than 17,800 N (4000 lb) at optimum asphalt content, values far exceeding the minimum Marshall design criteria for an acceptable mix of 5338 N (1200 lb) and 8006 N (1800 lb) for specimens compacted using 50 and 75 standard Marshall hammer blows compactive effort, respectively.

Because the Marshall stability values are meaningful only as a guide in mix design for passing or failing mixes, its usefulness in evaluating the effect of modifiers on mix permanent deformation performance is limited. Pradhan [2], on the basis of an evaluation of field experimental test sections in Montana, found no consistent correlation between the pavement in situ rut depth measurements and laboratory Marshall stability values for the modified and unmodified mixes evaluated in this study. Field sections containing Kraton-modified mixes were observed to have the lowest rut depths even though they had significantly higher air void contents compared with field sections containing unmodified mixes.

VII. USE OF UNCONVENTIONAL AND STATE-OF-THE-ART TEST METHODS FOR PERFORMANCE EVALUATION OF MODIFIED MIXES

Several test methods are available with the potential for quantitative evaluation of asphalt-aggregate mix permanent deformation performance. Test methods

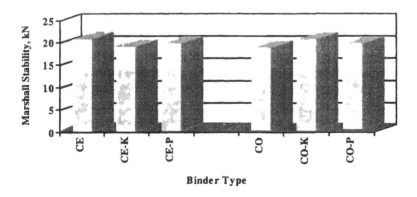

Figure 4   Marshall stability value versus binder type for modified and unmodified binders—optimum binder content.

included in this study were (a) axial loading tests, (b) direct or simple shear loading tests, and (c) partially confined constant-height repeated shear loading tests. Specimen configurations used for each of the test types are as follows: (a) 101.6 mm (4 in.) diameter by 203.2 mm (8 in.) high for the axial loading test, (b) 101.6 mm (4 in.) diameter by 63.5 mm (2.5 in.) high for simple shear loading test, and (c) 152.4 mm (6 in.) diameter by 50.8 mm (2 in.) high for the constant-height simple shear test. For the axial and simple shear tests, the specimens were compacted using the California kneading compactor. Specimens for the constant-height simple shear test were cored from a slab compacted using a rolling wheel compactor [10]. An asphalt content of 5.0% by weight of aggregate was used for all mixes. Two test temperatures, 40°C and 60°C, were used. Results for the mix permanent deformation response for each test type are discussed in the following sections.

A. Axial Repeated Load Test Results

The uniaxial loading test performed on cylindrical specimens has been one of the most extensively used tests for measuring the permanent deformation propensity of asphalt-aggregate mixes. This test can be conducted in several different forms: (a) as a creep test with a constant static loading for the entire duration of the test; (b) as a repeated load test, where the load is applied for a short time followed by a rest period until the next load application; and (c) as a dynamic load test, where the load is applied as a continuous sinusoidal function.

The use of the axial repeated test, in all of its different forms, has been reported by many researchers [11–17]. However, these investigations were limited to tests on conventional asphalt concrete mixes. In other studies the axial load test has been used to evaluate asphalt-aggregate mixes containing modified asphalt [7,18–20].

These latter studies have clearly demonstrated the advantages of axial repeated loading tests versus axial creep loading tests. Axial creep test results [7,19,20] have shown a lack of sensitivity in performance between asphalt-aggregate mixes with modified and unmodified asphalt binders. Accordingly, in this study, only the axial repeated load test has been utilized for mix evaluation.

The axial repeated load test was conducted using a stress level of 207 kPa (30 psi) with a loading time of 0.1 s and 1 Hz frequency. The axial stress, the elastic strain (recoverable), and the plastic strain were obtained as a function of the number of load repetitions from the repeated load test. Table 5 shows average test results for two response variables: (a) the initial resilient modulus at load repetition 100 computed as the ratio of the applied stress and the elastic (recoverable) strain, and (b) the natural logarithm of the number of load repetitions necessary to reach a threshold value of plastic (permanent) axial strain of 0.2%.

Figure 5 shows a plot of the log repetitions to reach a plastic strain value of 0.2% as a function of the asphalt binder type at 40 and 60°C. It can be noted

Table 5 Axial Repeated Load Test Results

Asphalt type	Percent air voids 40°C	Percent air voids 60°C	Ln N at 0.2% plastic strain 40°C	Ln N at 0.2% plastic strain 60°C	Resilient modulus (MPa) 40°C	Resilient modulus (MPa) 60°C
Unmodified Cenex (CE)	3.5	3.4	5.910	2.368	436	205
Kraton-modified Cenex (CE-K)	4.0	4.6	6.348	4.192	454	209
Polybilt-modified Cenex (CE-P)	3.2	2.7	8.852	5.057	667	265
Unmodified Conoco (CO)	3.1	3.0	9.779	5.410	638	253
Kraton-modified Conoco (CO-K)	3.6	3.6	7.929	4.517	407	110
Polybilt-modified Conoco (CO-P)	3.2	3.2	10.158	4.069	563	337

from this figure that for Cenex-modified binders, there is a consistent trend of improved resistance to plastic deformation with modification of the base Cenex asphalt at both temperatures. Polybilt-modified Cenex shows better performance, followed by the Kraton-modified Cenex and the unmodified Cenex, respectively.

Mixes containing unmodified Conoco asphalt perform better than mixes containing either the modified or unmodified Cenex asphalt. However, in general, modification for mixes containing Conoco asphalt results in reduced performance compared with that of mixes containing unmodified Conoco asphalt.

It should be noted that, as shown in Table 5, there are slight differences in air void contents of different mixes that will affect the permanent deformation performance, especially for mixes containing Kraton-modified Cenex. However, even though these mixes show 0.5 to 1% higher voids, they still outperform mixes containing unmodified Cenex. Therefore, in general, the axial repeated load test results indicate that Cenex asphalt is benefited by the addition of modifiers used in this investigation whereas Conoco is probably not. For the Cenex asphalt, the rank order of performance based on the number of load repetitions is the following:

$$CE\text{-}P > CE\text{-}K > CE$$

However, this rank order is not in agreement with that obtained using the binder tests.

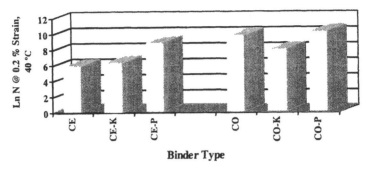

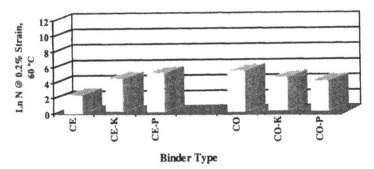

Figure 5   Axial repeated load test results, log repetitions to 0.2% plastic strain versus binder type, 40 and 60°C.

B.   Shear Creep and Repeated Load Test Results

Although, direct (simple) shear load tests have been popular and widely used in geotechnical engineering for the study of granular materials, adoption of these tests in characterizing the permanent deformation of asphalt mixes has been fairly recent. At the time this project was initiated, very little work was available concerning the use of shear creep and repeated load tests in evaluation of the asphalt-aggregate mixes containing modified or unmodified binders. Accordingly, in this study, both test methods were evaluated.

The shear creep and shear repeated loading tests were conducted using a shear stress level of 24 kPa (3.5 psi) and a vertical stress of 24 kPa (3.5 psi).

For the shear creep test, the shear load was held constant for a duration of 3600 s or until specimen failure. The shear stress and the creep strain were obtained as functions of time for this test. For the shear repeated load test, shear stress was applied for a duration of 0.1 s at 1 Hz frequency. The shear stress, elastic shear strain (recoverable), and plastic shear strain were obtained as functions of the number of load repetitions.

1. Shear Creep Load Test Results

Table 6 shows average test results for two response variables: (a) the shear creep modulus at 3600 s loading time computed as the ratio of applied shear stress and the resulting shear creep strain and (b) the shear creep strain at 3600 s loading time.

Because higher shear creep strain values equate to lower resistance to permanent deformation, to facilitate easy comparison with other test results the inverse shear creep strain is plotted as a function of the binder type shown in Figure 6 at 40 and 60°C. For mixes containing Cenex asphalt, modification, in general, results in improvement of permanent deformation resistance, with Kraton modification yielding higher resistance followed by the Polybilt modification. It should be noted that this ranking:

CE-K > CE-P > CE

is more in line with the ranking obtained from the asphalt binder tests. However, this is not in agreement with the axial repeated load test results.

Table 6 Shear Creep Loading Test Results

Asphalt type	Percent air voids		Creep strain at 3600 s		Creep modulus at 3600 s (MPa)	
	40°C	60°C	40°C	60°C	40°C	60°C
Unmodified						
Cenex (CE)	3.7	3.2	6.76	9.55	551	362
Kraton-modified						
Cenex (CE-K)	3.6	3.7	4.31	7.20	876	485
Polybilt-modified						
Cenex (CE-P)	3.2	3.5	5.56	10.10	688	372
Unmodified						
Conoco (CO)	3.3	3.0	3.25	5.40	1199	644
Kraton-modified						
Conoco (CO-K)	3.7	3.5	3.60	5.95	1094	624
Polybilt-modified						
Conoco (CO-P)	3.4	3.5	3.45	6.76	1102	585

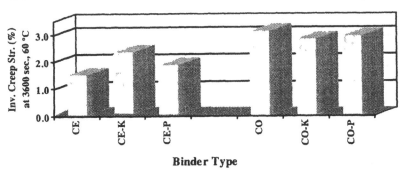

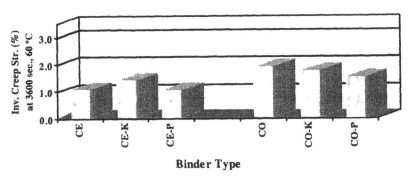

Figure 6   Shear creep load test results, inverse percent creep strain versus binder type, 40 and 60°C.

As was the case for the axial repeated load test, mixtures containing unmodified Conoco asphalt perform better than mixes containing the modified or unmodified Cenex asphalt. However, in general, modification of mixes containing Conoco asphalt results in reduced performance compared with mixes containing unmodified Conoco asphalt. The shear creep test results indicate that Cenex asphalt is benefited by addition of modifiers used in this study, whereas Conoco asphalt is not.

2.   Shear Repeated Load Test Results

Table 7 shows the average test results for two response variables: (a) the initial resilient modulus at load repetition 100 computed as the ratio of the shear stress and shear elastic (recoverable) strain and (b) the natural logarithm of load

Table 7 Shear Repeated Load Test Results

	Percent air voids		Ln N at 0.15% plastic strain		Resilient modulus (MPa)	
Asphalt type	40°C	60°C	40°C	60°C	40°C	60°C
Unmodified						
Cenex (CE)	3.3	3.2	5.00	2.26	82	10
Kraton-modified						
Cenex (CE-K)	4.4	4.0	5.70	3.20	63	17
Polybilt-modified						
Cenex (CE-P)	2.4	2.6	9.80	3.05	164	13
Unmodified						
Conoco (CO)	3.4	3.1	9.32	3.10	84	13
Kraton-modified						
Conoco (CO-K)	3.4	4.0	7.57	2.76	93	13
Polybilt-modified						
Conoco (CO-P)	3.3	3.6	10.13	2.72	64	13

repetitions necessary to reach a threshold value of shear plastic (permanent) strain of 0.15%.

Figure 7 shows a plot of the log repetitions to reach a shear plastic strain value of 0.15% as a function of the asphalt binder type at 40 and 60°C. The trends in the results obtained from this test are very similar to those for the axial repeated load test. Mixes containing Cenex asphalt are generally benefited by modification, whereas mixes containing Conoco asphalt are not. Once again, mixes containing unmodified Conoco asphalt outperform all other mixes. For mixes containing Cenex asphalt, Polybilt modification results in performance equal to or better than that seen with Kraton modification.

C. Constant-Height Shear Repeated Load Test Results

A problem with the simple shear creep and shear repeated load test is the dilation (volume increase) of the test specimen upon the application of the shear load. During the Strategic Highway Research Program (SHRP) a shear test device was developed at the University of California, Berkeley, under research contract SHRP A-003A [21]. Figure 8 shows one of the latest versions of the simple shear test device, which is capable of applying shear and axial loads to a test specimen simultaneously.

Axial loading is applied to a test specimen to prevent volume increase during the shear repeated load test so that the specimen height remains constant, hence the name constant-height shear repeated load test.

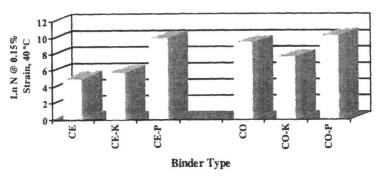

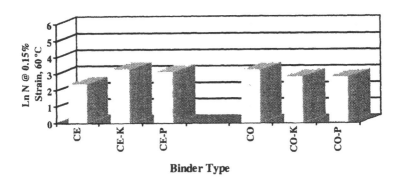

Figure 7 Shear repeated load test results, log repetitions to 0.15% plastic strain versus binder type, 40 and 60°C.

Results of this test for the modified and unmodified mixes are presented in Table 8. Figure 9 shows the average log repetitions necessary to reach a threshold shear plastic strain value of 2% as a function of binder type. Because of unavailability of Polybilt-modified Conoco binder, mixes containing this binder were not evaluated.

Unlike previous tests, in which mix ranking changes with test temperature, results of the constant-height shear repeated load test show consistency in ranking at both temperatures of 40 and 60°C. Figure 9 shows that Kraton modification increases permanent deformation resistance of mixes containing both asphalt types, Cenex and Conoco. All other performance tests indicated that Conoco

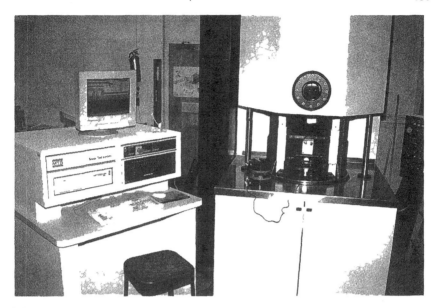

Figure 8 Simple shear test device.

asphalt is not benefited by modification. Ranking based on the constant-height shear repeated load test is in agreement with the ranking of the binders based on the ring and ball and kinematic viscosity test:

$$CE\text{-}K > CE\text{-}P > CE$$
$$CO\text{-}K > CO$$

It should be noted, however, that at 60°C the differences in performance between mixes containing unmodified and Polybilt-modified Cenex and unmodified

Table 8 Constant-Height Shear Repeated Load Test Results

	Percent air voids		Ln N at 2% plastic strain	
Asphalt type	40°C	60°C	40°C	60°C
Unmodified Cenex (CE)	2.1	2.1	8.71	4.33
Kraton-modified Cenex (CE-K)	3.0	2.9	14.96	6.84
Polybilt-modified Cenex (CE-P)	1.9	0.7	11.03	4.16
Unmodified Conoco (CO)	3.0	2.9	11.48	4.54
Kraton-modified Conoco (CO-K)	4.4	3.1	13.90	4.66

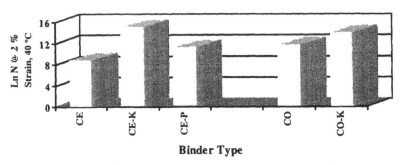

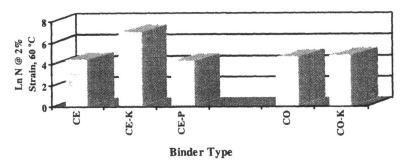

Figure 9   Constant-height shear repeated load test results, log repetitions to 2% plastic strain versus binder type, 40 and 60°C.

and Kraton-modified Conoco are small. For mixes containing modified and unmodified Cenex binders, the mix ranking based on the constant-height shear repeated load test is very similar to that based on the shear creep test at both temperatures.

In order to study the sensitivity of the constant-height shear repeated load test to asphalt source, three Kraton-modified mixes containing 120/150 pen graded asphalts—Conoco, Montana Refinery, and Exxon—were tested at 40°C. Average test results for these mixes are presented in Table 9. Figure 10 shows that the log repetitions to reach a shear strain value of 2% is sensitive to the asphalt source, with mixes containing Conoco asphalt performing best, followed by mixes containing Montana and Exxon asphalts, respectively.

Table 9 Constant-Height Shear Repeated Load Test Results for Rolling Wheel Compacted Specimens

Asphalt type	Percent air voids	Ln N at 2% plastic strain (40°C)
Kraton-modified Cenex (CE-K)	6.0	7.33
Kraton-modified Montana (MO-K)	6.1	6.70
Kraton-modified Exxon (EX-K)	6.0	6.56

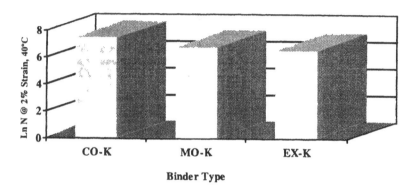

Figure 10 Constant-height shear repeated load test results, effect of asphalt source on log repetitions to 2% plastic strain, 40°C.

VIII. SUMMARY AND CONCLUSIONS

In this investigation a number of different tests have been used to evaluate the rutting propensity for mixes containing both conventional asphalts and modified binders. Test methods evaluated were the following:

1. Conventional mix evaluation tests methods:

(a) Hveem stabilometer test—stabilometer S value
(b) Marshall test—stability value

2. Unconventional mix evaluation methods:

(a) Unconfined axial repeated load test
(b) Simple shear creep and repeated load tests
(c) Constant-height simple shear repeated load test

Table 10 summarizes the ranking of the mixes containing the modified and unmodified binders. Results of these tests suggest the following conclusions:

1. Effects of modification on mix performance are not detected by the Hveem stabilometer. Similarly, binder modification produced no significant effects on Marshall stability values.
2. In general, results of both axial repeated load and simple shear load tests indicate that the permanent deformation resistance of mixes containing modified Cenex binders is improved over that of mixes containing the unmodified material. For mixes containing the Conoco binder, modification resulted in no improvement in mix performance compared with mixes containing the original unmodified asphalt.
3. Although both Cenex and Conoco are 120/150 pen grade asphalts, axial repeated load and simple shear creep and repeated load test results suggest that the permanent deformation resistance of mixes containing unmodified Conoco asphalt is equal to or better than that of mixes containing other modified or unmodified binders.
4. The axial repeated load test and simple shear repeated load test, generally, rank the binder in a similar manner, with Polybilt being a more effective modifier for the asphalts studied. However, temperature affects the rank order of the mixes.
5. Constant-height shear repeated load test ranks the mixes with Kraton being the more effective modifier at both temperatures. This trend is consistent with the ranking based on the binder tests.
6. Constant-height shear repeated load test results indicate that when comparing the performance of mixes containing different asphalts with same penetration grading and modified with Kraton, the influence of the modifier appears to be a function of asphalt source; that is, the performance ranking is (Conoco + Kraton) > (Montana + Kraton) > (Exxon + Kraton).

At the time when this investigation was completed, a comprehensive mix design and analysis system was available for the evaluation of rutting (permanent deformation) distress based on the constant-height shear repeated load test. SHRP technology also includes mix design and analysis systems for the evaluation of load-associated fatigue cracking [22] and low-temperature thermal cracking [23] in addition to the evaluation of the rutting performance.

ACKNOWLEDGMENTS

Funding for this research was provided by the California Department of Transportation. The research reported herein was conducted as a joint venture between

Table 10 Ranking of Modified and Unmodified Binders

Test	Test parameter	Ranking for Cenex asphalt			Ranking for Conoco asphalt		
		First	Second	Third	First	Second	Third
Binder tests							
Ring and ball (R&B)	R&B temperature	Kraton	Polybilt	Unmodified	Kraton	Polybilt	Unmodified
Penetration	Pen, dmm	Kraton	Polybilt	Unmodified	Kraton	Polybilt	Unmodified
Kinematic viscosity	Viscosity, cSt	Kraton	Polybilt	Unmodified	Kraton	Polybilt	Unmodified
Mix tests							
Hveem stabilometer	Hveem stability	Unmodified	Kraton	Polybilt	Kraton	Unmodified	Polybilt
Marshall	Marshall stability	Unmodified	Polybilt	Kraton	Kraton	Polybilt	Unmodified
Axial repeated load	Ln N at 0.2% plastic strain	Polybilt	Kraton	Unmodified	Unmodified	Kraton	Polybilt
Shear creep load	Inverse creep strain	Kraton	Polybilt	Unmodified	Unmodified	Kraton	Polybilt
Shear repeated load Constant height	Ln N at 0.15% plastic strain	Kraton	Polybilt	Unmodified	Unmodified	Kraton	Polybilt
Shear repeated load	Ln N at 2% plastic strain	Kraton	Polybilt	Unmodified	Kraton	Unmodified	Polybilt

the University of California, Berkeley, the Montana State University, the Montana Department of Highways, and the California Department of Transportation.

REFERENCES

1. J. D. Armijo, M. Pradhan, and M. Franchi, Effects of Commercial Modifiers on the Physical Properties of Montana Asphalts, report prepared for the Montana Department of Highways, Montana State University, 1989.
2. M. M. Pradhan, Investigation of Aggregate-Modified Asphalt Mixes Using Varied Materials and Molding Procedures with the Marshall Method, Doctoral dissertation, Montana State University, May 1995.
3. Annual Book of ASTM Standards, *Road and Paving Materials; Paving Management Technologies*, Vol. 04.03, *Construction*, American Society for Testing and Materials, Philadelphia (1995).
4. The Asphalt Institute, *Mix Design Methods for Asphalt Concrete and Other Hot-Mix Types*, Manual Series No. 2 (MS-2), 6th ed., Research Park Drive, Lexington, KY.
5. A. J. Tanco, Permanent Deformation Response of Conventional and Modified Asphalt-Aggregate Mixtures Under Simple and Compound Shear Loading Conditions, Dissertation, University of California at Berkeley, 1992.
6. A. A. Tayebali, J. Harvey, and C. L. Monismith, Performance Characteristics of Asphalt Concrete Mixes Containing Conventional and Modified Asphalt Binders, report prepared for the California Department of Transportation, Asphalt Research Program, Institute of Transportation Studies, University of California, Berkeley, July 1994.
7. A. A. Tayebali, J. L. Goodrich, J. B. Sousa, and C. L. Monismith, Relationships Between Modified Asphalt Binders Rheology and Binder-Aggregate Mixture Permanent Deformation Performance. *J. Asphalt Paving Technol.*, **60**, 121–159, (1991).
8. J. Armijo and M. Pradhan, Investigation of Conventional Aggregates and Modified Asphalt Mixes Using the Marshall Method, Partial Report on Permanent Deformation (Rutting) Characteristics of Binder-Aggregate Mixtures Containing Conventional and Modified Asphalt Binders, prepared by Montana State University for Montana Department of Highways, 1990.
9. J. Armijo and M. Pradhan, Findings of Investigation of Large Stone-Modified Asphalt Mixes Using Marshall Method, Part of Permanent Deformation (Rutting) Characteristics of Binder-Aggregate Mixtures Containing Conventional and Modified Asphalt Binders, Phase II, prepared by Montana State University for Montana Department of Highways, July 1991.
10. J. Harvey, University of California at Berkeley SHRP A-003A Asphalt Concrete Specimen Preparation Protocol, Version 3.0, Report No. TM-UCB-A-003A-91-2, prepared for the Strategic Highway Research Program, Asphalt Research Program, Institute of Transportation Studies, University of California, Berkeley, 1991.
11. P. J. van de Loo, Practical Approach to the Prediction of Rutting in Asphalt Pavements: The Shell Method, Transportation Research Record 616, Transportation Research Board, pp. 15–21, 1976.

12. C. L. Monismith, K. Inkabi, C. R. Freeme, and D. B. McLean, A Subsystem to Predict Rutting in Asphalt Concrete Pavement Structures, in *Proceedings, Fourth International Conference on Structural Design of Asphalt Pavements*, Vol. I, pp. 414–463 (1977).

13. F. R. P. Meyer and R. C. G. Haas, A Working Design Subsystem for Permanent Deformation in Asphalt Pavements, in *Proceedings, Fourth International Conference on the Structural Design of Asphalt Pavements*, Vol. 1, Ann Arbor, pp. 519–528 (1977).

14. W. J. Kenis, Predictive Design Procedures—A Design Method for Flexible Pavements Using the VESYS Structural Subsystem, in *Proceedings, Fourth International Conference on the Structural Design of Asphalts Pavements*, Vol. 1, Ann Arbor, pp. 101–130 (1977).

15. S. F. Brown and C. A. Bell, The Prediction of Permanent Deformation in Asphalt Pavements, in *Proceedings, Association of Asphalt Paving Technologists*, Vol. 53, pp. 415–439 (1979).

16. J. Uzan and R. Lytton, Structural Design of Flexible Pavements: A Simple Predictive System, paper presented at the 1982 Annual Meeting of the Transportation Research Board, Washington, DC, 1982.

17. R. Leahy, Permanent Deformation Characteristics of Asphalt Concrete, Ph.D. dissertation, University of Maryland, College Park, 1989.

18. Z. Yao and C. L. Monismith, Behavior of Asphalt Mixtures with Carbon Black Reinforcement, *Proceedings, Association of Asphalt Paving Technologists*, Vol. 55, (1986).

19. C. L. Monismith and A. Tayebali, Permanent Deformation (Rutting) Considerations in Asphalt Concrete Pavement Sections, *Proceedings, Association of Asphalt Paving Technologists*, Vol. 57, (1988).

20. C. Valkering, Rutting Resistance of Asphalt Mixes Containing Non-Conventional and Polymer-Modified Binders, *Proceedings, Association of Asphalt Paving Technologists*, Vol. 60, (1990).

21. Strategic Highway Research Program, Permanent Deformation Response of Asphalt Aggregate Mixes, SHRP Report No. SHRP-A-415, National Research Council, Washington, DC, 1994.

22. Strategic Highway Research Program, Fatigue Response of Asphalt Aggregate Mixes, SHRP Report No. SHRP-A-404, National Research Council, Washington, DC, 1994.

23. Strategic Highway Research Program, Final Report, Low Temperature Cracking—Test Selection, SHRP Report No. SHRP-A-401, National Research Council, Washington, DC, 1994.

23

Analytical Approach to Bitumen–Polymer Blends

Patrick Cogneau
Parc Industrial, Perwez, Belgium

I. INTRODUCTION

Over the past 15 years, bitumen–polymer mixes have been gaining in importance because of the very wide range of polymers that are compatible with bitumen, which means that today they can be used in the most diverse applications.

Our aim is to examine one of these mixes: bitumen-based waterproofing membranes modified by atactic polypropylene (APP). Several methods can be used to obtain an optimal form of these binary mixes: either investigating the properties required for the atactic polypropylene to provide a membrane of excellent quality when mixed with a defined bitumen, or the reverse, i.e., investigating the types of bitumen that provide the best qualities of membrane in the presence of judiciously selected APP. We followed the latter path, which enabled us to differentiate between several bitumens belonging to the same penetration class.

The so-called direct-distillation bitumens are obtained directly, by mixing bitumens of different penetrations or by fluxing. The crude oils from which these bitumens are derived come from the Middle East, the North Sea, Central America, or South America. The APP polymer is obtained by mixing various types of APP selected according to experience gathered mainly on the basis of infrared and differential thermal analysis techniques.

The bitumen and APP polymer are always mixed in the same proportions and under the same conditions. Then each mix is evaluated using different analytic techniques, each of which reveals the level of compatibility found.

The purpose of the present chapter is to establish the relationships between the parameters of generic composition and the working characteristics so that the latter can be predicted from the analysis of the bitumen. In order to establish certain correlations between the compositional aspects and the physicomechanical behavior, some modeling is required.

II. EXPERIMENTAL STUDY

A. Physicochemical Analysis of Bitumens

The experiments were performed on specimens, numbered 1 to 20 in Tables 1 and 2, of bitumens of the same penetration class but of different origins.

Table 1   Characteristics of Bitumens

Bitumen	Origin[a]	R & B[b] (°C)	Penetration at 25°C (dmm)	Viscosity at 135°C (mm²/s)	Ni (ppm)	V (ppm)
1	A + B?	43	101	295	34	117
2	C1	46	102	332	118	540
3	A + B	45	110	336	59	169
4	C2	43.5	118.5	305	83	410
5	A	41	118.5	234	20	54
6	C3 or C4	44.5	115	308	75	547
7	C3	44	106	310	78	575
8	C3 + C2	44	140	288	72	319
9	A	43.5	135	322	36	107
10	C2[c]	47	104	319	66	472
11	C3 + C2[c]	46	106	334	97	479
12	C3[c]	46.4	112	332	75	580
13	C4	48.6	110	350	79	587
14	C2 + X	49	103	320	70	375
15	A	45	111	281	60	187
16	C3 + C2	45.8	113	316	80	544
17	A	43.5	144	265	47	145
18	C3 + C2 + X	47.6	103	378	82	459
19	C5	45	125	350	75	570
20	A?	44	105	320	57	175

[a] A. from Middle East; B, from North Sea; C, from South America (numbers indicate different crudes); X, unknown.
[b] R & B. ring ang ball.
[c] Different composition.

Table 2 Generic Composition (Iatroscan Method)

Bitumen	% Asphalt	% Resins	% Aromatics	% Saturates	I_c
1	12.06	19.83	62.3	5.8	0.217
2	19.02	26.03	48.39	6.56	0.344
3	13.8	20	60.51	5.69	0.242
4	21.59	19.55	50.38	8.47	0.43
5	6.8	22.10	67	3.9	0.108
6	20.78	22.56	48.93	7.73	0.399
7	20.43	23.7	48.69	7.19	0.382
8	18.59	18.41	56.18	6.82	0.341
9	11.11	14.93	69.85	4.31	0.182
10	20.18	20.17	51.04	7.28	0.385
11	22.6	20.6	48.43	7.33	0.433
12	21.42	20.45	49.3	8.84	0.436
13	20.68	20.71	50.51	8.1	0.404
14	14.7	21.81	52.71	8.08	0.342
15	18.16	23.23	52.95	5.67	0.313
16	22.11	21.95	49.86	6.08	0.393
17	9.14	21.02	66.8	3.04	0.139
18	19.95	19.57	53.51	6.97	0.368
19	19.17	20.56	55.45	5.82	0.316
20	13.48	19.15	59.72	7.64	0.268

1. Standard Characteristics

A certain number of standard characteristics were determined for each specimen, including the following:

Needle penetration according to standard NFT.66.004
Ring and ball softening point as per standard NFT.66.008
Capillary viscosity at 135°C (Cannon-Fenske equipment)
Ni and V content (determined by plasma torch)

2. Generic Composition

Of the methods recommended for determining the generic composition of the bitumens, the one based on thin-layer adsorption chromatography connected to a flame ionization detector (FID), or the Iatroscan method, is a particularly suitable high-speed approach. One of its advantages is that there is no need for preliminary precipitation of asphaltenes. The bitumen, dissolved in tetrahydrofuran (2% w/v), is deposited on a series of 10 reusable chromarods (quartz rods coated with a thin layer of silica gel).

The constituents of the bitumen are separated by successive chromato-graphic elutions using n-heptane for separation of saturates; toluene and n-heptane (80:20) for separation of aromatics; and dichloromethane and methanol (95:5) for separation of resins. The various constituents (saturates, aromatics, resins, and asphaltenes) can be quantified by passing them through the flame of the detector. The generic composition results recorded in Table 2 have been plotted in a ternary diagram representing % aromatics, % asphaltenes + saturates, and % resins) (Figure 1).

The colloidal instability index I_c is defined by the following relationship:

$$I_c = \frac{\text{asphaltenes} + \text{saturates}}{\text{aromatics} + \text{resins}}$$

B. Bitumen–Atactic Polypropylene Associations

1. Choice of APP

The polymer batch is a mixture of different types of APP selected on the basis of our experience. It can be correlated with a basic formulation that has been tried and tested for the manufacture of waterproofing membranes.

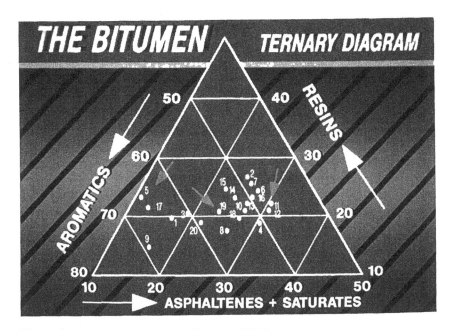

Figure 1   Generic composition by Iatroscan TH-10.

Mean characteristics are as follows:

Ring and ball softening point: (standard ASTM[a] D-36-26)	160°C
Needle penetration at 25°C (standard ASTM D 5–73) at 60°C	15–30 dmm
Cold bending test (according to UEATC directives)	−15/−25°C
Infrared, % isotacticity (standard ASTM E-204-72)	40–60%
Brookfield viscosity at 180°C, 100.000 to 700.000 cps (B 21 spindle, speed 5-10-20 rpm)	
Differential scanning calorimeter (DSC) (determination of the rate of crystallinity associated with the fusion peak—measurement of differential enthalpy)	
Peak 1	0.1–0.5 cal/g
Peak 2	3–5 cal/g

[a] ASTM, American Society for Testing and Materials.

2. Composition

The composition selected for the whole investigation was 70 parts bitumen to 30 parts of the APP batch.

3. Preparation of the Mixes

The bitumen–polymer mixes are produced by dispersion of the polymer in the bitumen, stirring it under heat. Thus, the mixes are prepared in a 1.2-L tank heated by thermal oil, and the mixer used is a Silverson L 4 R rotor-stator type with high-shear agitation and a speed of up to 6000 rpm.

The mixing temperature is fixed at 180°C; the maximum mixing time is 3 h and depends on the workability of the bitumen and its capacity for dispersion in the melted polymer.

C. Characterization of the Mixes (or Blends)

1. Standard Analyses

The characteristics of the blends were determined as follows:

Ring and ball (standard ASTM D-36-26), test carried out in glycerin. The temperature is recorded at 1 mm deformation and when the ball falls on the support.

Needle penetration at 25 and 60°C (standard D-5-73).

Brookfield viscosity at 180°C (viscometer type HB, spindle B 21, speed 5-10-20 rpm).

Cold bending test (according to UEATC directive): 4-mm-thick test specimen (12 × 2.5 cm), machine speed of 200 mm/min, 20 mm mandrel. The temperature indicated is that achieved before the test piece cracks.

Cold bending test after aging in an oven for 28 days at 80°C and for 6 months at 70°C.

2. Specific Analyses

Crushing Test. A pellet of the blend, 20.5 mm in diameter and 2 mm thick, is placed between sheets of absorbent filter paper. The assembled unit is placed under a weight of 100 g in an oven for 15 h at 135°C. Recording the deformation of the pellet and the oil migration makes it possible to evaluate the stability of the mix.

Demixing Test. This test is performed using an aluminum tube 10 cm in height and 3 cm in section into which the blend is poured and held for 63 h at 180°C. The softening point is then determined at the top and bottom parts of the tube (± 1 cm) and the temperature difference (ΔT) is recorded.

Visualization of the Mixes Under an Epifluorescence Microscope. The equipment consisted of an Olympus microscope, type BH-2 (exciter filter mount EY-455, exciter filter insertion slot B, dichroic mirror insertion port code B-0, barrier filter insertion slot 0-530), with magnification $125\times$ or $250\times$.

III. INTERPRETATION OF THE RESULTS

Bitumens emerge as compounds whose constitution and behavior are extremely complex. It is generally accepted that bitumens form a colloidal system represented by a suspension of micelles of asphaltenes peptized by resins in a medium of saturated and aromatic oils. The presence of polar groups at the surface of the colloids contributes to the saturation of the system. The viscoelastic behavior of the material can be explained on the basis of this colloidal model: interactions between the colloids lead to the formation of a temporary elastic network with reversible or partially reversible deformation depending on the deformation time. At low and elevated temperatures this elastic deformation is greatly reduced, first because the material will tend to behave like a glass and second because the bitumen will tend to behave like a Newtonian fluid.

The property of the polymers (APP) in the modified bitumen is their contribution to the formation of a less rigid and more permanent elastic network, thus extending the elastic properties of the modified bitumen in a range of temperatures distinctly wider than would be the case for bitumen on its own. The rheological properties therefore depend largely on the way in which the polymer and the asphaltenes are dispersed in the bitumen.

A. Micromorphology

Epifluorescence microscopy enables us to confirm that bitumens modified by APP are usually biphasic mixtures, one phase being made up of the polymer swollen by part of the bitumen oils (mainly the maltenes), the other comprising the constituents of the bitumen not involved in the solvation of the polymer (mainly the asphaltenes). Observation of the microstructure makes it possible to evaluate the quality of the bitumen–atactic polymer association and, consequently, the compatibility itself, i.e., the affinity of one constituent for another. If you examine the photographs in Figures 2 to 6, you can see that the structures are distinguishable by the fineness of the dispersion of the discontinuous phase in the continuous phase.

The comparative visual examination of the 20 prepared mixes has thus enabled us to classify them into five groups of decreasing dispersion quality, from a homogeneous biphase structure (fine dispersion) to a heterogeneous structure (coarse dispersion and/or partial separation of the phases).

The results given in Table 3 indicate that at least eight bitumens (belonging to groups 4 and 5) did not result in the formation of mixes with a homogeneous biphase structure. This poor behavior or incompatibility of the bitumen in the presence of the polymeric phase could be associated with the composition and

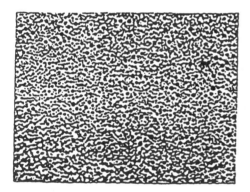

Figure 2   Epifluorescence microscopy category 1.

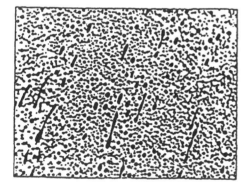

Figure 3   Epifluorescence microscopy category 2.

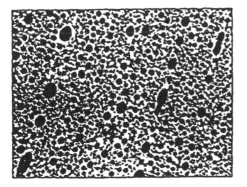

Figure 4 Epifluorescence microscopy category 3.

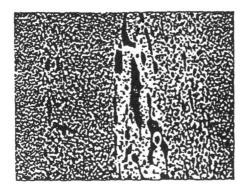

Figure 5 Epifluorescence microscopy category 4.

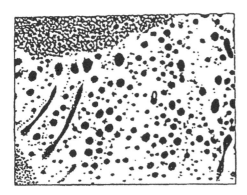

Figure 6 Epifluorescence microscopy category 5.

Table 3 Characteristics of Bitumen–APP Blend

Bitumen	Penetration (dmm) R & B (°C)	25°C	60°C	Viscosity at 180°C (cps)	Cold bending test (°C) Ref. new	Aged at 80°C 7 d	14 d	21 d	28 d	15 d	Aged at 70°C 1 mo	2 mo	4 mo	6 mo	Demixing (°C) R & B top	R & B bottom	ΔT	Crush test	Mic. epi. type
1	143/154	34	151	4426	−15	−7	−6	−7	−5	−11	−8	−7	−5	−4	149/155	72/98	56	0	4(−)
2	142/152	33	117	7233	−16	−13	−11	−13	−10	−12	−10	−10	−8	−5	145/153	78/117	36	30	4(+)
3	144/153	31	135	5658	−10	−6	−6	−3	−1	−8	−8	−2	+1	+1	148/155	74/93	62	0	5
4	144/154	35	122	10093	−20	−19	−18	−17	−18	−19	−17	−17	−16	−16	147/154	89/128	26	90	1
5	143/153	35	165	3002	−12	−11	−7	−3	−1	−7	−2	−1	+1	+3	145/154	57/75	79	0	5
6	143/153	32	125	10061	−19	−17	−17	−16	−14	−18	−15	−16	−14	−14	149/154	87/123	31	80	2
7	143/152	37	135	8373	−16	−15	−14	−13	−12	−14	−12	−13	−13	−11	148/154	91/124	30	50	3
8	143/153	35	124	6043	−15	−12	−10	−4	−6	−13	−10	−11	−8	−5	145/154	75/116	38	35	2(−)
9	130/152	33	132	3863	−15	−12	−7	−3	−1	−11	−6	−11	−1	−2	150/155	63/83	72	0	5
10	143/153	36	110	10592	−17	−15	−16	−15	−14	−14	−14	−13	−14	−14	150/152	94/121	31	80	2
11	142/152	36	114	11437	−22	−20	−19	−19	−18	−18	−17	−18	−16	−16	149/153	94/124	29	95	1
12	139/153	41	136	7398	−19	−19	−16	−16	−16	−15	−15	−16	−13	−11	148/153	90/122	31	75	1
13	142/154	38	137	12099	−21	−17	−14	−15	−17	−17	−18	−17	−17	−14	153/155	100/128	27	95	1
14	143/154	36	140	8132	−15	−16	−14	−8	−7	−14	−14	−13	−15	−11	153/155	98/124	31	55	4
15	145/154	29	91	6169	−5	−3	−3	−5	−6	−6	−7	−6	−5	−5	151/155	82/104	51	25	4
16	142/154	37	135	8965	−21	−20	−18	−17	−15	−18	−17	−18	−17	−17	142/154	86/125	29	65	3
17	143/154	40	186	2754	−14	−3	−1	+3	0	−5	−2	+1	0	+2	153/155	51/76	79	0	4
18	143/153	35	146	9371	−18	−16	−14	−15	−15	−16	−15	−14	−15	−14	146/153	78/122	31	30	3
19	142/153	34	165	5273	−18	−16	−13	−10	−8	−14	−14	−12	−10	−9	141/155	77/114	41	40	2(−)
20	141/153	28	127	7650	−17	−15	−14	−12	−10	−13	−10	−10	−11	−10	143/154	78/99	55	35	3

the properties of the maltenes (polarity, aromaticity, etc.). We note that a high level of aromatics provides poor compatibility between bitumen and APP.

B. Standard Characteristics (Table 3)

1. Ring-Ball Softening and Low-Temperature Flexibility (as-New Condition)

The modification of the bitumens leads to the formation of new materials whose plastic behavior temperature range is expanded upward, i.e., from 45–50°C to 150–155°C, as well as downward to −5 to −20°C. Our experience reveals that the value of the softening point depends little on the bitumen, varying mainly as a function of the choice of the APP and its level of crystallinity.

As regards the low-temperature flexibilities, greater variation is obtained in new products, clearly demonstrating the different behaviors of the bitumens and their ability to contribute (or not) to the formation of an elastic network in the polymer phase. Generally, the mixes for which the low-temperature flexibility is low are those whose dispersion is finest (bitumens, 4, 11, 12, and 13, for example).

2. Flexibility in the Aged Condition

Good quality of the flexibility results following aging is absolutely essential when a mix is being made for a waterproofing membrane, to ensure the service life of the membrane. Thus, in order to obtain a better interpretation of the behavior of the mixes over time, we used accelerated aging processes in an oven (28 days at 80°C and 6 months at 70°C). The analysis of the results indicates that certain bitumens (4, 11, 13, and 18, for example) behave favorably, losing 2 to 5 degrees in the aged condition, whereas the characteristics of others decline rapidly (1, 2, 5, and 17, for example) below a tolerable threshold.

3. Needle Penetration

The tests at 60°C, and to a lesser degree at 25°C, reveal notable variations. High penetrations may be expected for mixes whose structural states are heterogeneous or partially heterogeneous; the resistance to the penetration of the needle is not as great. Nevertheless, the results are difficult to interpret.

4. Brookfield Viscosity

In the fused state, the measured viscosities of the prepared mixes are between 2700 and 12,000 cps. On the one hand, it can be noted that the blends with low viscosities behave poorly in the aged condition (1, 5, and 17, for example) (and in certain cases the new products). The same bitumens present working problems (not very homogeneous dispersion). On the other hand, high-viscosity blends

(4, 6, 11, and 13, for example) have interesting flexibilities in the new and aged conditions.

In addition, a good relationship can be found between the viscosity of the mix and the colloidal instability index for the bitumen on its own (Table 3), again confirming that the rheological behavior depends very much on the colloidal structure of the bitumen and the affinity of the bitumen vis-à-vis the polymer.

C. Specific Characteristics

1. Crushing Test

We have chosen to illustrate this test (which is mainly visual) by representative examples of three extreme situations encountered (Figure 7). The first situation represents a stable mix, with little deformation and little oil migration (brown) on the absorbent paper (at the top). The second case shows migration of oil (brown), the appearance of black spots, and greater deformation. The third example indicates considerable deformation and very high migration. The mix is totally unstable and there is apparent separation of the constituents.

The comparative study of the results of the crushing tests on 20 mixes enables the allocation of a relative value to the rates of oil migration and deformation observed. Once again, we can see a good correlation between the crushing test and the other characteristics (flexibility, viscosity, etc.).

Figure 7 Results of the crushing test.

2. Demixing Test

The temperature differences (ΔT) measured between the top of the tube and its base are the result of separation, at elevated temperature, due to partial decantation of the polymer and the bitumen. A polymer-enriched polymeric phase demixes progressively, causing the deposition of a nonassociated bitumen-rich phase.

The ΔT value is therefore a measure of the thermal stability of the mix and indirectly a measure of its compatibility. The lower the compatibility between the bitumen and the polymer, the easier the separation between these two constituents and the higher the value of ΔT.

Mixes with low ΔT values (approximately 30°C) are those that retain good characteristics after aging and have a viscosity above +5000 cps. Blends with a high ΔT value also show mediocre performances with regard to other characteristics.

IV. MODELING

A weighting scale has been defined for each type of blend analysis described in this chapter to provide a means of defining the influence of the measured property on the quality of the membrane. Our aim is therefore to allocate a relative (and weighted) value to the performance attained by each mix for each of the characteristics that best reveals the level of compatibility found between the bitumen and a selected APP batch.

Performances are evaluated as follows (ponderation figures; see Appendix):

For Brookfield viscosity (score out of 50 points)
For microscope observation (score out of 50 points)
For the demixing test (score out of 100 points)
For the crushing test (score out of 100 points)
For the flexibility of fresh materials (score out of 100 points)
For flexibility after aging at 80 and 70°C (score out of 100 points)

Each property has been weighted and entered into a multilinear relationship so as to define each bitumen by a score, vis-à-vis a given APP.

Table 4 summarizes all our investigations, and Table 5 sets out the classification of the bitumens derived from them. This classification can be subdivided into three categories:

A, Compatible bitumens (vis-à-vis the APP)
B, Intermediate bitumens:
C, Incompatible bitumens (vis-à-vis the APP)

It is from bitumens of category A that the best mixing performances with APP will be obtained.

Table 4 Results of Ponderation

Bitumen	Demixing test (/100)	Crushing test (/100)	Viscosity (/50)	Dispersion and working (/50)	Cold bending fresh (/100)	Cold bending test (aged)			Total (/500)
						ΔT (a)	Ang. coeff. (b)	(a) + (b) (/100)	
1	20	0	5	5	20	0	5	5	55.00
2	40	30	15	10	40	10	10	20	155.00
3	0	0	15	0	0	10	0	10	25.00
4	100	90	45	50	90	50	45	95	470.00
5	0	0	0	0	0	0	0	0	0.00
6	70	80	45	40	80	35	35	70	385.00
7	70	50	30	25	40	35	20	55	270.00
8	40	35	15	30	20	10	5	15	155.00
9	0	0	0	0	20	0	0	0	20.00
10	70	80	45	40	60	50	30	80	375.00
11	85	95	50	50	100	35	45	80	460.00
12	70	75	15	50	80	20	35	55	345.00
13	100	95	50	50	100	20	40	60	455.00
14	70	55	30	10	20	20	10	30	215.00
15	20	25	15	10	0	50	5	55	125.00
16	85	65	40	25	100	35	45	80	395.00
17	0	0	0	0	10	0	0	0	10.00
18	70	30	40	25	70	35	35	70	305.00
19	20	40	5	30	70	10	10	20	185.00
20	20	35	30	25	60	20	10	30	200.00

Table 5 Bitumen Classification

Bitumen	Score
Category A	
4	470
11	460
13	455
16	395
6	385
10	375
12	345
Category B	
18	305
7	270
14	215
20	200
19	185
2	155
8	155
15	125
Category C	
1	55
3	25
9	20
17	10
5	0

V. CORRELATION BETWEEN GENERIC COMPOSITION AND COMPATIBILITY

The correlation between the table evaluating three categories of bitumens and the generic composition is illustrated by the ternary diagram in Figure 8. This figure shows the close relationship between the composition of the bitumen and its compatibility with the APP. We observe that the compatibility will be better, the closer the composition is to the defined domain.

It is therefore clearly apparent that a key parameter, which also governs the colloidal equilibrium of the bitumens, is the chemical structure of the constituents and their proportions. The quality of the bitumen will depend on the correct equilibrium between the different classes of constituents.

A. Bitumen Modified by Styrene-Butadiene-Styrene (SBS)

The first investigations into the modification of bitumen by SBS (results not recorded) seem to indicate the formation of a compatible domain situated in

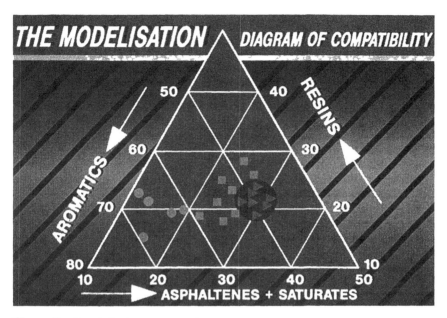

Figure 8 Correlation between generic composition by Iatroscan TH-10 and the performances on mixing with APP.

another slope of the ternary diagram (speckled zone), indicating the need for a bitumen richer in aromatic oils so as to produce better swelling of the polymer phases. Among the other physical properties of the bitumens (ring-ball softening, viscosity, infrared spectroscopy.* and Fraass point*). none is indisputably linked to the quality of the blend. It does emerge, however, that the least satisfactory bitumens are those with a low vanadium content, which are also the bitumens least rich in asphaltenes.

VI. CONCLUSIONS

The investigation described in this chapter demonstrates the relationships between the generic composition of bitumen and the rheological properties of the bitumen–polymer blend. The quality of the blend is evaluated semiquantitatively on the basis of crushing tests, viscosity tests, low-temperature flexibility tests (condition as new and after ageing), demixing tests, and micromorphologies. All these analyses and the evaluation of the performances of the mixes make it possible

---

* Result not mentioned.

to define the profile of an ideal bitumen for manufacture of the optimum membrane, taking due account of an assigned APP polymer.

APPENDIX: MODELING DETAILS OF PONDERATION

Performances are as follows.
For Brookfield viscosity (score out of 50 points):

Viscosity less than 4,000 cps:	0 pt
Viscosity between 4,000 and 5,500 cps:	5 pts
Viscosity between 5,500 and 7,500 cps:	15 pts
Viscosity between 7,500 and 8,500 cps:	30 pts
Viscosity between 8,500 and 10,000 cps:	40 pts
Viscosity between 10,000 and 11,000 cps:	45 pts
Viscosity greater than 11,000 cps:	50 pts

For microscope observation (score out of 50 points):

Class 1:	50 pts
Class 2:	40 pts
Class 3:	25 pts
Class 4:	10 pts
Class 5:	0 pt

For the demixing test (score out of 100 points):

ΔT less than 28°C:	100 pts
ΔT between 28 and 30°C:	85 pts
ΔT between 30 and 35°C:	70 pts
ΔT between 25 and 40°C:	40 pts
ΔT between 40 and 60°C:	20 pts
ΔT greater than 60°C:	0 pt

For flexibility on fresh material (score out of 100 points):

Flexibility less than −20:	100 pts
Flexibility equal to −20:	90 pts
Flexibility equal to −19:	80 pts
Flexibility equal to −18:	70 pts
Flexibility equal to −17:	60 pts
Flexibility equal to −16:	40 pts
Flexibility equal to −15:	20 pts
Flexibility equal to −14:	10 pts
Flexibility greater than −14:	0 pt

For flexibility after aging at 80 and 70°C (score out of 100 points). The weighting takes account of (a) the temperature difference recorded between flexibility in the as-new and aged (28 days or 6 months) conditions:

ΔT between 0 and 3°C:	50 pts
ΔT between 3 and 5°C:	35 pts
ΔT between 5 and 8°C:	20 pts
ΔT between 8 and 11°C:	10 pts
ΔT greater than 11°C:	0 pt

(b) the estimated angular coefficient of the temperature curve slope:

Flexibility less than -18:	50 pts
Flexibility between -18 and -17:	45 pts
Flexibility between -17 and -16:	40 pts
Flexibility between -16 and -15:	35 pts
Flexibility between -15 and -14:	30 pts
Flexibility between -14 and -13:	20 pts
Flexibility between -13 and -9:	10 pts
Flexibility between -9 and -5:	5 pts
Flexibility greater than -5:	0 pt

For crushing test (score out of 100 points) relative value (comparison of results):

INDEX

Adhesives, 249
Agrobased resin, 354-358
Aluminum pigment, 450, 451
  leafing, 452
  nonleafing, 453
Amorphous polyalphaolefins, 263
  blends, 271
Anacardium occidentale, 350
Antioxidants, 352, 358
Asphalt, 1, 2, 22
  accelerated aging of, 220
  acid in, 111, 114, 115
  aging, 13, 103
  aliphatic carbons, 14, 46, 48
  amorphous component, 63
  analytics, 372, 374, 504
  aromatic carbons, 23, 46
  asphaltene in, 88, 120, 260, 280,
    371, 375
  cements, 250
  classification, 371
  coatings, 441
  cracking, 49, 235
  crystalline component, 63
  cutbacks, 250
  emulsions, 250, 297
  fatigue, 49
  fluid, 250
  fumes from, 136

[Asphalt]
  gel, 260, 280
  hardening, 119
  heteroatoms, 12, 24, 91, 107, 371
  hydrogen bonding, 88
  low-temperature properties, 59
  maltene, 88, 121, 260
  masking agents for, 138
  metals, 30, 371
  micellar model of, 167
  mobile component, 12
  modifiers, 197, 340
  molecular,
    association, 50
    interaction, 12
    mobility, 31
    motion, 11, 20
  mopping, 253
  naphthalene aromatics in, 120,
    164
  neutral component, 95
  nonpolar, 69
  odor from, 136
  oxidation, 1, 6, 115, 119, 125, 349
  pavement failure, 59, 119
  polar aromatic in, 120, 164, 371,
    375
  polymer-like, 377
  properties with masks, 139

[Asphalt]
   rigid component, 12
   rolled carpet, 162
   roofing, 135
   rubber, 386
      history, 388
   rutting, 49, 235, 261, 371, 478
   saturates, 120, 375
   secondary structure, 377
   self-assembly, 1,7, 88
   shingles, 251
   sol, 260, 280
   sulfur content, 107
   thermoplastic, 369
   viscoelastic, 59, 197
   viscoelasticity, 12
Axial load test, 488

Beam, 203
Bending beam testing, 221, 226
Bitumen, 249, 259, 279, 337, 351,
      462
   mixing of, 284
Building products, 249, 346
Built-up-roofing, 135, 252, 297

Chromatography,
   inverse gas-liquid (IGLC), 104
   ion exchange, 69, 260
Coatings, 249, 442
   aluminum, 447
   applications, 456-458
   solvent-borne, 442-444
   water-borne, 444-447
Coordination polymerization,
      262
Corbett technique, 120, 121, 129
Differential scanning calorimetry,
      (DSC), 60, 64, 267, 325,
      376
   modulated, 60, 64, 86, 376
Direct tension testing, 220
Dynamic mechanical analysis, 237,
      269, 294, 325

Elastomer, 197, 461
Ethylene vinyl acetate, 202, 236,
      286, 287
   polyvinyl chloride, 236

Fatigue mechanism, 150
Fatigue testing, 205
Federal Highway Administration
      (FHWA), 387
Fluorescence microscopy, 282
Fracture, 150
   adhesive, 150
   cohesive, 150
   mechanics, 180

Gel permeation chromatography,
      (GPC), 7, 120, 127, 260, 268,
      376
Gilsonite, 249
Glass transition, 59, 62, 85, 267,
      272, 324, 376
   dopant effects, 96

Highways, 260
   traffic, 384
Hot mix asphalt, 384
Hveem stabillmeter test, 484

Iatroscan, 375
Infrared
   analysis, 112, 123
   spectroscopy, 401
Inverse gas-liquid chromatography,
      104
Ion exchange chromatography, 69,
      260
Low-temperature flexibility, 261

Marshal stability test, 486
Mass spectroscopy, 119
Melt viscosity, 264
Microcrack, 149, 187
   growth, 149, 200
   healing, 149

[Microcrack]
  healing index, 163, 188
Microstructures, 465, 466, 509, 510
Modification mechanism, 464
Modified crack layer model, 198,
  206, 211

Newtonian flow, 260
Nuclear magnetic resonance (NMR),
  2, 22, 119, 376
  benzylic peaks, 4
  carbon, 13, 22
  chemical shifts, 14
  cross-polarization, 17, 22
  hydrogen, 13
  magic angle, 17, 22
  relaxation time, 14, 20, 24, 28
  solid-state, 13, 22

Oil,
  crude, 370
    origin, 370

Paris' rule, 174
Pavement
  durability, 104, 112, 197
  recycling, 349, 351
Penetration, 265
Phase inversion, 377
  retardation, 379
Plastomer, 197, 461
Polybutadiene, 399
Polychloroprene latex, 297
  anionic, 298
  cationic, 303
Polyethylene, 236, 261, 319
Polymers, 236, 249, 261, 307, 349,
  373
  compatibility, 330, 463
  crystallinity, 322
  epoxy, 253
  ethylene, 313

[Polymers]
  modification, 254
  agrobased resin, 360
  APAO, 275
  epoxy, 254
  mechanism, 378
  phenolic, 256
  polyester, 255
  polyolefin, 342
  polysulfide, 255
  polyurethane, 255
  rubber, 256, 293, 343
  thermoplastic olefin, 257
  thermoplastic, 257, 370
  thermoset, 370
  modifying, 250
  network, 372
  stability, 379
  organic, 312
  phenolic, 253
  physical properties, 321
  polyester, 253
  polysulfide, 253
  polyurethane, 253
  properties, 316
  rubber, 236
  science, 308
  solubility, 329
  technology, 308, 337
  thermoplastic elastomers,
    254
  thermoplastic olefinics, 254
  tree, 315
  vinyl, 314
  viscoelasticity, 331
Polypropylene, 236, 252, 261,
  371, 503
  copolymers, 252, 263, 264,
    370, 371
Potentiometric titration, 106
Pressure aging vessel, 110,
  241

Rheology, 237, 269, 280, 281
   measurement, 239
Road binder, 286, 297
Roofing, 249
   sheets, 291
Roof membrane, 251
   atactic polypropylene (APP),
      252
   modbit, 251
   styrene-butadiene-styrene (SBS),
      252
   rolling thin-film oven test
      (RTFOT), 110, 241
Rubber,
   crumb,383
   application examples,
      418
   applications, 417
   common elastomers in,
      403
   composition, 396
   ground, 395
   manufacture, 390
   specification, 435
   natural, 399
   oxidation, 399
   styrene-butadiene, 261

Scanning electron microscopy, 198,
   205, 209, 215
Shale oil, 24
Shear creep test, 490
Shear modulus, 155
Shift factor, 162

Strategic Highway Research
   Program (SHRP), 1, 6, 22,
   217, 383
   performance grade, 38
Size exclusion chromatography,
   80
Solubility parameter, 197, 262,
   329
Static testing, 204
Styrene-butadiene-styrene (SBS),
   202, 217, 219, 224, 261,
   279, 286, 287, 370, 371
Styrene butadiene rubber, 261
Supercritical fluid extraction, 95

Thin film oven/pressure oxygen
   vessel (TFO/POV), 6
Thermal cracking, 49, 218, 228
   critical temperature, 228, 230
   model, 228
Thermogravimetric analysis, 400
Tire design, 385
   airplane, 396
   composition, 402
   durability, 385

Wax, 70, 77
Wheel
   French, 237
   Hambrug, 237
Williams, Landel, Ferry equation
   (WLF), 13, 285

Young's modulus, 155

Printed in the United States
by Baker & Taylor Publisher Services